D0757976

Progress in Mathematical Physics
Volume 40

Yaakov Friedman
with the assistance of Tzvi Scarr

Physical Applications of Homogeneous Balls

Birkhäuser
Boston • Basel • Berlin

SEP/AE
PHYS

Yaakov Friedman
Jerusalem College of Technology
Department of Mathematics
Jerusalem 91160
Israel
friedman@jct.ac.il

Tzvi Scarr
Jerusalem College of Technology
Department of Mathematics
Jerusalem 91160
Israel

AMS Subject Classifications: 51P05, 83A05, 22E70, 22E43, 17C90, 17C27, 78A35, 81P15 (Primary); 22E60, 81R25, 81R99, 17C40, 47L90, 47L50, 47L70, 22F30, 32A10, 37C85, 37C80, 81Q15, 81Q99, 57S20 (Secondary)

Library of Congress Cataloging-in-Publication Data
Friedman, Yaakov, 1948-
Physical applications of homogeneous balls / Yaakov Friedman, with the assistance of Tzvi Scarr.
p. cm. – (Progress in mathematical physics ; v. 40)
Includes bibliographical references and index.
ISBN 0-8176-3339-1 (acid-free paper)
1. Homogeneous spaces. 2. Mathematical physics–Mathematical models. 3. Special relativity (Physics) I. Scarr, Tzvi, II. Title. III. Series.

QC20.7.H63F75 2004
530.15'4-dc22 2004059980

ISBN 0-8176-3339-1 Printed on acid-free paper.

Printed in the United States of America. (TXQ/MV)

9 8 7 6 5 4 3 2 1 SPIN 10936436

www.birkhauser.com

SD 3/2/05 LL ag 3/10/05

To my wife Rachel

Contents

Preface

> Based on this century of experience, it is generally supposed that a final theory will rest on principles of symmetry.
>
> *Dreams of a Final Theory*
> Steven Weinberg
> 1979 Nobel Prize winner

This book introduces *homogeneous balls* as a new mathematical model for several areas of physics. It is widely known that the set of all relativistically admissible velocities is a ball in R^3 of radius c, the speed of light. It is also well known that the state space of a quantum system can be represented by positive trace-class operators on a Hilbert space that belong to the unit ball in the trace norm. In relativistic quantum mechanics, the Dirac bispinors belong to a ball in the space C^4. Is there something in common among these balls? At first glance, they look very different. Certainly, they do not represent commutative objects, for which the unit ball is a simplex. They cannot represent a binary algebraic operation, since for such an operation, we need an order on the space, and there is no order for the first and third examples. But as we will show, in all of the above situations, either the ball in question or its dual is *homogeneous*. Moreover, there is a triple structure which is uniquely constructed from either the homogeneity of the domain or the geometry of the dual ball.

Homogeneous balls could serve as a unifying language for different areas in physics. For instance, both the ball of relativistically admissible velocities in Special Relativity and the unit ball of operators on a Hilbert space, which is the dual of the state space in Quantum Mechanics, are homogeneous balls. In Special Relativity, the homogeneity of the velocity ball is an expression of the principle of Special Relativity and not artificially imposed. The surprising fact that the unit ball of the space of operators is a homogeneous ball was first discovered and utilized in solving the engineering problems involved in transatlantic telephone communication. But not much has been done in physics to take advantage of this structure.

In addition, some aspects of General Relativity may be described by the methods presented in the book. But in order to describe General Relativity efficiently, our model must be generalized. This generalization is obtained by weakening slightly one of the axioms of the triple product algebraic structure. At this stage, the generalized model needs to be developed further before it will be ready for applications.

Recall the definitions of a *homogeneous ball* and a *symmetric domain.* Let D be a domain in a real or complex Banach space. We denote by $Aut(D)$ the collection of all automorphisms (one-to-one smooth maps) of D. The exact meaning of "smooth" will vary with the context, but it will always mean either projective (preserving linear segments), conformal (preserving angles) or complex analytic. The unit ball D in a Banach space is one example of a bounded domain. It is called *homogeneous* if for any two points $z, w \in D$, there is an automorphism $\varphi \in Aut(D)$ such that $\varphi(z) = w$. A domain D is called *symmetric* if for any element $a \in D$, there is a symmetry $s_a \in Aut(D)$ which fixes only the point a. Any bounded symmetric domain can be realized as a homogeneous ball in a Banach space.

The theory of bounded symmetric domains as mathematical objects in their own right is highly developed (see [52], [62], [68] and [69]). However, these works are written at a high mathematical level and contain no physical applications.

The current text completely changes this situation. Not only do we develop the theory of homogeneous balls and bounded symmetric domains and their algebraic structure *informally*, but also our primary goal is to show how to construct an appropriate domain to model a given law of physics. The research physicist, and even the graduate student, can walk away with both an understanding of these domains *and* the ability to construct his own homogeneous balls. After seeing our new methodology applied to Special Relativity and Quantum Mechanics, the reader should be able to extrapolate our techniques to his own areas of interest.

In Chapter 1, we show how the principle of relativity leads to a symmetry on the space-time continuum. From this symmetry alone, we derive the Lorentz transformations and show that the set D_v of all relativistically admissible velocities is a homogeneous ball and a bounded symmetric domain with respect to the group $Aut_p(D_v)$ of projective automorphisms. We derive the formula for Einstein velocity addition and explore its geometric properties. We study the Lie algebra $aut_p(D_v)$ and show that relativistic dynamics is described by elements of this algebra. This observation provides an efficient tool for solving relativistic dynamic equations, regardless of initial conditions. As an example, we obtain explicit solutions for the relativistic evolution equation for a charged particle in an electric field E, a magnetic field B and an electromagnetic field E, B in which E and B are parallel.

In Chapter 2, we show that the ball D_s of all relativistically admissible *symmetric* velocities is a bounded symmetric domain with respect to the group $Aut_c(D_s)$ of conformal automorphisms and is a Cartan factor of type 4, called the *spin factor.* This enables us to express the non-commutativity and the non-associativity of Einstein velocity addition as well as the non-transitivity of parallelism among inertial frames in Special Relativity. The

Lie algebra $aut_c(D_s)$ is described in terms of the spin triple product. We describe relativistic evolution using elements of $aut_c(D_s)$. Utilizing the fact that the evolution equation for symmetric velocities of a charged particle in a constant, uniform electromagnetic field E, B, with $E \cdot B = 0$, becomes a one-dimensional complex analytic differential equation, we obtain *explicit* solutions for this evolution.

In Chapter 3, we study the complex spin factor, which is the complex extension of the conformal ball from the previous chapter. The natural basis in this space satisfies a triple product analog of the Canonical Anticommutation Relations. We derive a spectral decomposition for elements of this factor and then represent it geometrically. The two types of tripotents (building blocks of the triple product) determine a duality on this object. This duality is crucial in obtaining different representations of the Lorentz group on the spin factor. The three-dimensional complex spin factor efficiently represents the electromagnetic field, and the Lorentz group acts on it by linear operators defined directly by the triple product. We show that the properties of the field are related to the algebraic structure of its representation.

The four-dimensional complex spin factor admits several representations of the Lorentz group. The operators representing the generators of this group belong to a spin factor of dimension 6. If we use the representation provided by one type of tripotents, we obtain the usual representation of this group on four-vectors. These four-vectors form the invariant subspaces of the spin factor under this representation. If we switch the representation to the second type, the invariant subspaces are the Dirac bispinors with the proper action of the Lorentz group on them. This reveals the connection between the spin 1 and the spin 1/2 representations.

In Chapter 4, we study classical homogeneous unit balls of subspaces of operators on a Hilbert space. Since these operators are not necessarily self-adjoint, we first study some relevant results for non–self-adjoint operators. Based on ideas from transmission line theory, we show that such a ball is a symmetric domain with respect to the analytic automorphisms. Here we study the connection between the geometric properties of such domains and their JC^*-triple structure.

Chapter 5 consists of general results about homogeneous unit balls, bounded symmetric domains, and the Jordan triple product associated with them. Since these domains are homogeneous with respect to the analytic maps on a complex Banach space, we introduce and study some properties of such maps. From the study of the Lie group of analytic automorphisms of a bounded domain and its Lie algebra, we derive the Jordan triple product associated to the domain. We study the Peirce decomposition (which occurs also in earlier chapters) on JB^*-triples and their duals. We explore how the geometry inherited by the state space from the measuring process allows one to define grid bases on the set of observables. This justifies the use of homogeneous balls and bounded symmetric domains in modeling quantum mechanical phenomena.

Chapter 6 includes a complete classification of atomic JB^*-triples. We show how to build convenient bases, called grids, for such spaces. These grids are constructed from basic elements of the triple structure which may be interpreted as compatible observables. These grids span the full non-commutative object. Our methodology reveals why there are six different fundamental domains (called factors) for the same algebraic structure. This explains how apparently unrelated models in physics, corresponding to different types of factors, can have common roots. Furthermore, the mystery of the occurrence of two exceptional factors of dimensions 16 and 27 is explained.

In this book, the reader will find the answer to the following questions:

1. Does the principle of relativity imply the existence of an invariant speed and the preservation of an interval? (Answer: Section 1.2)
2. Why is there time contraction in the transformations from inertial system K to K', while space contraction is obtained in the transformations from K' to K? (Answer: Section 1.1.4)
3. The relative velocity between two inertial systems can be considered as a linear map between time displacement and space displacement. What is the adjoint of this map? (Answer: end of Section 1.2)
4. What geometry is preserved in the transformation of the ball of relativistically admissible velocities from one inertial system to another? (Answer: Section 1.4)
5. What is the connection between the relativistic dynamic equation and the Lie algebra of the velocity transformations? (Answer: Section 1.5)
6. If one has found a solution to the relativistic dynamic equation with a given initial condition, how can one obtain a solution which satisfies a different initial condition? (Answer: Section 1.5.5)
7. The relativistic evolution equation in the plane is not analytic. How can it be made analytic? What are the analytic solutions for a constant field in this case? (Answer: Sections 2.5 and 2.6)
8. How are the Canonical Anticommutation Relations related to the basis in a spin factor? (Answer: Section 3.1.2)
9. How can n Canonical Anticommutation Relations be represented in a space of dimension $2n$ (and not the usual space of dimension 2^n)? (Answer: Section 3.1.2)
10. What is the group of automorphisms of the spin factor? (Answer: Section 3.1.3)
11. Why, in quantum mechanics, do we use expressions like $a = \hat{x} + i\hat{p}_x$ and $J_+ = J_x + iJ_y$? (Answer: Section 3.3.7)
12. How can one represent the transformations of the electromagnetic field strength as operators of the triple product in the spin factor? (Answer: Section 3.5.4)
13. How can one represent four-vectors and Dirac bispinors on the same object? What is the relationship between these two representations? (Answer: Sections 3.5 and 3.6)

14. How can non–self-adjoint operators produce real numbers (similar to the eigenvalues of self-adjoint operators) and filtering projections? (Answer: Section 4.1)
15. Most balls of spaces of operators on a Hilbert space are homogeneous with respect to analytic maps. How can signal transformations in a lossless transmission line be used to demonstrate this homogeneity? (Answer: Section 4.2)
16. How can one derive an algebraic product from the geometry of a bounded homogeneous domain? (Answer: Section 5.3.5)
17. What is the algebraic non-commutative structure built on geometry alone? (Answer: Sections 5.3.5 and 5.4)
18. Can the homogeneity of the ball of observables be derived from the geometry of the state space induced by the measuring process? (Answer: section 5.7)
19. Why are there exactly six different types of bounded symmetric domains, or equivalently, JB^*-triple factors? (Answer: Section 6.3.1)
20. What is the principle difference between the spin domain and the domains in spaces of operators? (Answer: Section 6.3)
21. What is the bridge between the classical and the exceptional domains? (Answer: Sections 6.2.3 and 6.3.7)

For the most part, this book represents the results of more than 30 years of the author's research. During this period, the theory of homogeneous, bounded and unbounded symmetric domains has progressed significantly. We do not cover here all major topics of this area, but concentrate more on the aspects that seem currently ripe for physical applications.

I want to thank my research collaborators: Jonathan Arazy, with whom we started this project during our Ph.D. program, Bernard Russo with whom we worked together for more than 20 years, Thomas Barton, Truong Dang, Ari Naimark and Yuriy Gofman. Tzvi Scarr assisted me in writing this book. I want to thank Alexander Friedman and Hadar Crown for technical assistance. I want to thank Uziel Sandler, Mark Semon and Alex Gelman for helpful comments. This work was supported in part by a research grant from the Jerusalem College of Technology.

The book is dedicated to my wife Rachel, for without her encouragement over the last 30 years, I would not have been able to achieve the results presented in the book.

Jerusalem College of Technology

Yaakov Friedman
June 2004

List of Figures

List of Tables

1 Relativity based on symmetry

In this chapter, we will derive the Lorentz transformations without assuming the constancy of the speed of light. We will use only the principle of special relativity and the symmetry associated with it. We will see that this principle allows only Galilean or Lorentz space-time transformations between two inertial systems. In the case of the Lorentz transformations, we obtain the conservation of an interval and a certain speed. From known experiments, this speed is c, the speed of light in a vacuum.

The Einstein velocity-addition formula is also obtained. From this, it follows that the ball of all relativistically admissible velocities is a bounded symmetric domain. The Lie algebra of the automorphism group of this domain consists of the generators of boosts and rotations. The relativistic dynamics and the dynamics of a charged particle in an electromagnetic field are given by elements of this Lie algebra.

Our methodology in special relativity is outlined in the following steps, which we apply to two inertial systems:

Step 1 Choice of the parameters for the purpose of obtaining simpler (ideally, *linear*) transformations between the two systems
Step 2 Identification of *symmetry* in the basic principle of the area of application (in this case, the principle of special relativity)
Step 3 Choice of reference frames which preserve the symmetries
Step 4 Choice of inputs and outputs which reflect the description of the system
Step 5 Derivation of the explicit form of the symmetry operator
Step 6 Identification of invariants
Step 7 Construction of an appropriate (bounded) symmetric domain for the area of application
Step 8 Derivation of the equation of evolution based on the algebraic structure of the Lie algebra of the domain

1.1 Space-time transformation based on relativity

In this section, we derive the space-time transformation between two inertial systems, using only the isotropy of space and symmetry, both of which follow from the principle of relativity. The transformation will be defined uniquely up to a constant e, which depends only on the process of synchronization of clocks inside each system. If $e = 0$, the transformations reduce to Galilean.

1.1.1 Step 1 - Choice of the Parameters

We begin with two "systems", for example, an airplane flying at 30,000 feet and an observer standing on the ground. We assume that the airplane is flying in uniform motion with constant velocity, that there is no turbulence, etc. Passengers in the airplane feel themselves at rest. When they put their cup of coffee on the fold-down tray in front of them, it doesn't slide or move. When they drop a penny, it falls "straight" to the floor. This is a manifestation of the principle of special relativity, which states that (Pauli [59], page 4) "there exists a triply infinite set of reference systems moving rectilinearly and uniformly relative to one another, in which the phenomena occur in an identical manner."

Typically, there are events which are observable from both systems. Lake Michigan can be observed by both Observer A who is standing on its edge and also by Observer B who is flying over it (Figure 1.1). Certainly, Observers A and B will not observe Lake Michigan in the same way. To Observer A, Lake Michigan is next to him and standing still, while to Observer B, it is below him and moving. Each observer sets up a system of axes and scales in order to measure the position in space and time of each event. Imagine a kingfisher flying above the lake. It swoops down, snatches a fish from the lake, and takes off again. Let's take the snatch as our event. Each observer has a different set of four numbers to describe the location of this event in space-time. The connection between these two sets of four numbers is the *space-time transformation* between the two systems.

Why have we chosen space and time as the parameters with which to describe events? Why not velocity and time? Why not position and momentum? Why is the space-time description more convenient for transformations between inertial systems?

The advantage is *linearity*. Newton's First Law states that an object moves with constant velocity if there are no forces acting on it or if the sum of all forces on it is zero. Such a motion is called *free motion* and is described by straight lines in the space-time continuum, as shown in Figure 1.2. Conversely, any line (except lines with constant t) in the space-time continuum represents free motion. A system is called an *inertial system* if an object moves with constant velocity when there are no forces acting on it. By the definition of an inertial system, free motion will be observed as free motion in *any* inertial system. This means that the space-time transformations will map lines to lines. Thus the space-time description of events leads to *linear* transformations. For an example of how the choice of parameters with which to describe events affects the linearity of the transformations, see Figure 1.3.

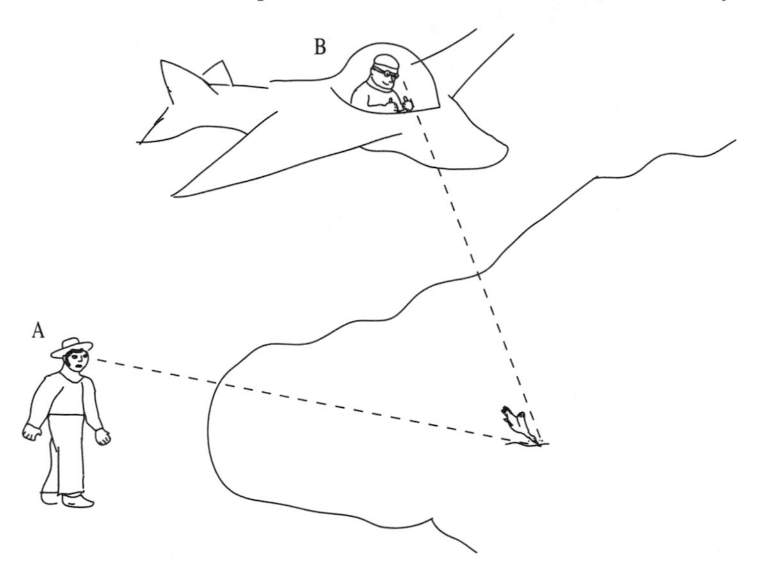

Fig. 1.1. The space-time transformation between two systems is the connection between the space-time coordinates of the same event (snatches of a fish) observed and described by two observers in the two systems. Above, Observer A is standing (at rest) on the edge of Lake Michigan, and Observer B is flying over it with constant velocity.

We restrict ourself to inertial systems with the same space origin at time $t = 0$. By a well-known theorem in mathematics, a transformation between

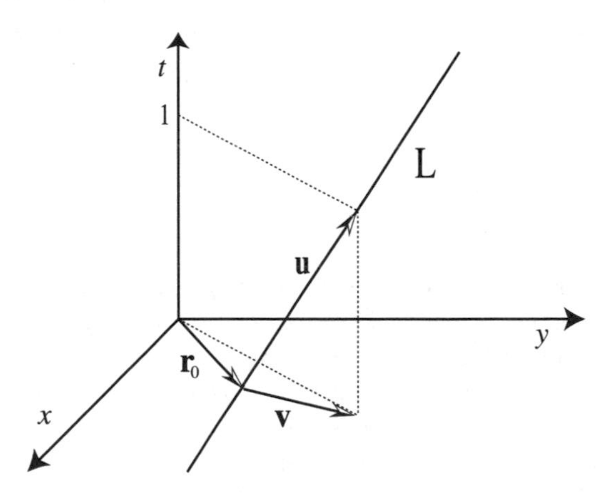

Fig. 1.2. Lines in space-time and free motion. Two space coordinates, x and y, and time are displayed. The line L intersects the plane $t = 0$ at $\mathbf{r}_0$, the position of an object at time $t = 0$. The direction of L is given by a vector $\mathbf{u} = (1, \mathbf{v})$, where $\mathbf{v}$ is the constant velocity of the object. The line $L = \{(t, \mathbf{r}_0 + \mathbf{v}t) : t \in R\}$ is the world-line of this free motion.

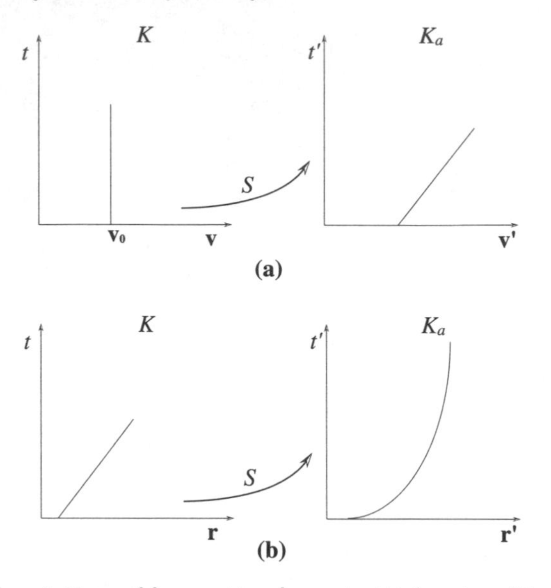

Fig. 1.3. Two descriptions of free motion for an inertial system K and a system K_a whose acceleration with respect to K is $\mathbf{a}$. (a) In the velocity-time description, the constant velocity $\mathbf{v}_0$ in K is represented by a line $L = \{(t, \mathbf{v}_0) : t \in R\}$ in K and also by a line $L = \{(t, \mathbf{v}_0 - \mathbf{a}t) : t \in R\}$ in K_a. (b) In the space-time description, the constant velocity $\mathbf{v}_0$ in K is represented by a line $L = \{(t, \mathbf{r}_0 + \mathbf{v}_0 t) : t \in R\}$ in K, while in K_a, it is represented by a parabola $(t, \mathbf{r}_0 + \mathbf{v}_0 t - 0.5\mathbf{a}t^2) : t \in R\}$. Hence, the space-time transformation between K and K_a cannot be linear.

two vector spaces which maps lines to lines and the origin to the origin is linear. Thus, the space-time transformation between our two systems is a *linear* map. After choosing space axes in each system, we can represent this transformation by a matrix.

1.1.2 Step 2 - Identification of symmetry inherent in principle of special relativity

Albert Einstein formulated the *principle of special relativity* ([21], p.25): "If K is an inertial system, then every other system K' which moves uniformly and without rotation relatively to K, is also an inertial system; the laws of nature are in concordance for all inertial systems." Observation of the same event from these two systems defines the space-time transformation between the systems. By the principle of special relativity, this transformation will depend only on the choice of the space axes, the measuring devices (consisting of rods and clocks) and the relative position in time between these systems. The relative position in time between two inertial systems is described by their *relative velocity*. We denote by $\mathbf{v}$ the relative velocity of K' with respect to K and by $\mathbf{v}'$ the relative velocity of K with respect to K'. If we choose the measuring devices in each system to be the same and choose the axes in such a way that the coordinates of $\mathbf{v}$ are equal to the coordinates of $\mathbf{v}'$, then the space-time transformation S from K to K' will be equal to the space-time transformation S' from K' to K. Since, in general, $S' = S^{-1}$, in this case we will have $S^2 = I$. Such an operator S is called a *symmetry*. Thus, the principle of special relativity implies that with an appropriate choice of axes and measuring devices, the space-time transformation S between two inertial systems is a symmetry.

1.1.3 Step 3 - Choice of reference frames

Following Einstein, the space axes in special relativity are chosen as in Figure 1.4. If we assume that the interval $ds^2 = (cdt)^2 - d\mathbf{r}^2$ is conserved, the

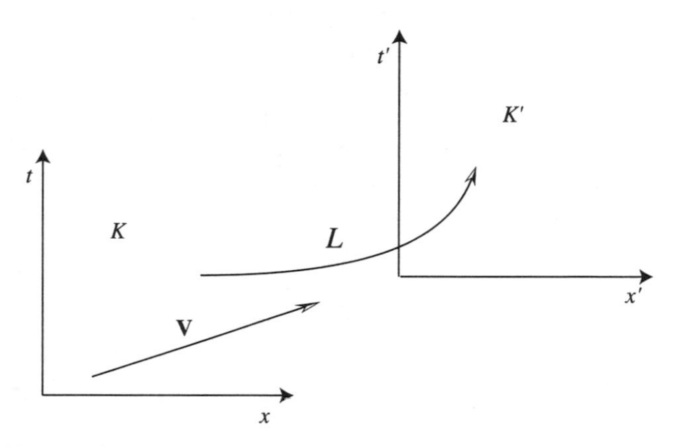

Fig. 1.4. The usual Lorentz space-time transformations between two inertial systems K and K', moving with relative velocity $\mathbf{v}$. The space axes are chosen to be parallel. The Lorentz transformation L transforms the space-time coordinates $(t, \mathbf{r})$ in K of an event to the space-time coordinates $(t', \mathbf{r}')$ in K' of the same event.

resulting space-time transformation between systems is called the *Lorentz transformation.* In the case $\mathbf{v} = (v, 0, 0)$, the Lorentz transformation L is given by

$$\begin{aligned} t' &= \frac{1}{\sqrt{1-v^2/c^2}}\left(t - \frac{vx}{c^2}\right), \\ x' &= \frac{1}{\sqrt{1-v^2/c^2}}(x - vt), \\ y' &= y, \\ z' &= z. \end{aligned} \tag{1.1}$$

Note that the assumption that the system K' is moving with velocity $\mathbf{v}$ with respect to the system K implies that system K is moving with velocity $-\mathbf{v}$ with respect to K'. And this apparently minor lack of symmetry means that the Lorentz transformation L' from system K' to system K will be different from L. In fact, we have

$$\begin{aligned} t &= \tfrac{1}{\sqrt{1-v^2/c^2}}(t' + \tfrac{vx'}{c^2}), \\ x &= \tfrac{1}{\sqrt{1-v^2/c^2}}(x' + vt'), \\ y &= y', \\ z &= z'. \end{aligned} \tag{1.2}$$

We would like to arrange things so that the two transformations L and L' are the *same*! It certainly would help if K were *also* moving with velocity $\mathbf{v}$ with respect to system K'.

We will synchronize the two systems by observing events from each system and comparing the results. System 1 begins with the following configuration. There is a set of three mutually orthogonal space axes and a system of rods. In this way, each point in space is associated with a unique vector in R^3. In addition, there is a clock at each point in space, and all of the clocks are synchronized to each other by some synchronization procedure. System 2 has the same setup, only we do not assume that the rods of system 1 are identical to the rods of system 2, nor do we assume that the clock synchronization procedure in system 2 is the same as that of system 1.

First, we synchronize the origins of the frames. Produce an event E_0 at the origin O of system 1 at time $t = 0$ on the system 1 clock positioned at O. This event is observed at some point O' in system 2, and the system 2 clock at O' shows some value $t' = t'_0$. Translate the origin of system 2 to the point O' (without rotating). Subtract t'_0 from the system 2 clock at O'. Synchronize all of the system 2 clocks to this clock. This completes the synchronization of the origins.

Next, we will adjust the x-axis of each system. Note that system 2 is moving with some (perhaps unknown) constant velocity $\mathbf{v}$ with respect to system 1 and that the origin O' of system 2 was at the point O of system 1 at time $t = 0$. Therefore, the point O' will always be on the line $\mathbf{v}t$ in system 1. Rotate the axes in system 1 so that the new negative x-axis coincides with the ray $\{\mathbf{v}t : t > 0\}$. Similarly, system 1 is moving with some constant velocity $\mathbf{w}$ with respect to system 2, and the origin O of system 1 was at the point O' of system 2 at time $t' = 0$. Therefore, the point O will always be on the line $\mathbf{w}t$ in system 2. Rotate the axes in system 2 so that the new negative x'-axis coincides with the ray $\{\mathbf{w}t : t > 0\}$. The two x-axes now coincide as lines and point in opposite directions. We are finished manipulating the axes and clocks of system 1 and will henceforth refer to system 1 as the inertial frame K. However, it still remains to manipulate system 2, as we must adjust the y'- and z'-axes of system 2 to be parallel and oppositely oriented to the corresponding axes of K.

To adjust the y'-axis of system 2, produce an event E_1 at the point $\mathbf{r} = (0, 1, 0)$ of K. This event is observed in system 2 at some point $\mathbf{r}'$. Rotate the space axes of system 2 around the x'-axis so that $\mathbf{r}'$ will lie in the new

x'-y' plane and have a negative y' coordinate y'_1. After this rotation, the z-axis of K and the z'-axis of system 2 will be parallel. We need to make sure that they have opposite orientations. Produce an event E_2 at the point $\mathbf{r} = (0, 0, 1)$ of K. This event is observed in system 2 at some point $\mathbf{r}'$. If the z' coordinate of $\mathbf{r}'$ is positive, reverse the direction of the z'-axis. This completes the adjustment of the space axes of the two systems. See Figure 1.5.

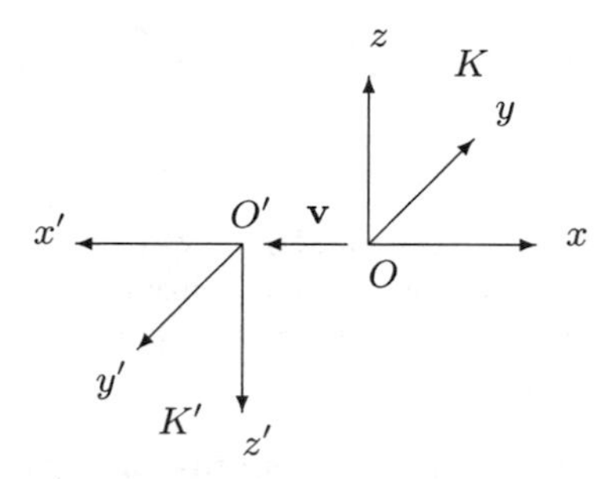

Fig. 1.5. Two symmetric space reference frames. The relative velocity of the inertial system K' with respect to K is $\mathbf{v}$. The coordinates of $\mathbf{v}$ in K are equal to the coordinates (in K') of the relative velocity of the system K with respect to K'.

It remains to redefine the space and time units of system 2 to match those of K. The new space unit of system 2 is defined to be y'_1 times the previous space unit. In order to adjust the time unit of system 2, we will measure the speed $|\mathbf{v}|$ of system 2 with respect to K and the speed $|\mathbf{v}'|$ of K with respect to system 2. To calculate $|\mathbf{v}|$, produce an event E_3 at O' at any time $t' > 0$. This event is observed in K at some point $\mathbf{r} = (x_0, 0, 0)$, and the clock at this point shows time t_0. The relative speed of system 2 with respect to K is $|\mathbf{v}| = |x_0|/t_0$. The calculation of $|\mathbf{v}'|$ is symmetric. Produce an event E_4 at O at any time $t > 0$. This event is observed in system 2 at some point $\mathbf{r}' = (x'_0, 0, 0)$, and the clock at this point shows time t'_0. The relative speed of K with respect to system 2 is $|\mathbf{v}'| = |x'_0|/t'_0$. Finally, the time unit in system 2 is chosen as $|\mathbf{v}'|/|\mathbf{v}|$ times the previous unit. With this choice of units, the speeds $|\mathbf{v}|$ and $|\mathbf{v}'|$ are equal. System 2 will henceforth be called K'.

The transformations from system K to system K' will now be mathematically *identical* to the transformations from system K' to system K. In other words, the space-time transformation S from system K to system K' will be a symmetry operator.

The space-time coordinates of K and K' will be denoted $\begin{pmatrix} t \\ \mathbf{r} \end{pmatrix}$ and $\begin{pmatrix} t' \\ \mathbf{r}' \end{pmatrix}$, respectively. These coordinates will be considered as a 4×1 matrix. By the above synchronization procedure, the frames have the same origin and the two clocks at each origin are synchronized at time $t = 0$. Moreover, the space axes are reversed as in Figure 1.5 . Note that with this choice of axes, the velocity coordinates of O' in K are equal to the velocity coordinates of O in K'. Thus, the transformation is fully symmetric with respect to K and K' (see Figure 1.6). We will denote the space-time transformation from K to K'

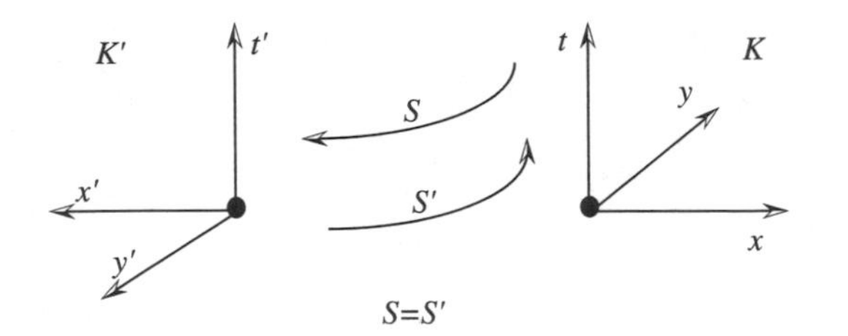

Fig. 1.6. The time t and two space axes x and y of systems K and K' are displayed. With our choice of the axes, the space-time map S from K to K' is identical to the space-time map S' from K' to K, and, thus, S is a symmetry.

by $S_{\mathbf{v}}$, since it is a symmetry and depends only on the velocity $\mathbf{v}$ between the systems.

1.1.4 Step 4 - Choice of inputs and outputs

The space-time transformation between two inertial systems can be considered as a "two-port linear black box" transformation with two inputs and two outputs. There are two ways to define the inputs and outputs for such a transformation.

Cascade connection

The first one, called the *cascade connection*, takes time and space of one of the systems, say $\begin{pmatrix} t \\ \mathbf{r} \end{pmatrix}$ of K, as input, and gives time and space of the second system, say $\begin{pmatrix} t' \\ \mathbf{r}' \end{pmatrix}$ of K', as output (see Figure 1.7) [1]

The cascade connection is the one usually used in special relativity.

We represent the linear transformation induced by the cascade connection by a 4×4 matrix E, which we decompose into four block matrix components E_{ij}, as follows:

$$\begin{pmatrix} t' \\ \mathbf{r}' \end{pmatrix} = E \begin{pmatrix} t \\ \mathbf{r} \end{pmatrix} = \begin{pmatrix} E_{11} & E_{12} \\ E_{21} & E_{22} \end{pmatrix} \begin{pmatrix} t \\ \mathbf{r} \end{pmatrix}. \tag{1.3}$$

[1] We use a circle instead of the usual box notation in order that the connection between any two ports will be displayed *inside* the box (see Figure 1.8).

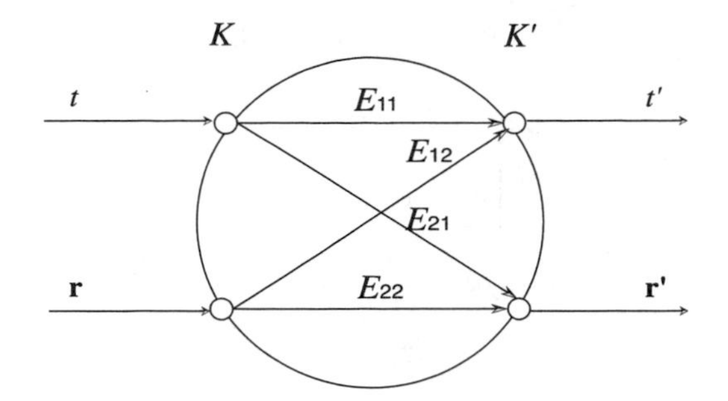

Fig. 1.7. The cascade connection for space-time transformations. The circle represents a black box. One side has two input ports: the time t and the space $\mathbf{r}$ coordinates of an event in system K. The other side has two output ports: the time t' and the space $\mathbf{r}'$ coordinates of the same event in system K'. The linear operators E_{ij} represent the functional connections between the corresponding ports.

To understand the meaning of the blocks, assume that the system K is the airplane. Let t be the time between two events (say crossing two lighthouses) measured by a clock at rest at $\mathbf{r} = 0$ on the airplane. The time difference t' of the same two events measured by synchronized clocks at the two lighthouses (in system K', the earth) will be equal to $t' = E_{11}t$. If we denote the distance between the lighthouses by $\mathbf{r}'$, then $\mathbf{r}' = E_{21}t$, and E_{21} is the so-called *proper velocity* of the plane. Generally, the proper velocity $\mathbf{u}$ of an object (the airplane) in an inertial system is the ratio of the space displacement $d\mathbf{r}$ in this system (the earth) divided by the time interval, called the proper time interval $d\tau$, measured by the clock of the object (on the plane). Thus,

$$\mathbf{u} = \frac{d\mathbf{r}}{d\tau}. \tag{1.4}$$

Hybrid connection

The second type of connection, called the *hybrid connection*, uses time of one of the systems, say t of K, and the space coordinates $\mathbf{r}'$ of the second system K', as input, and gives $\begin{pmatrix} t' \\ \mathbf{r} \end{pmatrix}$ as output (see Figure 1.8). Usually we use relative *velocity* (not relative proper velocity) to describe the relative position between inertial systems. To define the relative position of system K' with respect to K, we consider an event that occurs at O', corresponding to $\mathbf{r}' = 0$, at time t, and express its position $\mathbf{r}$ in K. If we denote by $\mathbf{v}$ the uniform velocity of system K' with respect to K, then

$$\mathbf{r} = \mathbf{v}t. \tag{1.5}$$

Note our use of the hybrid connection. In this section we will use the hybrid connection in order to be consistent with the description of relative position

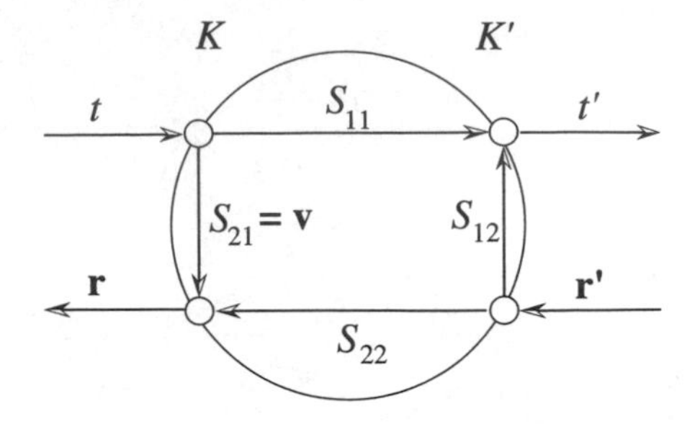

Fig. 1.8. The hybrid connection for space-time transformations. The circle represents a black box. The two input ports are the time t of an event, as measured in system K, and its space coordinates $\mathbf{r}'$, as measured in system K'. The two output ports are the time t' of the same event, calculated in system K', and its space $\mathbf{r}$ coordinates, calculated in K. The linear operators S_{ij} represent the functional connections between the corresponding ports. For instance, to define the map S_{21}, we consider an event that occurs at O', corresponding to input $\mathbf{r}' = 0$, at time t in K. Then $S_{21}t$ represents the space displacement of O' in K during time t, which is, by definition, the relative velocity $\mathbf{v}$ of system K' with respect to system K.

between the systems and because we will be interested later in velocities (rather than proper velocities).

Thus, for the transformation $S_{\mathbf{v}}$, we choose the inputs to be the scalar t, the time of the event in K, and the three-dimensional vector $\mathbf{r}'$ describing the position of the event in K'. Then our outputs are the scalar t', the time of the event in K', and the three-dimensional vector $\mathbf{r}$ describing the position of the event in K. As above with respect to the cascade connection, here we also decompose the 4×4 matrix $S_{\mathbf{v}}$ into block components:

$$\begin{pmatrix} t' \\ \mathbf{r} \end{pmatrix} = S_{\mathbf{v}} \begin{pmatrix} t \\ \mathbf{r}' \end{pmatrix} = \begin{pmatrix} S_{11} & S_{12} \\ S_{21} & S_{22} \end{pmatrix} \begin{pmatrix} t \\ \mathbf{r}' \end{pmatrix} \tag{1.6}$$

(see Figure 1.8).

The transformation between cascade and hybrid connections

Note that the matrices E and $S_{\mathbf{v}}$ describing the space-time transformations between two inertial systems using the cascade and hybrid connections, respectively, are related by some transformation Ψ. To define this transformation, note that equation (1.3) can be rewritten as a scalar equation

$$t' = E_{11}t + E_{12}\mathbf{r} \tag{1.7}$$

and a vector equation

$$\mathbf{r}' = E_{21}t + E_{22}\mathbf{r}. \tag{1.8}$$

The matrix E_{22} is invertible, since from its physical meaning it is one-to-one and onto. By multiplying (1.8) on the left by E_{22}^{-1}, we get

$$\mathbf{r} = -E_{22}^{-1}E_{21}t + E_{22}^{-1}\mathbf{r}'. \tag{1.9}$$

Substituting this expression for $\mathbf{r}$ into (1.7), we get

$$t' = (E_{11} - E_{12}E_{22}^{-1}E_{21})t + E_{12}E_{22}^{-1}\mathbf{r}'. \tag{1.10}$$

Thus

$$\begin{pmatrix} t' \\ \mathbf{r} \end{pmatrix} = \begin{pmatrix} E_{11} - E_{12}E_{22}^{-1}E_{21} & E_{12}E_{22}^{-1} \\ -E_{22}^{-1}E_{21} & E_{22}^{-1} \end{pmatrix} \begin{pmatrix} t \\ \mathbf{r}' \end{pmatrix}, \tag{1.11}$$

implying that

$$\begin{pmatrix} S_{11} & S_{12} \\ S_{21} & S_{22} \end{pmatrix} = \begin{pmatrix} E_{11} - E_{12}E_{22}^{-1}E_{21} & E_{12}E_{22}^{-1} \\ -E_{22}^{-1}E_{21} & E_{22}^{-1} \end{pmatrix}. \tag{1.12}$$

Define a transformation Ψ by

$$\Psi \begin{pmatrix} E_{11} & E_{12} \\ E_{21} & E_{22} \end{pmatrix} = \begin{pmatrix} E_{11} - E_{12}E_{22}^{-1}E_{21} & E_{12}E_{22}^{-1} \\ -E_{22}^{-1}E_{21} & E_{22}^{-1} \end{pmatrix}. \tag{1.13}$$

This transformation is called the *Potapov–Ginzburg transformation.* Then,

$$\begin{pmatrix} S_{11} & S_{12} \\ S_{21} & S_{22} \end{pmatrix} = \Psi \begin{pmatrix} E_{11} & E_{12} \\ E_{21} & E_{22} \end{pmatrix}. \tag{1.14}$$

A symmetric argument shows that

$$\begin{pmatrix} E_{11} & E_{12} \\ E_{21} & E_{22} \end{pmatrix} = \Psi \begin{pmatrix} S_{11} & S_{12} \\ S_{21} & S_{22} \end{pmatrix}. \tag{1.15}$$

It is easy to check that $S_{\mathbf{v}}$ is a symmetry (that is, $S_{\mathbf{v}}^2 = I$) if and only if $E = \Psi(S_{\mathbf{v}})$ is a symmetry.

The meaning of the operators in the hybrid connection

We explain now the meaning of the four linear maps S_{ij}. To define the maps S_{21} and S_{11}, consider an event that occurs at O', corresponding to $\mathbf{r}' = 0$, at time t in K. Then $S_{21}(t)$ expresses the position of this event in K, and $S_{11}(t)$ expresses the time of this event in K'. Obviously, S_{21} describes the relative velocity of K' with respect to K, and

$$S_{21}(t) = \mathbf{v}t, \tag{1.16}$$

while $S_{11}(t)$ is the time shown by the clock positioned at O' of an event occurring at O' at time t in K and is given by

$$S_{11}(t) = \alpha t \tag{1.17}$$

for some constant α.

To define the maps S_{12} and S_{22}, we will consider an event occurring at time $t = 0$ in K in space position $\mathbf{r}'$ in K'. Then $S_{12}(\mathbf{r}')$ will be the time of this event in K', and $S_{22}(\mathbf{r}')$ will be the position of this event in K. Note that $S_{12}(\mathbf{r}')$ is also the time difference of two clocks, both positioned at time $t = 0$ at $\mathbf{r}'$ in K', where the first one was synchronized to the clock at the common origin of the two systems within the frame K', and the second one was synchronized to the clock at the origin within the frame K. Thus S_{12} describes the *non-simultaneity* in K' of simultaneous events in K with respect to their space displacement in K', following from the difference in synchronization of clocks in K and K'. Since S_{12} is a linear map from R^3 to R, it is given by

$$S_{12}(\mathbf{r}') = \mathbf{e}^T\mathbf{r}' \tag{1.18}$$

for some vector $\mathbf{e} \in R^3$, where $\mathbf{e}^T$ denotes the transpose of $\mathbf{e}$. Note that $\mathbf{e}^T\mathbf{r}'$ is the dot product of $\mathbf{e}$ and $\mathbf{r}'$. See Figure 1.9 for the connection between the time of events in two inertial systems.

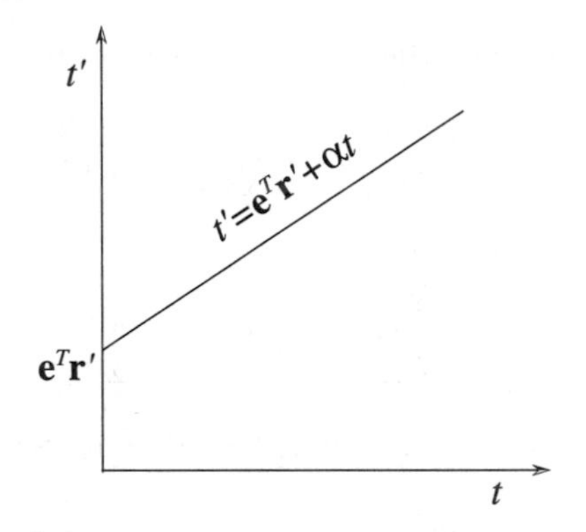

Fig. 1.9. The times t' and t of an event at space point $\mathbf{r}'$ in system K'. The difference in timings is caused both by the difference in the rates of clocks (time slowdown) in each system and by the different synchronization of the clocks positioned at different space points.

Finally, the map S_{22} describes the transformation of the space displacement in K of simultaneous events in K with respect to their space displacement in K', and it is given by

$$S_{22}(\mathbf{r}') = A\mathbf{r}' \tag{1.19}$$

for some 3×3 matrix A.

1.1.5 Step 5 - Derivation of the explicit form of the symmetry operator

Our black box transformation can now be described by a 4×4 matrix $S_{\mathbf{v}}$ with block matrix entries from (1.16), (1.17), (1.18) and (1.19):

$$\begin{pmatrix} t' \\ \mathbf{r} \end{pmatrix} = S_{\mathbf{v}} \begin{pmatrix} t \\ \mathbf{r}' \end{pmatrix} = \begin{pmatrix} \alpha & \mathbf{e}^T \\ \mathbf{v} & A \end{pmatrix} \begin{pmatrix} t \\ \mathbf{r}' \end{pmatrix}. \tag{1.20}$$

If we now interchange the roles of systems K and K', we will get a matrix $S'_{\mathbf{v}}$:

$$\begin{pmatrix} t \\ \mathbf{r}' \end{pmatrix} = S'_{\mathbf{v}} \begin{pmatrix} t' \\ \mathbf{r} \end{pmatrix} = \begin{pmatrix} \alpha' & \mathbf{e}'^T \\ \mathbf{v}' & A' \end{pmatrix} \begin{pmatrix} t' \\ \mathbf{r} \end{pmatrix}. \tag{1.21}$$

But the principle of relativity implies that switching the roles of K and K' is nonrecognizable. Hence

$$\alpha = \alpha', \;\; \mathbf{e}^T = \mathbf{e}'^T, \;\; \mathbf{v} = \mathbf{v}', \;\; A = A'.$$

By combining (1.20) and (1.21), we get $S_{\mathbf{v}}^2 = I$, implying that $S_{\mathbf{v}}$ is a symmetry operator. Hence,

$$\begin{pmatrix} \alpha & \mathbf{e}^T \\ \mathbf{v} & A \end{pmatrix} \begin{pmatrix} \alpha & \mathbf{e}^T \\ \mathbf{v} & A \end{pmatrix} = \begin{pmatrix} 1 & \mathbf{0}^T \\ \mathbf{0} & I \end{pmatrix}, \tag{1.22}$$

where I is the 3×3 identity matrix. Equation (1.22) is equivalent to the following four equations:

$$\alpha^2 + \mathbf{e}^T \mathbf{v} = 1, \tag{1.23}$$

$$\alpha \mathbf{e}^T + \mathbf{e}^T A = \mathbf{0}^T, \tag{1.24}$$

$$\alpha \mathbf{v} + A\mathbf{v} = \mathbf{0}, \tag{1.25}$$

and

$$\mathbf{v}\,\mathbf{e}^T + A^2 = I. \tag{1.26}$$

Note that since space is isotropic and the configuration of our systems has one unique divergent direction $\mathbf{v}$, the vector $\mathbf{e}$ is collinear to $\mathbf{v}$. Thus

$$\mathbf{e} = e\mathbf{v} \tag{1.27}$$

for some constant e. Since the choice of direction of the space coordinate system in the frame is free, the constant e depends only on $|\mathbf{v}|$ and not on

$\mathbf{v}$. Finally, from (1.18) and (1.27), it follows that this constant has units $(\text{length}/\text{time})^{-2}$.

By using (1.23) and (1.27), we obtain $\alpha = \pm\sqrt{1 - e|\mathbf{v}|^2}$. To choose the appropriate sign for α, we use the fact that the transformation is continuous in $\mathbf{v}$ and that for $\mathbf{v} = 0$ we have $\alpha = 1$. Thus,

$$\alpha = \sqrt{1 - e|\mathbf{v}|^2}. \tag{1.28}$$

Note that by use of (1.27), the operator $\mathbf{v}\mathbf{e}^T$ acts on an arbitrary vector $\mathbf{u} \in R^3$ as follows:

$$\mathbf{v}\,\mathbf{e}^T\mathbf{u} = e\mathbf{v}\,\mathbf{v}^T\mathbf{u} = e|\mathbf{v}|^2 \frac{\mathbf{v}^T\mathbf{u}}{|\mathbf{v}|^2}\mathbf{v} = e|\mathbf{v}|^2 P_{\mathbf{v}}\mathbf{u}, \tag{1.29}$$

where $P_{\mathbf{v}}$ denotes the orthogonal projection onto the direction of $\mathbf{v}$. Now from (1.26), we get

$$A^2 = I - e|\mathbf{v}|^2 P_{\mathbf{v}} = \alpha^2 P_{\mathbf{v}} + (I - P_{\mathbf{v}}).$$

Using once more the continuity in $\mathbf{v}$, we get

$$S_{22} = A = -\alpha P_{\mathbf{v}} - (I - P_{\mathbf{v}}). \tag{1.30}$$

Thus, the space-time transformation between the two inertial frames K and K' is

$$\begin{pmatrix} t' \\ \mathbf{r} \end{pmatrix} = S_{\mathbf{v}} \begin{pmatrix} t \\ \mathbf{r}' \end{pmatrix} = \begin{pmatrix} \alpha & e\mathbf{v}^T \\ \mathbf{v} & -\alpha P_{\mathbf{v}} - (I - P_{\mathbf{v}}) \end{pmatrix} \begin{pmatrix} t \\ \mathbf{r}' \end{pmatrix}, \tag{1.31}$$

with α defined by (1.28) as $\alpha = \sqrt{1 - e|\mathbf{v}|^2}$ (see Figure 1.10).

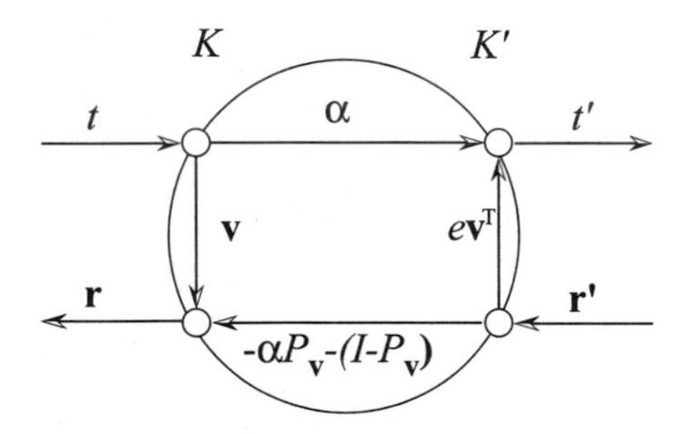

Fig. 1.10. The hybrid connection for space-time transformations. The circle represents a black box. The two input ports are the time t of an event, as measured in system K, and its space coordinates $\mathbf{r}'$, as measured in system K'. The two output ports are the time t' of the same event, calculated in system K', and its space $\mathbf{r}$ coordinates, calculated in K. The explicit form of the linear operators representing the functional connections between the corresponding ports is shown.

If we choose $\mathbf{v} = (v, 0, 0)$ and write $\mathbf{r} = (x, y, z)$ and $\mathbf{r}' = (x', y', z')$, then the above matrix becomes

$$S_{\mathbf{v}} = \begin{pmatrix} \alpha & ev & 0 & 0 \\ v & -\alpha & 0 & 0 \\ 0 & 0 & -1 & 0 \\ 0 & 0 & 0 & -1 \end{pmatrix}. \tag{1.32}$$

To compare this result with the usual space-time transformations in special relativity, we have to recalculate our result for the cascade connection. To obtain $\begin{pmatrix} t' \\ \mathbf{r}' \end{pmatrix}$ as a function of $\begin{pmatrix} t \\ \mathbf{r} \end{pmatrix}$, we use the map Ψ from (1.15) and obtain

$$\begin{pmatrix} t' \\ \mathbf{r}' \end{pmatrix} = \Psi(S_{\mathbf{v}}) \begin{pmatrix} t \\ \mathbf{r} \end{pmatrix} = \gamma(\mathbf{v}) \begin{pmatrix} 1 & -e\mathbf{v}^T \\ \mathbf{v} & -P_{\mathbf{v}} - \alpha(I - P_{\mathbf{v}}) \end{pmatrix} \begin{pmatrix} t \\ \mathbf{r} \end{pmatrix}, \tag{1.33}$$

where $\gamma(\mathbf{v}) = 1/\sqrt{1 - |\mathbf{v}|^2/c^2}$. This defines an explicit form for the operators of the space-time transformations using the cascade connection (see Figure 1.11). For the particular case $\mathbf{v} = (v, 0, 0)$, we get

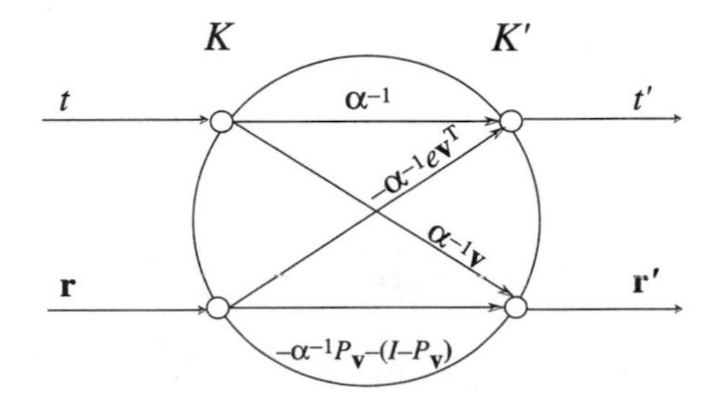

Fig. 1.11. The cascade connection for space-time transformations. The circle represents a black box. The two input ports are the time t of an event and its space coordinates $\mathbf{r}$, as measured in system K, and the two output ports are the time t' of the same event and its space $\mathbf{r}'$ coordinates, calculated in system K'. The explicit form of the linear operators representing the functional connections between the corresponding ports is shown.

$$\begin{aligned} t' &= \gamma(\mathbf{v})(t - evx), \\ x' &= \gamma(\mathbf{v})(vt - x), \\ y' &= -y, \\ z' &= -z, \end{aligned} \tag{1.34}$$

which are the usual Lorentz transformations (with space reversal) when $e = 1/c^2$. If $e = 0$, then $\alpha = 1$, and the transformations are the Galilean transformations.

1.2 Step 6 - Identification of invariants

In this section and the two which follow, we will show that the principle of relativity alone implies that an interval is conserved, that the ball D_v of all relativistically admissible velocities is conserved and that D_v is a bounded symmetric domain with respect to the projective maps. The symmetry of this ball, resulting from the above space-time transformations, determines the so-called *symmetric velocity.*

1.2.1 Eigenvectors of $S_{\mathbf{v}}$

As mentioned above, the space-time transformation between the systems K and K' is a symmetry transformation. Such a symmetry is a reflection with respect to the set of fixed points. We now want to determine the events fixed by $S_{\mathbf{v}}$, meaning that in both systems the event will have the same coordinates. It follows from (1.31) that such an event must satisfy

$$\begin{pmatrix} t' \\ \mathbf{r} \end{pmatrix} = \begin{pmatrix} t \\ \mathbf{r}' \end{pmatrix} = \begin{pmatrix} \alpha & e\mathbf{v}^T \\ \mathbf{v} & -\alpha P_{\mathbf{v}} - (I - P_{\mathbf{v}}) \end{pmatrix} \begin{pmatrix} t \\ \mathbf{r}' \end{pmatrix}. \tag{1.35}$$

This can be rewritten as

$$t = \alpha t + e\langle \mathbf{v} | \mathbf{r}' \rangle \tag{1.36}$$

and

$$\mathbf{r}' = \mathbf{v}t - (\alpha P_{\mathbf{v}} + (I - P_{\mathbf{v}}))\mathbf{r}'. \tag{1.37}$$

By multiplying the previous equation by $(I - P_{\mathbf{v}})$, we get $(I - P_{\mathbf{v}})\mathbf{r}' = 0$. Hence, $(1 + \alpha)\mathbf{r}' = \mathbf{v}t$, implying that

$$\frac{\mathbf{r}'}{t} = \frac{\mathbf{v}}{1+\alpha} = \frac{\mathbf{r}}{t} := \mathbf{w}_1. \tag{1.38}$$

Note that if $(t, \mathbf{r}')$ satisfies (1.38), then by use of (1.28), it also satisfies (1.36). The meaning of this is that all of the events fixed by the transformation $S_{\mathbf{v}}$ lie on a straight world-line through the origin of both frames at time $t = 0$, moving with velocity $\mathbf{w}_1$ defined by (1.38) in both frames (see Figures 1.13 and 1.14).

The velocity $\mathbf{w}_1$ will be called the *symmetric velocity* between the systems K and K'. Note that $\mathbf{w}_1$ is equal to its hybrid velocity $\mathbf{r}'/t$, defined above as the space displacement measured in one inertial frame divided by the time interval measured in the second frame.

In addition to the mathematical meaning of the symmetric velocity, we can give it the following physical interpretation. Place two objects of equal mass (test masses) at the origin of each inertial system. The center of mass

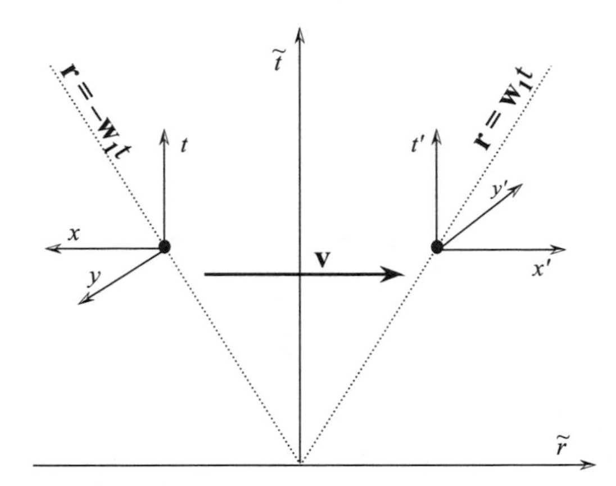

Fig. 1.12. The symmetric velocity $\pm\mathbf{w}_1$ is the velocity of each system with respect to their center, while $\mathbf{v}$ is the velocity of one system with respect to the other.

of the two objects will be called the *center of the two inertial systems*. The symmetric velocity is the velocity of each system with respect to the center of the systems (see Figure 1.12).

Let us now find the events represented by $\begin{pmatrix} t \\ \mathbf{r}' \end{pmatrix}$, with $\mathbf{r}'$ in the direction of $\mathbf{v}$, which are the -1 eigenvectors of $S_{\mathbf{v}}$. By modifying equations (1.35), (1.36) and (1.37) accordingly, we get

$$\frac{\mathbf{r}'}{t} = \frac{\mathbf{v}}{\alpha - 1} = \frac{\mathbf{r}}{t} := \mathbf{w}_{-1}. \tag{1.39}$$

The relative position of the 1 and -1 eigenvectors of $S_{\mathbf{v}}$ differs for the two cases $\alpha < 1$, which by (1.28) corresponds to $e > 0$, and $\alpha > 1$, corresponding to $e < 0$. In Figure 1.13, we show the position of the eigenspaces in the case $\alpha < 1$, and in Figure 1.14, we show the position of the eigenspaces in the case $\alpha > 1$.

1.2.2 Unique speed and interval conservation

The symmetry $S_{\mathbf{v}}$ becomes an isometry if we introduce an appropriate inner product. Under this inner product, the 1 and -1 eigenvectors of $S_{\mathbf{v}}$ will be orthogonal. Figure 1.15 shows the action of a general symmetry S in two cases: (a) when the eigenspaces are not orthogonal and (b) when the eigenspaces are orthogonal. Only in the second case are the lengths of intervals preserved.

The new inner product is obtained by leaving the inner product of the space components unchanged and introducing an appropriate weight μ for the time component. This change of time scale maps $t \to \mu t$ and velocity

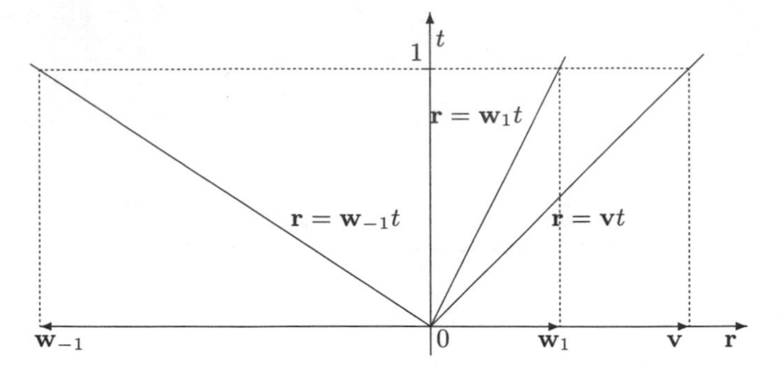

Fig. 1.13. Eigenspaces of the symmetry if $\alpha < 1$. A two-dimensional section of space-time is presented: the time direction and one dimension of space, in the direction of $\mathbf{v}$. In this case, by changing the scale of the time t, we could make the world-lines corresponding to velocities $\mathbf{w}_1$ and $\mathbf{w}_{-1}$ orthogonal.

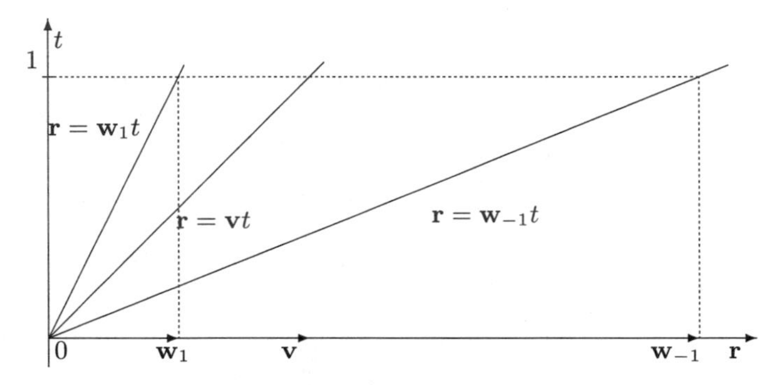

Fig. 1.14. Eigenspaces of the symmetry if $\alpha > 1$. A two-dimensional section of space-time is presented: the time direction and one dimension of space, in the direction of $\mathbf{v}$. In this case, by changing the scale of the time t, we cannot make the world-lines corresponding to velocities $\mathbf{w}_1$ and $\mathbf{w}_{-1}$ orthogonal, since the angle between them will always be less than 90°.

$\mathbf{w} \rightarrow \mathbf{w}/\mu$. Thus $\mathbf{w}t$ is unchanged. The orthogonality of the eigenvectors means that

$$\langle \begin{pmatrix} \mu t \\ \mathbf{w}_1 t \end{pmatrix} \mid \begin{pmatrix} \mu t \\ \mathbf{w}_{-1} t \end{pmatrix} \rangle = t^2(\mu^2 + \langle \mathbf{w}_1 | \mathbf{w}_{-1} \rangle) = 0. \tag{1.40}$$

By use of (1.38), (1.39) and (1.28), this becomes

$$\mu^2 + \frac{|\mathbf{v}|^2}{(1+\alpha)(\alpha-1)} = \mu^2 + \frac{|\mathbf{v}|^2}{\alpha^2 - 1} = \mu^2 - \frac{1}{e} = 0. \tag{1.41}$$

If $e > 0$, corresponding to $\alpha < 1$, the orthogonality of the 1 and -1 eigenvectors of $S_\mathbf{v}$ is achieved (see Figure 1.13) by setting

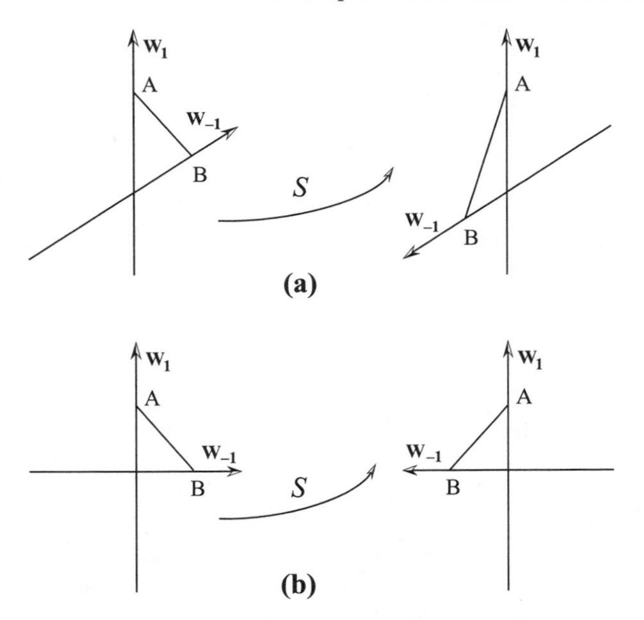

Fig. 1.15. The action of a general symmetry S for two cases. (a) the 1-eigenspace $\mathbf{w}_1$ and the -1-eigenspace $\mathbf{w}_{-1}$ are not orthogonal. (b) these eigenspaces are orthogonal. Only in the second case is the length of an interval [A,B] preserved.

$$\mu = \frac{1}{\sqrt{e}}. \tag{1.42}$$

In this case, $S_{\mathbf{v}}$ becomes an isomctry with respect to the inner product with weight μ, implying that

$$(\mu t)^2 + |\mathbf{r}'|^2 = (\mu t')^2 + |\mathbf{r}|^2, \tag{1.43}$$

or, equivalently,

$$(\mu t')^2 - |\mathbf{r}'|^2 = (\mu t)^2 - |\mathbf{r}|^2. \tag{1.44}$$

The previous equation implies that our space-time transformation from K to K' conserves the *relativistic interval*

$$ds^2 = (\mu dt)^2 - |d\mathbf{r}|^2, \tag{1.45}$$

with μ defined by (1.42). See Figure 1.16 for the meaning of the interval.

In particular, the transformation $S_{\mathbf{v}}$ maps zero interval world-lines to zero interval world-lines. Since zero interval world-lines correspond to uniform motion with unique speed μ, for any relativistic space-time transformation between two inertial systems with $e > 0$, there is a speed μ defined by (1.42) which is conserved. Obviously, the cone $ds^2 > 0$, corresponding to the positive *Lorentz cone*, is also preserved under this transformation.

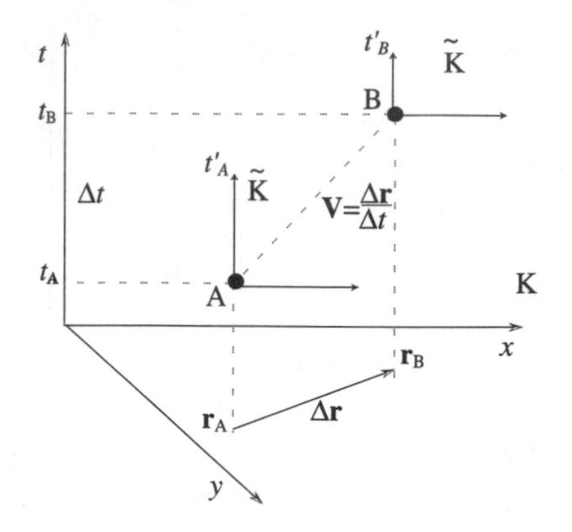

Fig. 1.16. The meaning of the relativistic interval. Two events are depicted: A, with space-time coordinates $(t_A, \mathbf{r}_A)$, and B, with space-time coordinates $(t_B, \mathbf{r}_B)$ in system K. An inertial system $\widetilde{K}$ is chosen with the space origin at $\mathbf{r}_A$ at time t_A and at $\mathbf{r}_B$ at time t_B. The relative velocity of $\widetilde{K}$ with respect to K is $\mathbf{v} = (\mathbf{r}_B - \mathbf{r}_A)/(t_B - t_A) = \Delta\mathbf{r}/\Delta t$. The time of event A in $\widetilde{K}$ is t'_A, and the time of event B is t'_B. The interval between the events A and B is $ds^2 = (\mu(t'_B - t'_A))^2 = (\mu\Delta t)^2 - |\Delta\mathbf{r}|^2$. If conservation of the speed of light is assumed, then $\mu = c$.

Let us now show that $e = e_v$ is independent of the relative velocity $\mathbf{v}$ between the frames K and K'. To do this, consider Figure 1.17, in which two intermediate inertial systems $\widetilde{K}$ and $\widetilde{K}'$ have been added to the configuration of Figure 1.5.

Note that the systems $\widetilde{K}$ and $\widetilde{K}'$ are simply space-reversed, and thus the transformation between them is $S = S_0$. The space-time transformation $S_{\mathbf{v}}$ can now be decomposed as follows:

$$K \xrightarrow{S_{\mathbf{w}_1}} \widetilde{K} \xrightarrow{S} \widetilde{K}' \xrightarrow{S_{\mathbf{w}_1}} K'$$

and, hence,

$$S_{\mathbf{v}} = S_{\mathbf{w}_1} S S_{\mathbf{w}_1}. \tag{1.46}$$

From our discussion, we know that the speed $\mu = 1/\sqrt{e_{\mathbf{w}_1}}$ is preserved by $S_{\mathbf{w}_1}$, and, since S preserves every speed, it follows that $S_{\mathbf{v}}$ also preserves this speed, implying that $e_v = e_{\mathbf{w}_1}$. By a standard argument, this implies that e is independent of $\mathbf{v}$.

Several experiments at the end of the 19th century showed that the speed of light is the same in all inertial systems. Thus

$$\mu = c \quad \text{and} \quad e = \frac{1}{c^2},$$

where c is the *speed of light* in a vacuum, and we have shown that the space-time transformations between two inertial systems are the Lorentz transformations.

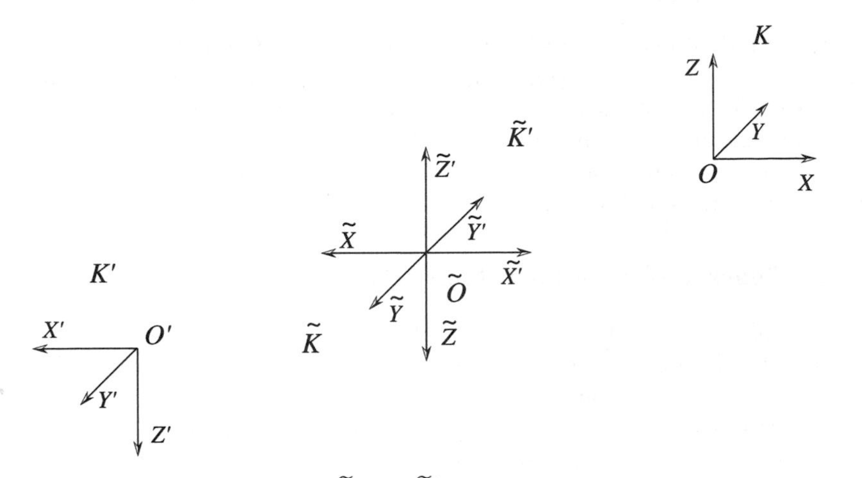

Fig. 1.17. The inertial systems $\widetilde{K}$ and $\widetilde{K}'$ are introduced between systems K and K'. The origin $\widetilde{O}$ is at the center of the systems K and K'. The system $\widetilde{K}$ is space-reversed to K, and $\widetilde{K}'$ is space-reversed to K'.

Note that the operator $S_{\mathbf{v}}$ becomes self-adjoint with respect to the inner product defined by (1.40) with $\mu = c$. Moreover, we can now calculate the adjoint of the relative velocity $\mathbf{v}$ as a linear operator from time t to space displacement $\mathbf{r}$. Since

$$\langle \mathbf{v}|\mathbf{r}\rangle t = \langle \mathbf{v}t|\mathbf{r}\rangle = \langle t|\mathbf{v}^*\mathbf{r}\rangle = \mu^2 t\mathbf{v}^*\mathbf{r},$$

we get

$$\mathbf{v}^*(\mathbf{r}) = \frac{1}{c^2}\langle \mathbf{v}|\mathbf{r}\rangle = e\langle \mathbf{v}|\mathbf{r}\rangle = \mathbf{e}^T\mathbf{r} = S_{12}(\mathbf{r}).$$

This shows that the *adjoint to the relative velocity* between two inertial systems K' and K is the operator that describes the *non-simultaneity* in K' of simultaneous events in K displaced at a distance $\mathbf{r}$.

Theoretically, there is also the possibility that $e = 0$. In this case, the space-time transformations defined by (1.31) become the Galilean transformations. In the next section, we will show that the case $e < 0$ leads to physically absurd results, leaving only two possibilities for relativistic space-time transformations: the Galilean and Lorentz transformations.

1.3 Relativistic velocity addition

We begin this section by deriving a general formula for relativistic velocity addition. Using this formula, we then show that $e < 0$ leads to physically absurd results. Thus, we will assume that $e > 0$, implying that the space-time transformations between the inertial systems are the Lorentz transformations. In this case, the velocity addition coincides with the well-known Einstein velocity addition. We will study the mathematical properties of this velocity addition and its effect on the ball D_v of relativistically admissible velocities. We end the section by showing that the symmetric velocity is the relativistic half of its corresponding velocity.

1.3.1 General formula for velocity addition

In special relativity, the addition of two velocities $\mathbf{v}$ and $\mathbf{u}$ is defined as follows. Let K_1 and K_2 be two inertial systems, with space axes parallel (not reversed), where the relative velocity of K_2 with respect to K_1 is $\mathbf{v}$. Consider an object moving with uniform velocity $\mathbf{u}$ in K_2. If this object is observed in K_1, it is moving with uniform velocity $\mathbf{v} \oplus \mathbf{u}$, called the *relativistic sum of the velocities* $\mathbf{v}$ and $\mathbf{u}$ (see Figure 1.18).

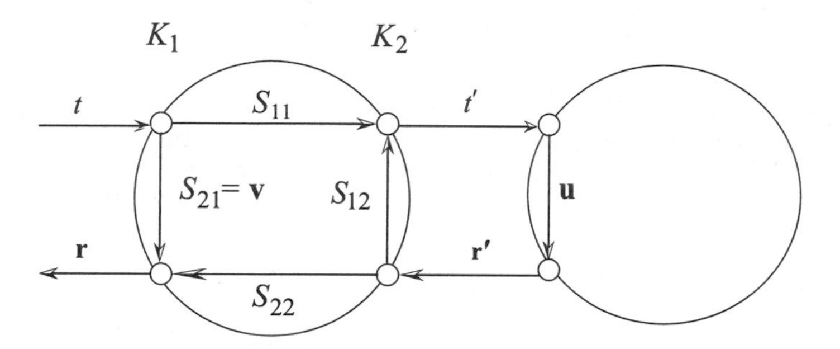

Fig. 1.18. The velocity addition $\mathbf{v} \oplus \mathbf{u}$ is the velocity in K_1 of an object moving with uniform velocity $\mathbf{u}$ in K_2, where K_2 moves relative to K_1 with velocity $\mathbf{v}$ and the space axes of the two systems are parallel (not reversed).

To derive an explicit formula for the velocity addition, we will associate with our moving object an inertial system K_3 with axes parallel to the axes of K_2. The following diagram (Figure 1.19), showing the hybrid connection between these three systems, will be used to derive the above-mentioned formula. From (1.31), with sign modification due to non-reversal of space frames, it follows that the operators in this diagram are

$$\alpha_1 = \sqrt{1 - e|\mathbf{v}|^2}, \quad \alpha_2 = \sqrt{1 - e|\mathbf{u}|^2}, \tag{1.47}$$

$$\mathbf{e}_1 = -e\mathbf{v}^T, \quad \mathbf{e}_2 = -e\mathbf{u}^T, \tag{1.48}$$

$$A_1 = \alpha_1 P_{\mathbf{v}} + I - P_{\mathbf{v}}, \tag{1.49}$$

and

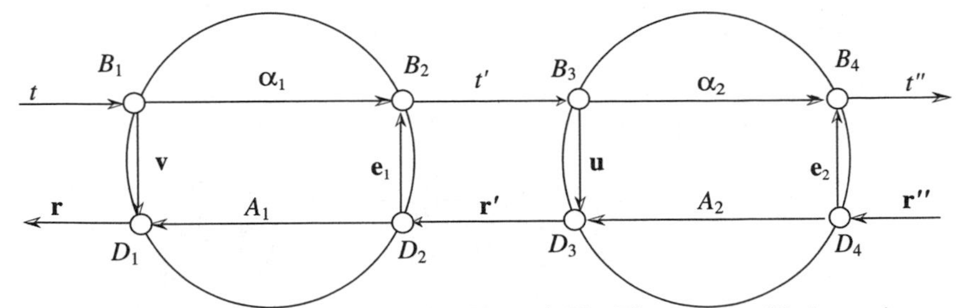

Fig. 1.19. Three inertial systems K_1, K_2 and K_3. The system K_1 is moving parallel to K_2, and K_2 is moving parallel to K_3. The circles represent two space-time transformations: $S_{\mathbf{v}}$ from K_1 to K_2 and $S_{\mathbf{u}}$ from K_2 to K_3. The ports B_1 and D_2 are inputs to $S_{\mathbf{v}}$, the ports B_2 and D_1 are outputs of $S_{\mathbf{v}}$, the ports B_3 and D_4 are inputs to $S_{\mathbf{u}}$, and the ports B_4 and D_3 are outputs of $S_{\mathbf{u}}$. The operators inside the transformations are similar to the ones in Figure 1.10, page 14.

$$A_2 = \alpha_2 P_{\mathbf{u}} + I - P_{\mathbf{u}}. \tag{1.50}$$

To define the velocity addition $\mathbf{v} \oplus \mathbf{u}$, we use inputs $\mathbf{r}'' = 0$ at port D_4 and $\triangle t$ at port B_1. The output at port D_1 will be $(\mathbf{v} \oplus \mathbf{u})\triangle t$. This output is combined from the following passes through the diagram:
$B_1 \longrightarrow D_1$,
$B_1 \longrightarrow B_2 \longrightarrow B_3 \longrightarrow D_3 \longrightarrow D_2 \longrightarrow D_1$,
$B_1 \longrightarrow B_2 \longrightarrow B_3 \longrightarrow D_3 \longrightarrow D_2 \longrightarrow B_2 \longrightarrow B_3 \longrightarrow D_3 \longrightarrow D_2 \longrightarrow D_1$,
and so on.

By substituting the transformations for these passes and using the formula for the sum of a geometric series, we get

$$\begin{aligned}(\mathbf{v} \oplus \mathbf{u})\triangle t &= \mathbf{v}\triangle t + A_1 \mathbf{u} \alpha_1 \triangle t + A_1 \mathbf{u}\, \mathbf{e}_1 \mathbf{u} \alpha_1 \triangle t + A_1 \mathbf{u}\, \mathbf{e}_1 \mathbf{u}\, \mathbf{e}_1 \mathbf{u} \alpha_1 \triangle t + \cdots \\ &= \mathbf{v}\triangle t + A_1 \mathbf{u}(1 - \mathbf{e}_1 \mathbf{u})^{-1} \alpha_1 \triangle t,\end{aligned}$$

or

$$(\mathbf{v} \oplus \mathbf{u})\triangle t = \mathbf{v}\triangle t + A_1 \mathbf{u}(1 + e\langle \mathbf{v}|\mathbf{u}\rangle)^{-1} \alpha_1 \triangle t. \tag{1.51}$$

Using (1.47), (1.48) and (1.49), we get the velocity-addition formula

$$\mathbf{v} \oplus \mathbf{u} = \mathbf{v} + (\alpha^2 P_{\mathbf{v}} + \alpha(I - P_{\mathbf{v}})) \frac{\mathbf{u}}{1 + e\langle \mathbf{v}|\mathbf{u}\rangle}, \tag{1.52}$$

where $\alpha = \sqrt{1 - e|\mathbf{v}|^2}$.

Note that in (1.51), the second velocity summand is obtained by the following three corrections of $\mathbf{u}$. The first correction, described by α_1, is due to the time slowdown of the clocks in K_2 with respect to K_1. The second correction, $\triangle t' \to \triangle t' + e\langle \mathbf{v}|\mathbf{u}\triangle t'\rangle$, is needed to correct the difference in the settings of two clocks at a distance $\mathbf{u}\triangle t'$, synchronized in K_2, with the clocks at the same points synchronized in K_1. Finally, the last correction, described by A_1, expresses the space contraction from K_2 to K_1.

1.3.2 Non-negativity of e.

Consider now the case when the velocities $\mathbf{v}$ and $\mathbf{u}$ are parallel. Since, in this case, $\mathbf{u} = P_{\mathbf{v}}\mathbf{u}$ and $\langle \mathbf{v}|\mathbf{u}\rangle \mathbf{v} = |\mathbf{v}|^2\mathbf{u}$, from (1.52) we get

$$\mathbf{v} \oplus \mathbf{u} = \mathbf{v} + (1 - e|\mathbf{v}|^2)\frac{\mathbf{u}}{1 + e\langle \mathbf{v}|\mathbf{u}\rangle} = \frac{\mathbf{v} + \mathbf{u}}{1 + e\langle \mathbf{v}|\mathbf{u}\rangle}. \tag{1.53}$$

If $e < 0$, this implies that

$$|\mathbf{v} \oplus \mathbf{u}| > |\mathbf{v}| + |\mathbf{u}|,$$

and, thus, the magnitude of $\mathbf{v} \oplus \mathbf{u}$ can become arbitrarily large. For a fixed $\mathbf{v}$, the length of the vector $\mathbf{v} \oplus \mathbf{u}$ in the direction of $\mathbf{v}$, as a function of $\mathbf{u}$, is shown in Figure 1.20.

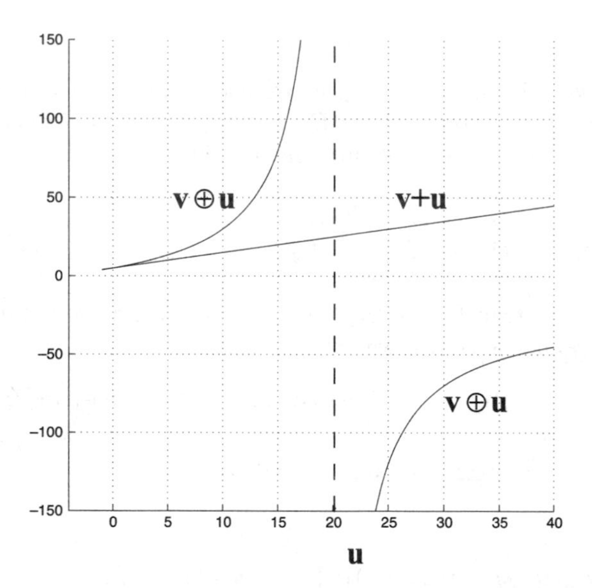

Fig. 1.20. Velocity addition with negative e. In the figure, $e = -0.01$ and $\mathbf{v} = (5, 0, 0)$. For small $\mathbf{u} = (u, 0, 0)$, the relativistic sum $\mathbf{v} \oplus \mathbf{u}$ is slightly larger then $\mathbf{v}+\mathbf{u}$, but when u approaches 20, the sum goes to infinity, and later it even changes direction.

This leads to physically absurd results: the sum is not defined for any $\mathbf{u}$ such that $\langle \mathbf{v}|\mathbf{u}\rangle = -e^{-1}$, and for large $\mathbf{u}$, the direction of the sum is opposite that of each summand. Therefore, we will assume from now that $e > 0$.

1.3.3 Velocity addition in special relativity

We will assume from now on that $e = 1/c^2$, and, thus, the space-time transformations $S_{\mathbf{v}}$ given by (1.31) become

$$\begin{pmatrix} t' \\ \mathbf{r} \end{pmatrix} = \begin{pmatrix} \alpha & \mathbf{v}^T/c^2 \\ \mathbf{v} & -\alpha P_{\mathbf{v}} - (I - P_{\mathbf{v}}) \end{pmatrix} \begin{pmatrix} t \\ \mathbf{r}' \end{pmatrix}, \tag{1.54}$$

with

$$\alpha = \sqrt{1 - |\mathbf{v}|^2/c^2}. \tag{1.55}$$

In this case, the relativistic addition formula (1.52) becomes

$$\mathbf{v} \oplus_E \mathbf{u} = \mathbf{v} + (\alpha^2 P_{\mathbf{v}} + \alpha(I - P_{\mathbf{v}})) \frac{\mathbf{u}}{1 + \langle \mathbf{v}|\mathbf{u}\rangle/c^2}, \tag{1.56}$$

which is known as the *Einstein velocity-addition* formula.

For some calculations, it is convenient to use a different form for Einstein velocity-addition. The alternative formula follows from (1.56) by substituting $P_{\mathbf{v}}(\mathbf{u}) = \frac{\langle \mathbf{u}|\mathbf{v}\rangle \mathbf{v}}{|\mathbf{v}|^2}$ and $\alpha^2 = 1 - |\mathbf{v}|^2/c^2$. With these substitutions, we get

$$\mathbf{v} \oplus_E \mathbf{u} = \frac{1}{1 + \langle \mathbf{v}|\mathbf{u}\rangle/c^2} (\mathbf{v} + \alpha \mathbf{u} + \frac{1}{(1+\alpha)c^2} \langle \mathbf{v}|\mathbf{u}\rangle \mathbf{v}). \tag{1.57}$$

The space-time transformations between frames K_1 and K_3 can also be obtained by use of the diagram in Figure 1.21. However, since the notion

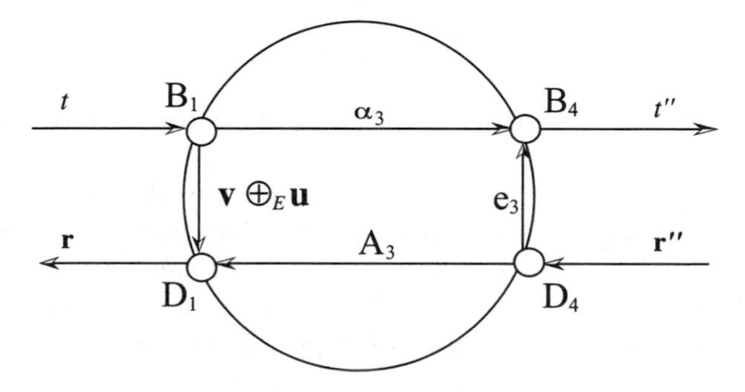

Fig. 1.21. Diagram of the connection between systems K_1 and K_3

of parallelism is not transitive in special relativity, the assumption that the space axes of K_2 were chosen to be parallel to those of K_1 and that the space axes of K_3 were chosen to be parallel to those of K_2, does not imply that the space axes of K_3 are parallel to those of K_1. An explicit expression for the non-transitivity of parallelism in special relativity will be given in section 2.2.4 of Chapter 2, page 64. Thus, the space contraction operator A_3 and the vector $\mathbf{e}_3$ will depend on a rotation which will make the axes of K_3 parallel to the axes of K_1. Nevertheless, we can calculate the time contraction α_3 using the same argument we used for deriving the expression for $\mathbf{v} \oplus_E \mathbf{u}$, obtaining

$$\alpha_3 = \frac{\alpha_2 \alpha_1}{1 - \langle \mathbf{e}_1|\mathbf{u}\rangle} = \sqrt{1 - \frac{|\mathbf{v}|^2}{c^2}} \sqrt{1 - \frac{|\mathbf{u}|^2}{c^2}} \frac{1}{1 + \langle \mathbf{v}|\mathbf{u}\rangle/c^2}. \tag{1.58}$$

1.3.4 Examples of velocity addition

We consider now two special cases of Einstein velocity addition.

Velocity addition of perpendicular vectors

If the velocity $\mathbf{u}$ is perpendicular to the velocity $\mathbf{v}$, meaning that $\langle \mathbf{v}|\mathbf{u}\rangle = 0$, then from (1.57) we have

$$\mathbf{v} \oplus_E \mathbf{u} = \mathbf{v} + \alpha_v \mathbf{u}, \tag{1.59}$$

where $\alpha_v = \sqrt{1 - |\mathbf{v}|^2/c^2}$. In this case, there is no correction for the difference in synchronization of clocks in the two systems, and there is no space contraction. The only correction which we have to perform on $\mathbf{u}$ is that due

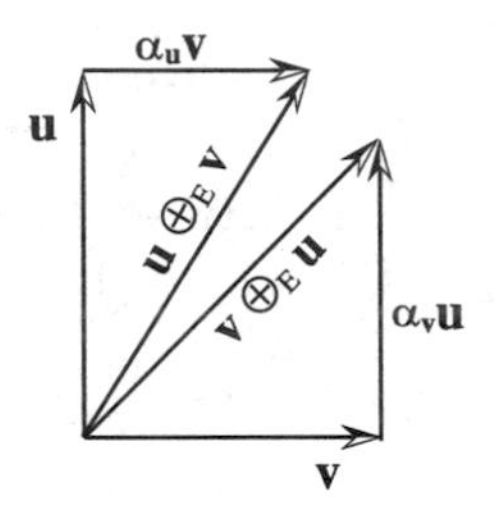

Fig. 1.22. Velocity addition with $\mathbf{u}$ perpendicular to $\mathbf{v}$. Note that $\mathbf{u} \oplus_E \mathbf{v} \neq \mathbf{v} \oplus_E \mathbf{u}$.

to the slowdown of the clocks in K_2 with respect to the clocks in K_1. Note that here the velocity addition is not commutative, since $\mathbf{u} \oplus_E \mathbf{v} = \mathbf{u} + \alpha_u \mathbf{v}$, with $\alpha_u = \sqrt{1 - |\mathbf{u}|^2/c^2}$, and differs from $\mathbf{v} \oplus_E \mathbf{u}$ calculated by (1.59) (see Figure 1.22). For an expression which quantifies the non-commutativity of velocity addition, see Section 2.2.2 of Chapter 2, page 62.

Velocity addition of parallel vectors

Consider now the case when $\mathbf{u}$ is parallel to $\mathbf{v}$. Let

$$I_{\mathbf{v}} = \{\mathbf{u} \in D_v : \mathbf{u} = \lambda \mathbf{v},\ \lambda \in R\},$$

the set of velocities parallel to $\mathbf{v}$. For any $\mathbf{u} \in I_{\mathbf{v}}$, we use (1.53) to get

$$\mathbf{v} \oplus_E \mathbf{u} = \frac{\mathbf{v} + \mathbf{u}}{1 + \langle \mathbf{v}|\mathbf{u}\rangle/c^2}. \tag{1.60}$$

Note that in this case we have $\mathbf{v} \oplus_E \mathbf{u} = \mathbf{u} \oplus_E \mathbf{v}$, so the addition is commutative. Moreover, it could be shown that *only* in this case is the Einstein velocity addition commutative. Denote the direction of $\mathbf{v}$ by $\mathbf{j} = \mathbf{v}/|\mathbf{v}|$. Then any $\mathbf{u} \in I_{\mathbf{v}}$ can be written as

$$\mathbf{u} = uc\mathbf{j}, \tag{1.61}$$

where $|u| = |\mathbf{u}|/c$ is the relative length of $\mathbf{u}$ with respect to the speed of light. Now we can rewrite (1.60) as

$$\mathbf{v} \oplus_E \mathbf{u} = \frac{v+u}{1+vu} c\mathbf{j}. \tag{1.62}$$

The commutative addition on $I_{\mathbf{v}}$ is connected to the usual addition of real numbers in the following way. Recall that the hyperbolic tangent function is defined by $\tanh(x) = (e^x - e^{-x})/(e^x + e^{-x})$ (see Figure 1.23) and that

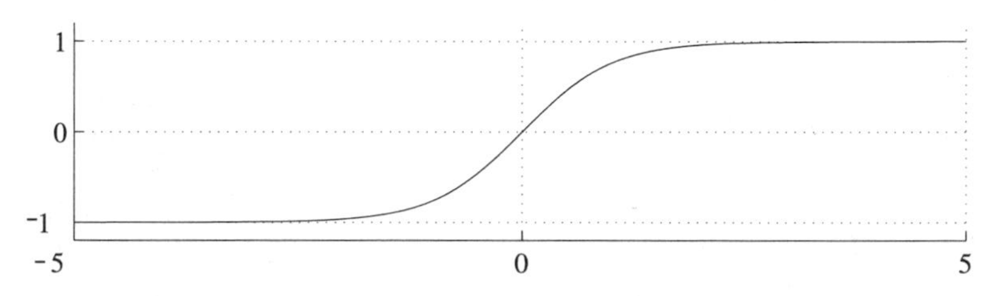

Fig. 1.23. The hyperbolic tangent *tanh* function. This function maps the real line R onto the open interval $(-1, 1)$ and often serves as a connecting function between the addition of the real numbers and the addition of the *commutative* subspaces of *bounded* symmetric domains.

$$\tanh(a+b) = \frac{\tanh a + \tanh b}{1 + \tanh a \, \tanh b}. \tag{1.63}$$

Combining (1.62) with (1.63), we get, for any real numbers a and b,

$$\tanh(a)c\mathbf{j} \oplus_E \tanh(b)c\mathbf{j} = \tanh(a+b)c\mathbf{j}. \tag{1.64}$$

By introducing the map $\phi : R \to I_{\mathbf{v}}$, defined by $\phi(a) = \tanh(a)c\mathbf{j}$, we have the following commutative diagram:

$$\begin{array}{ccc} R \times R & \xrightarrow{\;+\;} & R \\ {\scriptstyle \phi\times\phi}\downarrow & & \downarrow{\scriptstyle \phi} \\ I_{\mathbf{v}} \times I_{\mathbf{v}} & \xrightarrow{\;\oplus_E\;} & I_{\mathbf{v}} \end{array}$$

By using this diagram and the fact that $\phi^{-1}(\mathbf{u}) = \tanh^{-1}(|\mathbf{u}|/c)$, the velocity addition in $I_{\mathbf{v}}$ is given by

$$\mathbf{u} \oplus_E \mathbf{w} = \tanh(\tanh^{-1}(|\mathbf{u}|/c) + \tanh^{-1}(|\mathbf{w}|/c))c\mathbf{j}, \tag{1.65}$$

for any $\mathbf{u}, \mathbf{w} \in I_{\mathbf{v}}$, where $\mathbf{j} = \mathbf{v}/|\mathbf{v}|$.

Given an arbitrary $\mathbf{u}$ (not necessarily in $I_{\mathbf{v}}$), we decompose it as $\mathbf{u} = \mathbf{u}_1 + \mathbf{u}_2$, where $\mathbf{u}_1 = P_{\mathbf{v}}\mathbf{u} \in I_{\mathbf{v}}$ and $\mathbf{u}_2 = (1 - P_{\mathbf{v}})\mathbf{u}$. Then, from (1.56), we have

$$\mathbf{v} \oplus_E \mathbf{u} = (\mathbf{v} \oplus_E \mathbf{u}_1) + \delta \mathbf{u}_2, \tag{1.66}$$

where the constant $\delta = \frac{\alpha}{1+\langle \mathbf{v}|\mathbf{u}_1\rangle/c^2}$ depends on $\mathbf{u}_1$. Consider the disc $\Delta_{\mathbf{u}_1}$ obtained from the intersection of the plane $\mathbf{v}_x = \mathbf{u}_1$ (which is perpendicular to $\mathbf{v}$) with D_v. Note that all $\mathbf{u}$ ending on $\Delta_{\mathbf{u}_1}$ have the same $\mathbf{u}_1$ component and differ only in the $\mathbf{u}_2$ component. Thus, all vectors $\mathbf{v} \oplus_E \mathbf{u}$ end on the disc $\Delta_{\mathbf{v}\oplus_E \mathbf{u}_1}$ (see Figure 1.24).

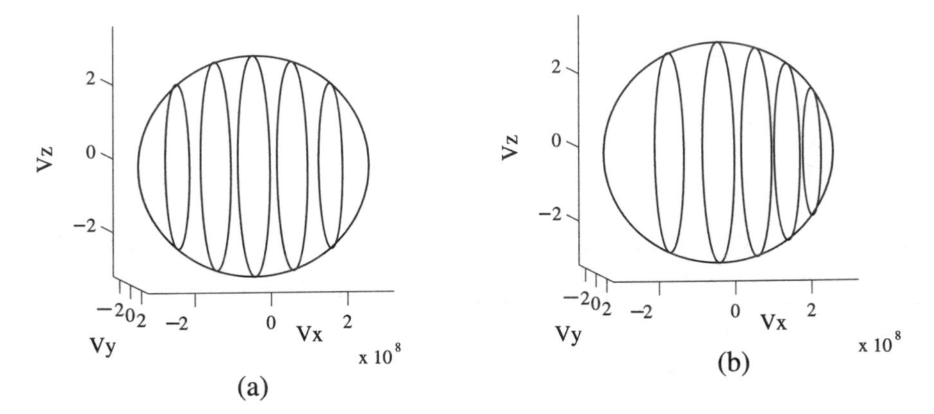

Fig. 1.24. (a) A set of five uniformly spread discs Δ_j obtained by intersecting the three-dimensional velocity ball D_v of radius $c = 3 \cdot 10^8$m/s with $y - z$ planes at $x = 0, \pm\, 10^8, \pm 2 \cdot 10^8$m/s. (b) The images of these Δ_j under the map $\varphi_{\mathbf{v}}(\mathbf{u}) = \mathbf{v} \oplus_E \mathbf{u}$, with $\mathbf{v} = (10^8, 0, 0)$m/s. Note that $\varphi_{\mathbf{v}}(\Delta_j)$ is also a disc in D_v, perpendicular to $\mathbf{v}$ and moved in the direction of $\mathbf{v}$. On each disc Δ_j, the map $\varphi_{\mathbf{v}}$ acts as multiplication by a constant in the $\mathbf{u}_2$ component.

The connection between a velocity and its symmetric velocity

As mentioned on page 16, the symmetry $S_{\mathbf{v}}$ fixes only the velocity $\mathbf{w}_1$. We want to find the connection between $\mathbf{w}_1$ and $\mathbf{v}$. From the definition (1.38) of the symmetric velocity $\mathbf{w}_1$ of $\mathbf{v}$, we have

$$\mathbf{w}_1 = \frac{\mathbf{v}}{1 + \sqrt{1 - |\mathbf{v}|^2/c^2}}, \tag{1.67}$$

which is a vector in the same direction as $\mathbf{v}$. Thus $\mathbf{w}_1 \in I_{\mathbf{v}}$ and has length $|\mathbf{w}_1| = \frac{|\mathbf{v}|}{1+\sqrt{1-|\mathbf{v}|^2/c^2}}$ (see Figure 1.25).

Note that $\langle \mathbf{v}|\mathbf{w}_1\rangle = |\mathbf{v}||\mathbf{w}_1|$ and

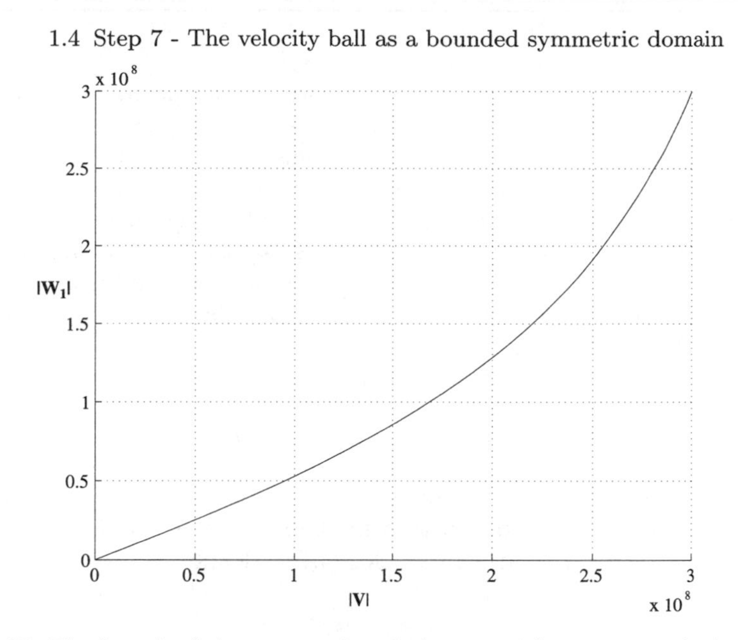

Fig. 1.25. The length of the symmetric velocity $\mathbf{w}_1$ with respect to the length of $\mathbf{v}$. For small velocities, the length of the symmetric velocity is approximately half the length of the corresponding velocity, but for speeds close to the speed of light $3 \cdot 10^8$m/s, they are almost the same.

$$\mathbf{v} = \mathbf{w}_1 + \sqrt{1 - |\mathbf{v}|^2/c^2}\mathbf{w}_1. \tag{1.68}$$

This implies that

$$|\mathbf{v}|^2 - 2\langle \mathbf{v}|\mathbf{w}_1\rangle + |\mathbf{w}_1|^2 = |\mathbf{w}_1|^2 - |\mathbf{v}|^2|\mathbf{w}_1|^2/c^2, \tag{1.69}$$

and

$$|\mathbf{v}| = \frac{2|\mathbf{w}_1|}{1 + |\mathbf{w}_1|^2/c^2}. \tag{1.70}$$

From this, it follows that

$$\mathbf{v} = \frac{2\mathbf{w}_1}{1 + |\mathbf{w}_1|^2/c^2} = \mathbf{w}_1 \oplus_E \mathbf{w}_1. \tag{1.71}$$

Thus, the symmetric velocity $\mathbf{w}_1$ is the *relativistic half of the velocity* $\mathbf{v}$, as we should expect (see Figure 1.12).

1.4 Step 7 - The velocity ball as a bounded symmetric domain

1.4.1 The symmetry on D_v

Recall the definition of a *symmetric domain.* Let D be a domain in a real or complex Banach space. We denote by $Aut(D)$ the collection of all automorphisms (one-to-one smooth maps) of D. The exact meaning of "smooth" will vary with the context, but it will always mean either projective (preserving linear segments), conformal (preserving angles), or complex analytic. We will sometimes denote the particular automorphism group under discussion by $Aut_p(D)$, $Aut_c(D)$ and $Aut_a(D)$ in order to indicate the type of smoothness. A domain D is called *symmetric* if for any element $a \in D$, there is a symmetry $s_a \in Aut(D)$ fixing only the point a. It is easy to show that a domain D is a symmetric domain if it has a symmetry about one point and is homogeneous in the sense that for any two points $z, w \in D$, there is an automorphism $\varphi \in Aut(D)$ such that $\varphi(z) = w$.

We show now that the set D_v, defined by

$$D_v = \{\mathbf{v} : \mathbf{v} \in R^3, \ |\mathbf{v}| < c\}, \tag{1.72}$$

representing all relativistically admissible velocities in an inertial frame K, is a bounded symmetric domain with respect to the projective automorphisms of D_v. Let $\mathbf{a} \in D_v$ be an arbitrary velocity. We define

$$\mathbf{v} = \mathbf{a} \oplus_E \mathbf{a}. \tag{1.73}$$

From (1.71), it follows that $\mathbf{a}$ is the symmetric velocity of $\mathbf{v}$, and, thus, from Section 2.1, the line $(t, \mathbf{a}t)$ in space-time is fixed by $S_{\mathbf{v}}$ and also by the map $E_{\mathbf{v}} = \Psi(S_{\mathbf{v}})$ defined by (1.33).

The map $E_{\mathbf{v}}$ induces a map $s_{\mathbf{a}}$ of the velocity ball as follows. Any point $\mathbf{u}$ in the velocity ball D_v can be identified as the intersection of a line $L = \{(t, \mathbf{u}t) : t \in R\}$ through the origin in space-time of K with the plane $\Pi = \{(1, \mathbf{r}) \in K : \mathbf{r} \in R\}$ (see Figure 1.26). Let K' be an inertial system moving

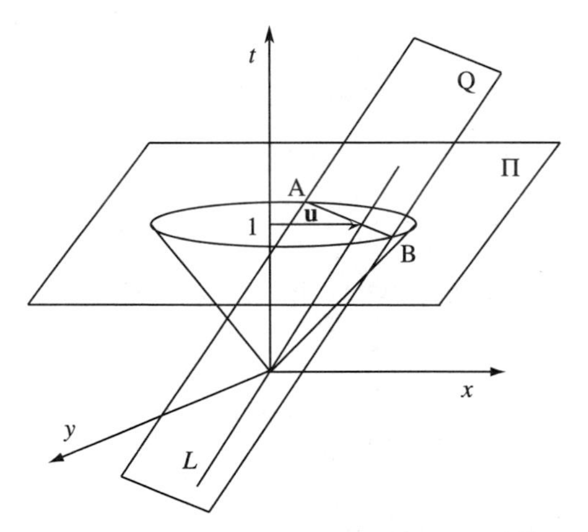

Fig. 1.26. The velocity ball D_v in space-time. Time and two dimensions of space are displayed. The velocity $\mathbf{u}$ is the intersection of a line $L = \{(t, \mathbf{u}t) : t \in R\}$ through the origin with the plane $\Pi = \{(1, \mathbf{r}) \in K : \mathbf{r} \in R\}$. The segment $[AB]$ in D_v is the intersection of D_v with a two-dimensional plane Q through the origin.

with relative velocity $\mathbf{v}$ with respect to K, whose space axes are reversed to those of K.

Under the space-time transformation $E_{\mathbf{v}}$ between systems K and K', the line $L : (t, \mathbf{u}t)$ in K is mapped to a line through the origin in K'. From the definition of Einstein velocity addition, this line is $L' : (t', (\mathbf{v} \oplus_E (-\mathbf{u}))t')$ in K' (the minus sign came from the space reversal). We define $s_{\mathbf{a}}(\mathbf{u})$ to be the intersection of this line with the plane $\Pi = \{(1, \mathbf{r}) \in K : \mathbf{r} \in R\}$. From (1.56), the transformation $s_{\mathbf{a}}$ is given by

$$s_{\mathbf{a}}(\mathbf{u}) = \mathbf{v} + (\alpha^2 P_{\mathbf{v}} + \alpha(I - P_{\mathbf{v}}))\frac{-\mathbf{u}}{1 - \langle \mathbf{v}|\mathbf{u}\rangle/c^2}, \tag{1.74}$$

with $\alpha = \sqrt{1 - |\mathbf{v}|^2/c^2}$, which is the Einstein velocity sum of the relative velocity $\mathbf{v}$ of the systems with $-\mathbf{u}$ (and not $\mathbf{u}$, due to the space reversal). To visualize $s_{\mathbf{a}}$, decompose the velocity $\mathbf{u}$ into $\mathbf{u} = \mathbf{u}_1 + \mathbf{u}_2$, where $\mathbf{u}_1 = P_{\mathbf{v}}\mathbf{u}$ and $\mathbf{u}_2 = (1 - P_{\mathbf{v}})\mathbf{u}$. Then, from (1.74), we get

$$s_{\mathbf{a}}(\mathbf{u}) = (\mathbf{v} \oplus_E (-\mathbf{u}_1)) + \delta(-\mathbf{u}_2), \tag{1.75}$$

where the constant $\delta = \frac{\alpha}{1 - \langle \mathbf{v}|\mathbf{u}_1\rangle/c^2}$ depends only on $\mathbf{u}_1$. The first term $\mathbf{v} \oplus_E (-\mathbf{u}_1)$ is depicted in Figure 1.27. The second term represents reversal and stretching of the component of $\mathbf{u}$ perpendicular to $\mathbf{v}$.

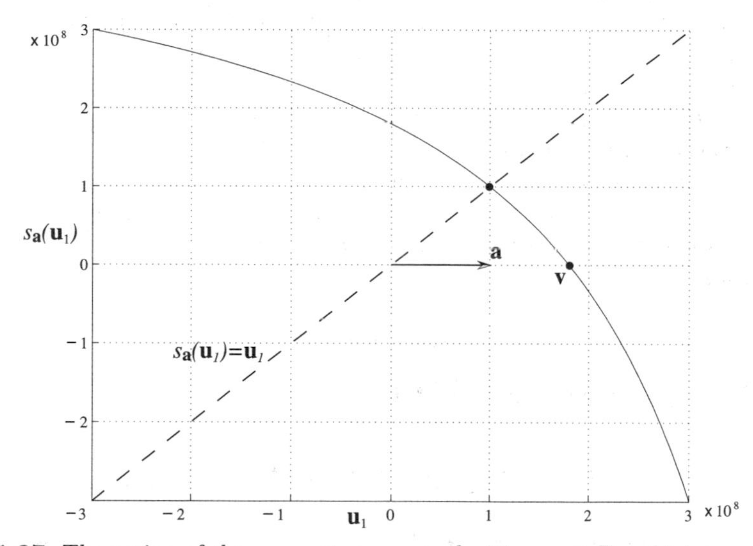

Fig. 1.27. The action of the symmetry $s_{\mathbf{a}}$ on velocities $\mathbf{u}_1 \in D_v$ which are parallel to $\mathbf{a}$, where $\mathbf{a} = 10^8 m/s$. Note that only the point $\mathbf{a}$ is fixed and the graph intersects the $\mathbf{u}$ axis at $\mathbf{v} = \mathbf{a} \oplus_E \mathbf{a}$.

We will show that $s_{\mathbf{a}}$ is a projective map and a symmetry fixing only $\mathbf{a}$. Note that any segment in D_v is obtained from the intersection of D_v with a two-dimensional plane Q through the origin in space-time. The plane Q is mapped by $E_{\mathbf{v}}$ to a two-dimensional plane in space-time $(t', \mathbf{r}')$ in K'. Thus a segment of D_v is mapped by $s_{\mathbf{a}}$ to a segment, implying that $s_{\mathbf{a}}$ is a projective map. As mentioned on page 16, the symmetry $S_{\mathbf{v}}$ fixes only the line associated with $\mathbf{a}$, the symmetric velocity of $\mathbf{v}$. Therefore $s_{\mathbf{a}}$ fixes only $\mathbf{a}$. By use of (1.74) and the definition of α, it is easy to show that $s_{\mathbf{a}}(s_{\mathbf{a}}(\mathbf{u})) = \mathbf{u}$, implying that $s_{\mathbf{a}}^2 = I$, and thus $s_{\mathbf{a}}$ is a symmetry.

1.4.2 The group $Aut_p(D_v)$ of projective automorphisms of D_v

We denote by $Aut_p(D_v)$ the set of all projective automorphisms of the domain D_v. This set is a group, since the composition of two projective automorphisms is a projective automorphism, and the inverse of a projective automorphism (which always exists) is a projective automorphism. Note that for any $\mathbf{a} \in D_v$, the map $\varphi_{\mathbf{a}}$ defined by

$$\varphi_{\mathbf{a}}(\mathbf{u}) = \mathbf{a} \oplus_E \mathbf{u}, \tag{1.76}$$

where $\mathbf{a} \oplus_E \mathbf{u}$ is defined by either (1.56) or (1.57), is an element of $Aut_p(D_v)$. The fact that $\varphi_{\mathbf{a}}$ is a projective map follows from the same argument which showed that $s_{\mathbf{a}}$ is projective. It is obvious that for any velocity $\mathbf{u} \in D_v$ in the system K_2, which is moving parallel to K_1 with relative velocity $\mathbf{a}$, there is a unique corresponding velocity $\varphi_{\mathbf{a}}(\mathbf{u}) \in D_v$ in K_1. Conversely, every velocity in K_1 corresponds to a unique velocity in K_2. Thus, the map $\varphi_{\mathbf{a}} : D_v \to D_v$ is one-to-one and onto (see Figure 1.28).

Next, we characterize the elements of $Aut_p(D_v)$. Let ψ be any projective automorphism of D_v. Set $\mathbf{a} = \psi(0)$ and $U = \varphi_{\mathbf{a}}^{-1}\psi$. Then U is a projective map that maps $0 \to 0$ and is thus a linear map which can be represented by a 3×3 matrix. Since U maps D_v onto itself, it is an isometry and represented by an orthogonal matrix. Since $\psi = \varphi_{\mathbf{a}} U$, the group $Aut_p(D_v)$ of all projective automorphisms is defined by

$$Aut_p(D_v) = \{\varphi_{\mathbf{a}} U : \mathbf{a} \in D_v,\ U \in O(3)\}. \tag{1.77}$$

We write $\varphi_{\mathbf{a},U}$ instead of $\varphi_{\mathbf{a}} U$, and, from (1.57), we have

$$\varphi_{\mathbf{a},U}(\mathbf{u}) = \frac{1}{1 + \langle \mathbf{a} | U\mathbf{u} \rangle / c^2} \left(\mathbf{a} + \alpha U\mathbf{u} + \frac{\langle \mathbf{a} | U\mathbf{u} \rangle \mathbf{a}}{(1+\alpha)c^2}\right), \tag{1.78}$$

for $\mathbf{u} \in D_v$. The group $Aut_p(D_v)$ is a real Lie group of dimension 6, since any element of the group is determined by an element $\mathbf{a}$ of the 3-dimensional open ball of radius c in R^3 and an element U of the 3-dimensional orthogonal group $O(3)$.

By a *one-parameter subgroup* $g(s)$ of $Aut_p(D_v)$, we mean a map $g : R \to Aut_p(D_v)$, such that for any $s_1, s_2 \in R$, we have

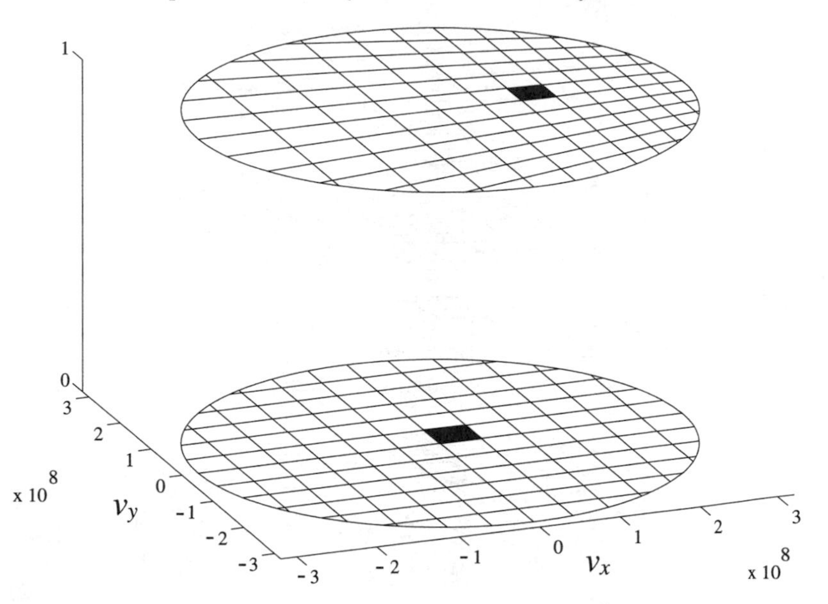

Fig. 1.28. The velocity ball transformation $\varphi_{\mathbf{a}}$, for $\mathbf{a} = 10^8$m/s. On the zero level we see a two-dimensional section of the velocity ball D_v of radius $c = 3 \cdot 10^8$m/s, with a rectangular grid. On level one we see the image of this ball under the map $\varphi_{\mathbf{a}}$. One cell of the grid has been darkened along with its image to help visualize the effect of the transformation. Note how the grid moves in the positive direction of the v_x axis.

$$g(s_1 + s_2) = g(s_1)g(s_2) = g(s_2)g(s_1). \tag{1.79}$$

Any physically meaningful evolution generates a one-parameter subgroup of transformations of the state space of the system. This subgroup is *commutative*, since the evolution of the system during the time interval $s_1 + s_2$ is independent of the way we partition this interval. Note, however, that the full group $Aut_p(D_v)$ is *not* commutative. This means that the set of possible evolution equations is restricted to those stemming from the commutative subgroups of $Aut_p(D_v)$.

For any $\mathbf{a} \in D_v$, the one-parameter subgroup generated by $\varphi_{\mathbf{a}}$ is obtained as follows. Denote the direction of $\mathbf{a}$ by $\mathbf{j} = \mathbf{a}/|\mathbf{a}|$ and define $k = \tanh^{-1}(|\mathbf{a}|/c)$. For any real s, define $\mathbf{b}(s) = \tanh(sk)c\mathbf{j}$. Then $\mathbf{b}(1) = \mathbf{a}$, and, from (1.64), it follows that for any real s_1, s_2, we have

$$\mathbf{b}(s_1 + s_2) = \mathbf{b}(s_1) \oplus_E \mathbf{b}(s_2) = \mathbf{b}(s_2) \oplus_E \mathbf{b}(s_1). \tag{1.80}$$

We call $g(s) = \varphi_{\mathbf{b}(s)}$ the one-parameter subgroup generated by $\varphi_{\mathbf{a}}$. See Figure 1.29 for an example.

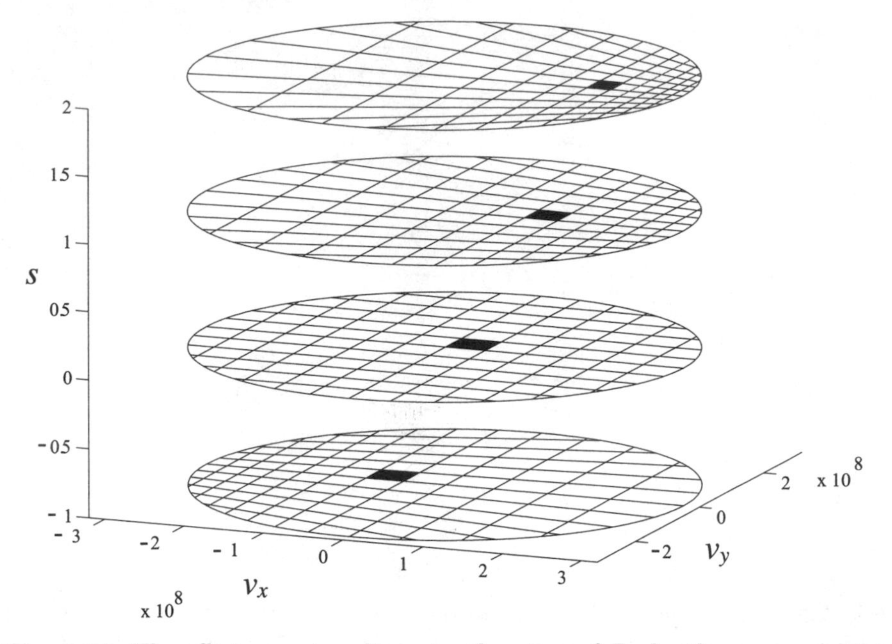

Fig. 1.29. The effect on a two-dimensional section of D_v by the one-parameter subgroup $g(s)$ generated by the map $\varphi_{\mathbf{a}}$ from Figure 1.28, for $s = -1, 0, 1, 2$. One cell of the grid has been darkened along with its images to help visualize the effect of the transformation. Note that $g(-1) = \varphi_{\mathbf{a}}^{-1} = \varphi_{-\mathbf{a}}$, $g(0) = I$—the identity, $g(1) = \varphi_{\mathbf{a}}$ and $g(2) = \varphi_{\mathbf{a}}^2 = \varphi_{\mathbf{a}\oplus_E \mathbf{a}}$.

1.4.3 The group $Aut_p(D_v)$ in two inertial systems

Consider two inertial systems K and K', with common origins at time $t = 0$. Denote by $\mathbf{a}$ the relative velocity of system K with respect to K', and by U the relative rotation between the axes of K and K'. Then any velocity $\mathbf{u} \in D_v$ in system K is observed in K' as $\mathbf{u}' = \varphi(\mathbf{u}) = \varphi_{\mathbf{a},U}(\mathbf{u})$ in D'_v. The map φ between the velocity balls D_v and D'_v induces a map between their projective automorphism groups $Aut_p(D_v)$ and $Aut_p(D'_v)$. Given $\psi \in Aut_p(D_v)$, define $\widetilde{\psi} \in Aut_p(D'_v)$ by

$$\widetilde{\psi} = \varphi\psi\varphi^{-1}. \tag{1.81}$$

The map $\widetilde{\psi}$ is called *the conjugate of ψ with respect to φ*. Thus, the transformation of the automorphism groups of the velocity balls between two inertial systems is given by conjugation, and the following diagram is commutative:

$$\begin{array}{ccc} D_v & \xrightarrow{\varphi_{\mathbf{a},U}} & D'_v \\ \psi \downarrow & & \downarrow \widetilde{\psi} \\ D_v & \xrightarrow{\varphi_{\mathbf{a},U}} & D'_v \end{array}$$

1.5 Step 8 - Relativistic dynamics

It is well known that a force generates a velocity change, or acceleration. There are two types of forces. The first type generates changes in the *magnitude* of the velocity and can be considered a velocity boost. An example is the force of an electric field on a charged particle. The second type of force generates a change in the *direction* of the velocity — a rotation or, equivalently, acceleration in a direction perpendicular to the velocity of the object. An example is a magnetic field acting on a moving charge. Thus a force can be considered as a generator of velocity change. During the time evolution, the velocity of an object cannot leave the velocity ball D_v. Therefore, it is natural to assume that the generator of a relativistic evolution is an element of the Lie algebra $aut_p(D_v)$, which consists of the generators of the group $Aut_p(D_v)$ generated by velocity addition.

1.5.1 The generators of $Aut_p(D_v)$

The elements of a Lie algebra are, by definition, the tangent space of the identity of the group. To define the elements of $aut_p(D_v)$, consider differentiable curves $g(s)$ from a neighborhood I_0 of 0 into $Aut_p(D_v)$, with $g(0) = \varphi_{0,I}$, the identity of $Aut_p(D_v)$. Any such $g(s)$ has the form

$$g(s) = \varphi_{\mathbf{a}(s),U(s)}, \tag{1.82}$$

where $\mathbf{a} : I_0 \to D_v$ is a differentiable function satisfying $\mathbf{a}(0) = \mathbf{0}$ and $U(s) : I_0 \to O(3)$ is differentiable and satisfies $U(0) = I$. We denote by δ the element of $aut_p(D_v)$ generated by $g(s)$. For any fixed $\mathbf{u} \in D_v$, $g(s)(\mathbf{u})$ is a smooth curve in D_v, with $g(0) = \mathbf{u}$, and $\delta(\mathbf{u})$ is a tangent vector to this line. Thus, the elements of $aut_p(D_v)$ are vector fields $\delta(\mathbf{u})$ on D_v defined by

$$\delta(\mathbf{u}) = \frac{d}{ds} g(s)(\mathbf{u})\Big|_{s=0}. \tag{1.83}$$

We now obtain the explicit form of $\delta(\mathbf{u})$. First, define

$$\mathbf{E} = \mathbf{a}'(0), \tag{1.84}$$

which is a vector in R^3, and $A = U'(0)$, which is a 3×3 skew-symmetric matrix (*i.e.*, $A^T = -A$). Combining (1.82) and (1.78), we get

$$g(s)(\mathbf{u}) = \varphi_{\mathbf{a}(s),U(s)}(\mathbf{u})$$

$$= \frac{1}{1 + \langle \mathbf{a}(s)|U(s)\mathbf{u}\rangle/c^2}(\mathbf{a}(s) + \alpha(s)U(s)\mathbf{u} + \frac{\langle \mathbf{a}(s)|U(s)\mathbf{u}\rangle \mathbf{a}(s)}{(1+\alpha(s))c^2}), \tag{1.85}$$

where $\alpha(s) = \sqrt{1 - |\mathbf{a}(s)|^2/c^2}$. A simple calculation shows that

$$\alpha(0) = 1, \quad \frac{d}{ds}\alpha(s)\Big|_{s=0} = 0. \tag{1.86}$$

Moreover,

$$\frac{1}{1 + \langle \mathbf{a}(s)|U(s)\mathbf{u}\rangle/c^2}\Big|_{s=0} = 1, \tag{1.87}$$

$$\frac{d}{ds}(\frac{1}{1 + \langle \mathbf{a}(s)|U(s)\mathbf{u}\rangle/c^2})\Big|_{s=0} = -c^{-2}\langle \mathbf{E}|\mathbf{u}\rangle, \tag{1.88}$$

$$\mathbf{a}(s) + \alpha(s)U(s)\mathbf{u} + \frac{\langle \mathbf{a}(s)|U(s)\mathbf{u}\rangle \mathbf{a}(s)}{(1+\alpha(s))c^2}\Big|_{s=0} = \mathbf{u} \tag{1.89}$$

and

$$\frac{d}{ds}(\mathbf{a}(s) + \alpha(s)U(s)\mathbf{u} + \frac{\langle \mathbf{a}(s)|U(s)\mathbf{u}\rangle \mathbf{a}(s)}{(1+\alpha(s))c^2})\Big|_{s=0} = \mathbf{E} + A\mathbf{u}. \tag{1.90}$$

Thus, by using the formula for the derivative of the product, we get

$$\delta(\mathbf{u}) = \frac{d}{ds}g(s)(\mathbf{u})\Big|_{s=0} = \mathbf{E} + A\mathbf{u} - c^{-2}\langle \mathbf{u}|\mathbf{E}\rangle \mathbf{u}. \tag{1.91}$$

Since A is skew-symmetric, it has the form

$$\begin{pmatrix} 0 & a_{12} & a_{13} \\ -a_{12} & 0 & a_{23} \\ -a_{13} & -a_{23} & 0 \end{pmatrix}, \tag{1.92}$$

and if we let $\mathbf{B} = \begin{pmatrix} a_{23} \\ -a_{13} \\ a_{12} \end{pmatrix}$, we have

$$A\mathbf{u} = \mathbf{u} \times \mathbf{B}, \tag{1.93}$$

where $\times$ denotes the vector product in R^3. Thus, the Lie algebra

$$aut_p(D_v) = \{\delta_{\mathbf{E},\mathbf{B}} : \mathbf{E}, \mathbf{B} \in R^3\}, \tag{1.94}$$

where $\delta_{\mathbf{E},\mathbf{B}} : D_v \longrightarrow R^3$ is the vector field defined by

$$\delta_{\mathbf{E},\mathbf{B}}(\mathbf{u}) = \mathbf{E} + \mathbf{u} \times \mathbf{B} - c^{-2}\langle \mathbf{u} \,|\mathbf{E}\rangle \mathbf{u}. \tag{1.95}$$

Note that any $\delta(\mathbf{u})$ is a polynomial in $\mathbf{u}$ of degree less than or equal to 2. (A term $f(\mathbf{u})$ is *linear* in $\mathbf{u}$ if $f(\mathbf{u}+\mathbf{v}) = f(\mathbf{u}) + f(\mathbf{v})$ and $f(k\mathbf{u}) = kf(\mathbf{u})$ for all $\mathbf{u}, \mathbf{v}$ in the domain of f and all $k \in R$. A term $f(\mathbf{u})$ is *quadratic* in $\mathbf{u}$ if $f(\mathbf{u}) = g(\mathbf{u}, \mathbf{u})$ for some bilinear form $g(\mathbf{u}, \mathbf{v})$). Note also that at any boundary point $\mathbf{u}$, $|\mathbf{u}| = c$, of D_v, the vector $\delta_{\mathbf{E},\mathbf{B}}(\mathbf{u})$ is tangent to D_v. To see this, note that $\mathbf{u} \times \mathbf{B}$ is perpendicular to $\mathbf{u}$ and therefore tangent to D_v. Moreover, since the projection of $\mathbf{E}$ onto the direction of $\mathbf{u}$ is $c^{-2}\langle \mathbf{u}|\mathbf{E}\rangle\mathbf{u}$, the vector $\mathbf{E} - c^{-2}\langle \mathbf{u}|\mathbf{E}\rangle\mathbf{u}$ has zero projection onto the direction of $\mathbf{u}$ and is also tangent to D_v. Thus, $\delta_{\mathbf{E},\mathbf{B}}(\mathbf{u})$ is tangent to D_v. Two examples are shown in Figures 1.30 and 1.31.

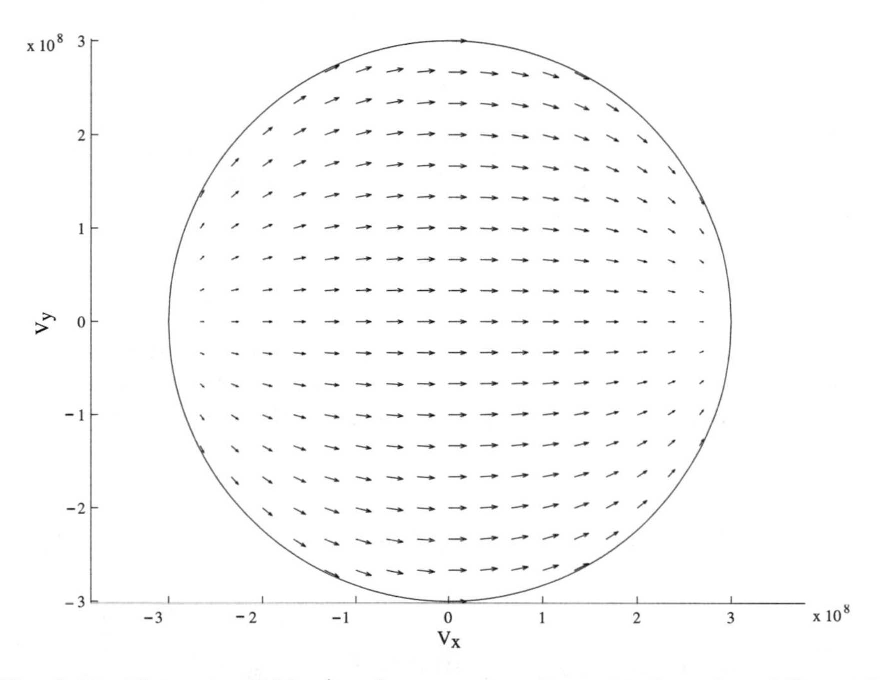

Fig. 1.30. The vector field $q/m \cdot \delta_{\mathbf{E},\mathbf{B}}$ on a two-dimensional section of D_v, with $q/m = 10^7 C/kg$, $\mathbf{E} = (2,0,0)V/m$ and $\mathbf{B} = 0$. Since $\mathbf{E}$ is in the positive direction of the v_x-axis, the field tends to move particles in this direction. However, near the edge of D_v, the vectors either shrink to zero magnitude or become nearly tangent to D_v, reflecting the fact that the flow generated by this field cannot leave D_v.

1.5.2 The Lie algebra of $Aut_p(D_v)$

To show that the set $aut_p(D_v)$ defined by (1.94) and (1.95) is a Lie algebra, it remains to check that this set is closed under the Lie bracket. Recall that the Lie bracket of two vector fields δ and ξ is defined as

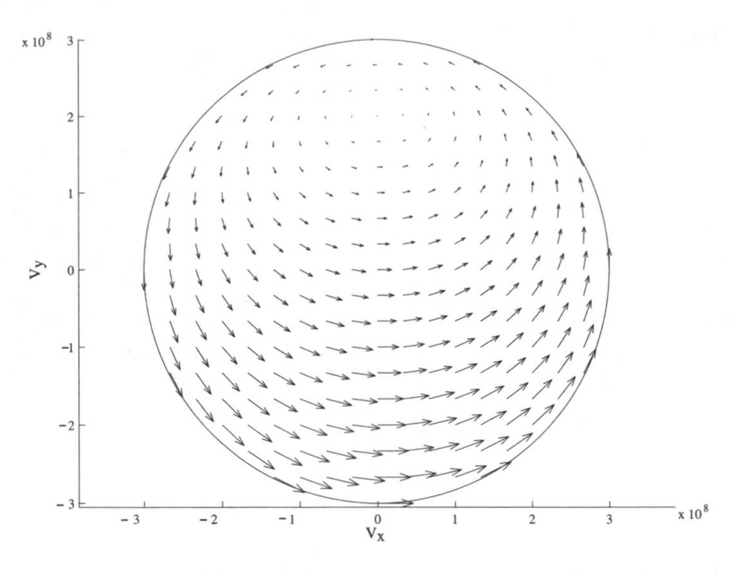

Fig. 1.31. The vector field $q/m \cdot \delta_{\mathbf{E},\mathbf{B}}$ on a two-dimensional section of D_v, with $q/m = 10^7 C/kg$, $\mathbf{E} = (2,0,0)V/m$ and $c\mathbf{B} = (0,0,3)V/m$. Here, the addition of a magnetic field $\mathbf{B}$ causes a rotation.

$$[\delta, \xi](\mathbf{u}) = \frac{d\delta}{d\mathbf{u}}(\mathbf{u})\xi(\mathbf{u}) - \frac{d\xi}{d\mathbf{u}}(\mathbf{u})\delta(\mathbf{u}), \tag{1.96}$$

where $\mathbf{u} \in D_v$ and $\frac{d\delta}{d\mathbf{u}}(\mathbf{u})\xi(\mathbf{u})$ denotes the derivative of δ at the point $\mathbf{u}$ in the direction of the vector $\xi(\mathbf{u})$. Let $\delta_{\mathbf{E},\mathbf{B}}$ and $\delta_{\widetilde{\mathbf{E}},\widetilde{\mathbf{B}}}$ be arbitrary elements of $aut_p(D_v)$. To show that $aut_p(D_v)$ is closed under the Lie bracket, we shall calculate $[\delta_{\mathbf{E},\mathbf{B}}, \delta_{\widetilde{\mathbf{E}},\widetilde{\mathbf{B}}}](\mathbf{u})$ and show that it has the form (1.95).

Note that

$$\frac{d\delta_{\mathbf{E},\mathbf{B}}}{d\mathbf{u}}(\mathbf{u})d\mathbf{u} = d\mathbf{u} \times \mathbf{B} - c^{-2}\langle d\mathbf{u}|\mathbf{E}\rangle\mathbf{u} - c^{-2}\langle \mathbf{u}|\mathbf{E}\rangle d\mathbf{u}. \tag{1.97}$$

Thus,

$$\begin{aligned}[\delta_{\mathbf{E},\mathbf{B}}, \delta_{\widetilde{\mathbf{E}},\widetilde{\mathbf{B}}}](\mathbf{u}) &= \delta_{\widetilde{\mathbf{E}},\widetilde{\mathbf{B}}}(\mathbf{u}) \times \mathbf{B} - c^{-2}\langle \delta_{\widetilde{\mathbf{E}},\widetilde{\mathbf{B}}}(\mathbf{u})|\mathbf{E}\rangle\mathbf{u} \\ &-c^{-2}\langle \mathbf{u}|\mathbf{E}\rangle\delta_{\widetilde{\mathbf{E}},\widetilde{\mathbf{B}}}(\mathbf{u}) - \delta_{\mathbf{E},\mathbf{B}}(\mathbf{u}) \times \widetilde{\mathbf{B}} \\ &+c^{-2}\langle \delta_{\mathbf{E},\mathbf{B}}(\mathbf{u})|\widetilde{\mathbf{E}}\rangle\mathbf{u} + c^{-2}\langle \mathbf{u}|\widetilde{\mathbf{E}}\rangle\delta_{\mathbf{E},\mathbf{B}}(\mathbf{u}). \end{aligned} \tag{1.98}$$

Using (1.95), the previous expression becomes a second-degree polynomial in $\mathbf{u}$, with constant term $\widetilde{\mathbf{E}} \times \mathbf{B} - \mathbf{E} \times \widetilde{\mathbf{B}}$, linear term

$$(\mathbf{u} \times \widetilde{\mathbf{B}}) \times \mathbf{B} - (\mathbf{u} \times \mathbf{B}) \times \widetilde{\mathbf{B}} - c^{-2}\langle \mathbf{u}|\mathbf{E}\rangle\widetilde{\mathbf{E}} + c^{-2}\langle \mathbf{u}|\widetilde{\mathbf{E}}\rangle\mathbf{E}, \tag{1.99}$$

and quadratic term

$$-c^{-2}\langle \mathbf{u} \times \widetilde{\mathbf{B}} | \mathbf{E} \rangle \mathbf{u} + c^{-2} \langle \mathbf{u} \times \mathbf{B} | \widetilde{\mathbf{E}} \rangle \mathbf{u}. \tag{1.100}$$

By using the identities

$$(\mathbf{u} \times \widetilde{\mathbf{B}}) \times \mathbf{B} - (\mathbf{u} \times \mathbf{B}) \times \widetilde{\mathbf{B}} = \mathbf{u} \times (\widetilde{\mathbf{B}} \times \mathbf{B}), \tag{1.101}$$

$$\langle \mathbf{u} | \widetilde{\mathbf{E}} \rangle \mathbf{E} - \langle \mathbf{u} | \mathbf{E} \rangle \widetilde{\mathbf{E}} = \mathbf{u} \times (\mathbf{E} \times \widetilde{\mathbf{E}}) \tag{1.102}$$

and

$$\langle \mathbf{u} \times \mathbf{B} | \widetilde{\mathbf{E}} \rangle = -\langle \mathbf{u} | \widetilde{\mathbf{E}} \times \mathbf{B} \rangle, \tag{1.103}$$

the expression (1.99) can be written as

$$\mathbf{u} \times (\widetilde{\mathbf{B}} \times \mathbf{B} + c^{-2}(\mathbf{E} \times \widetilde{\mathbf{E}})) \tag{1.104}$$

and (1.100) as

$$-c^{-2}\langle \mathbf{u} | \widetilde{\mathbf{E}} \times \mathbf{B} - \mathbf{E} \times \widetilde{\mathbf{B}} \rangle \mathbf{u}. \tag{1.105}$$

Thus, from (1.95), the expression for the *Lie bracket* in $aut_p(D_v)$ is

$$[\delta_{\mathbf{E},\mathbf{B}}, \delta_{\widetilde{\mathbf{E}},\widetilde{\mathbf{B}}}] = \delta_{\widetilde{\mathbf{E}} \times \mathbf{B} - \mathbf{E} \times \widetilde{\mathbf{B}},\, c^{-2}(\mathbf{E} \times \widetilde{\mathbf{E}}) - \mathbf{B} \times \widetilde{\mathbf{B}}}, \tag{1.106}$$

an element of $aut_p(D_v)$.

For example, $[\delta_{\mathbf{E},0}, \delta_{0,\mathbf{B}}] = -\delta_{\mathbf{E} \times \mathbf{B},0}$ is 0 if and only if $\mathbf{E}$ and $\mathbf{B}$ are parallel.

1.5.3 The commutation relations for the Lorentz group

We will now use (1.106) to derive the commutation relations for the Lorentz group. Recall that the Lorentz space-time transformations induce projective maps of the velocity ball and generate the Lie group $Aut_p(D_v)$. The generators of this group are of the form $\delta_{\mathbf{E},\mathbf{B}}$, for $\mathbf{E}, \mathbf{B} \in R^3$, and belong to the Lie algebra $aut_p(D_v)$ of $Aut_p(D_v)$. There is a basis for the generators consisting of the generators of rotations about the x, y, and z-axes and the boosts in the direction of these axes.

Let $\mathbf{i}, \mathbf{j}$ and $\mathbf{k}$ be unit vectors in the direction of the positive x, y and z axes, respectively. Then from (1.92) and (1.93) page 36, it follows that $\delta_{0,\mathbf{i}}$ acts on $\mathbf{v}$ like the momentum of rotation about the x-axis. This momentum is denoted by J_1. Thus, we can represent J_1 as a generator $\delta_{0,\mathbf{i}}$ of a projective map on the velocity ball D_v and denote it by $\pi_p(J_1)$, where the subscript p indicates that it generates a projective map. Similarly, we can represent the generators of rotation about the other axes, and we have

$$\pi_p(J_1) = \delta_{0,\mathbf{i}},\ \pi_p(J_2) = \delta_{0,\mathbf{j}},\ \pi_p(J_3) = \delta_{0,\mathbf{k}}. \tag{1.107}$$

From (1.106), it follows that

$$[\delta_{0,\mathbf{i}}, \delta_{0,\mathbf{j}}] = \delta_{0,-\mathbf{i}\times\mathbf{j}} = -\delta_{0,\mathbf{k}}. \tag{1.108}$$

This implies that $[\pi_p(J_1), \pi_p(J_2)] = -\pi_p(J_3)$ and similarly for the other pairs of generators. Since the same relations hold for the momentums of rotations about the axes, we get the first set of commutation relations

$$[J_1, J_2] = -J_3,\ \ [J_3, J_1] = -J_2,\ \ [J_2, J_3] = -J_1. \tag{1.109}$$

It is obvious that the generator of a boost in the x-direction is a multiple of $\delta_{\mathbf{i},0}$, which we denote by $\lambda\delta_{\mathbf{i},0} = \delta_{\lambda\mathbf{i},0}$. Similarly, the generator of a boost in the y-direction will be denoted by $\delta_{\lambda\mathbf{j},0}$. Then, from (1.106), we get

$$[\delta_{\lambda\mathbf{i},0}, \delta_{\lambda\mathbf{j},0}] = \delta_{0,c^{-2}\lambda^2\mathbf{i}\times\mathbf{j}} = c^{-2}\lambda^2\delta_{0,\mathbf{k}}.$$

In order to simplify the commutation relations, we will take $\lambda = c$. This suggests representing the boosts K_1, K_2 and K_3 in the directions x, y and z, respectively, by

$$\pi_p(K_1) = \delta_{c\mathbf{i},0},\ \ \pi_p(K_2) = \delta_{c\mathbf{j},0},\ \ \pi_p(K_3) = \delta_{c\mathbf{k},0}. \tag{1.110}$$

From the above discussion, we get

$$[\pi_p(K_1), \pi_p(K_2)] = \pi_p(J_3),\ [\pi_p(K_3), \pi_p(K_1)] = \pi_p(J_2),$$

$$[\pi_p(K_2), \pi_p(K_3)] = \pi_p(J_1),$$

and, for the boosts themselves,

$$[K_1, K_2] = J_3,\ \ [K_3, K_1] = J_2,\ \ [K_2, K_3] = J_1. \tag{1.111}$$

Direct use of (1.106) leads to the following commutation relations for the remaining pairs of momentums and boosts

$$[J_1, K_1] = 0,\ [J_1, K_2] = -K_3,\ [J_1, K_3] = K_2, \tag{1.112}$$

$$[J_2, K_1] = K_3,\ [J_2, K_2] = 0,\ [J_2, K_3] = -K_1, \tag{1.113}$$

$$[J_3, K_1] = -K_2,\ [J_3, K_2] = K_1,\ [J_3, K_3] = 0. \tag{1.114}$$

The commutation relations (1.109),(1.111),(1.112),(1.113) and (1.114) form the full set of commutation relations for the generators of the Lorentz group. The representation π_p, defined by (1.107) and (1.110), is a representation of this group into the projective maps on the velocity ball D_v.

1.5.4 Transformation of $aut_p(D_v)$ between two inertial systems

Consider two inertial systems K and K', with common origins at time $t = 0$ and space axes parallel each to other. Denote by $\mathbf{a}$ the relative velocity of system K with respect to K'. Let D_v be the velocity ball in K, and let D'_v be the velocity ball in K'. Let

$$\delta_{\mathbf{E},\mathbf{B}}(\mathbf{u}) = \mathbf{E} + \mathbf{u} \times \mathbf{B} - c^{-2}\langle \mathbf{u}\,|\mathbf{E}\rangle \mathbf{u}$$

be an arbitrary element of $aut_p(D_v)$, generated by some curve $g(s)$ into $Aut_p(D_v)$. This curve generates a curve $\widetilde{g}(s)$ into $Aut_p(D'_v)$, which, from (1.81), is

$$\widetilde{g}(s) = \varphi_{\mathbf{a}} g(s) \varphi_{\mathbf{a}}^{-1}. \tag{1.115}$$

This curve defines a generator $\widetilde{\delta}$, which is an element of $aut_p(D'_v)$ and thus of the form

$$\widetilde{\delta}_{\mathbf{E}',\mathbf{B}'}(\mathbf{u}') = \mathbf{E}' + \mathbf{u}' \times \mathbf{B}' - c^{-2}\langle \mathbf{u}'\,|\mathbf{E}'\rangle \mathbf{u}'. \tag{1.116}$$

We want to find the relationship between $\mathbf{E}, \mathbf{B}$ and $\mathbf{E}', \mathbf{B}'$.

To do this, we first rewrite (1.115) as

$$\widetilde{g}(s)\varphi_{\mathbf{a}}(\mathbf{u}) = \varphi_{\mathbf{a}} g(s)(\mathbf{u}) \tag{1.117}$$

for $\mathbf{u} \in D_v$. By differentiating this equation with respect to s and substituting $s = 0$, we get

$$\widetilde{\delta}_{\mathbf{E}',\mathbf{B}'}(\varphi_{\mathbf{a}}(\mathbf{u})) = \frac{d\varphi_{\mathbf{a}}}{d\mathbf{u}}(\mathbf{u})\delta_{\mathbf{E},\mathbf{B}}(\mathbf{u}). \tag{1.118}$$

Now we have to calculate $\frac{d\varphi_{\mathbf{a}}}{d\mathbf{u}}(\mathbf{u})d\mathbf{u}$. Using (1.56) and (1.102), we get

$$\frac{d\varphi_{\mathbf{a}}}{d\mathbf{u}}(\mathbf{u})d\mathbf{u} = \frac{d}{d\mathbf{u}}(\mathbf{a} \oplus_{\mathbf{E}} \mathbf{u})d\mathbf{u}$$

$$= \frac{d}{d\mathbf{u}}(\mathbf{a} + (\alpha^2 P_{\mathbf{a}} + \alpha(I - P_{\mathbf{a}}))\frac{\mathbf{u}}{1 + \langle \mathbf{a}|\mathbf{u}\rangle/c^2})d\mathbf{u}$$

$$= (\alpha^2 P_{\mathbf{a}} + \alpha(I - P_{\mathbf{a}}))(\frac{d\mathbf{u}}{1 + \langle \mathbf{a}|\mathbf{u}\rangle/c^2} - \frac{\mathbf{u}\langle \mathbf{a}|d\mathbf{u}\rangle/c^2}{(1 + \langle \mathbf{a}|\mathbf{u}\rangle/c^2)^2})$$

$$= (\alpha^2 P_{\mathbf{a}} + \alpha(I - P_{\mathbf{a}}))\frac{d\mathbf{u} - c^{-2}\mathbf{a} \times (\mathbf{u} \times d\mathbf{u})}{(1 + \langle \mathbf{a}|\mathbf{u}\rangle/c^2)^2}. \tag{1.119}$$

Substituting these expressions into (1.118) and comparing the terms constant in $\mathbf{u}$ (*i.e.*, setting $\mathbf{u} = 0$), we get

$$\widetilde{\delta}_{\mathbf{E}',\mathbf{B}'}(\mathbf{a}) = \mathbf{E}' + \mathbf{a} \times \mathbf{B}' - c^{-2}\langle \mathbf{a}\,|\mathbf{E}'\rangle \mathbf{a} = (\alpha^2 P_{\mathbf{a}} + \alpha(I - P_{\mathbf{a}}))\mathbf{E},$$

and, hence,

$$\mathbf{E} = (\alpha^{-2} P_{\mathbf{a}} + \alpha^{-1}(I - P_{\mathbf{a}}))(\mathbf{E}' + \mathbf{a} \times \mathbf{B}' - c^{-2}\langle \mathbf{a}\,|\mathbf{E}'\rangle \mathbf{a}). \tag{1.120}$$

In the particular case when the relative velocity of system K with respect to K' is $\mathbf{a} = (a, 0, 0)$, the previous equation yields

$$\begin{cases} \mathbf{E}_1 = & \mathbf{E}'_1, \\ \mathbf{E}_2 = & \alpha^{-1}(\mathbf{E}'_2 - a\mathbf{B}'_3), \\ \mathbf{E}_3 = & \alpha^{-1}(\mathbf{E}'_3 + a\mathbf{B}'_2), \end{cases} \tag{1.121}$$

where $\alpha = \sqrt{1 - a^2/c^2}$. This coincides with the usual formula for the transformation of an electric field from one system to another.

To obtain a similar formula for the transformation of a magnetic field, we will compare the terms of (1.118) which are linear in $\mathbf{u}$. This leads to

$$\widetilde{\mathbf{u}} \times \mathbf{B}' - c^{-2}\langle \mathbf{a}|\mathbf{E}'\rangle \widetilde{\mathbf{u}} - c^{-2}\langle \widetilde{\mathbf{u}}|\mathbf{E}'\rangle \mathbf{a}$$

$$= (\alpha^2 P_{\mathbf{a}} + \alpha(I - P_{\mathbf{a}}))(\mathbf{u} \times \mathbf{B} - 2c^{-2}\langle \mathbf{a}|\mathbf{u}\rangle \mathbf{E} - c^{-2}\mathbf{a} \times (\mathbf{u} \times \mathbf{E})), \tag{1.122}$$

where

$$\widetilde{\mathbf{u}} = (\alpha^2 P_{\mathbf{a}} + \alpha(I - P_{\mathbf{a}}))\mathbf{u} \tag{1.123}$$

is the linear term of $\varphi_{\mathbf{a}}(\mathbf{u})$. Assume now that $\mathbf{a} = (a, 0, 0)$. If we choose $\mathbf{u} = (0, 0, u_3)$, then, comparing the first component in this equation, we get

$$\mathbf{B}_2 = \alpha^{-1}(\mathbf{B}'_2 + c^{-2}a\mathbf{E}'_3), \tag{1.124}$$

and for the second component, we get

$$\mathbf{B}_1 = \mathbf{B}'_1. \tag{1.125}$$

If we choose $\mathbf{u} = (0, u_2, 0)$, then, comparing the first component of equation (1.122), we get

$$\mathbf{B}_3 = \alpha^{-1}(\mathbf{B}'_3 - c^{-2}a\mathbf{E}'_2). \tag{1.126}$$

This coincides with the usual formula

$$\begin{cases} \mathbf{B}_1 = & \mathbf{B}'_1, \\ \mathbf{B}_2 = & \alpha^{-1}(\mathbf{B}'_2 + c^{-2}a\mathbf{E}'_3), \\ \mathbf{B}_3 = & \alpha^{-1}(\mathbf{B}'_3 - c^{-2}a\mathbf{E}'_2) \end{cases} \tag{1.127}$$

for the transformation of a magnetic field from one system to another.

Thus, the elements of $aut_p(D_v)$ and the electromagnetic field strength transform between two inertial systems in the same way. We saw earlier that the space-time transformations between two inertial systems are linear and preserve the interval. Such transformations preserve the velocity ball D_v and are given by projective maps. So it is natural to ask "What is preserved by the action of the above transformations on $aut_p(D_v)$?"

To answer this question, we combine the two real-valued three-dimensional vectors $\mathbf{E}$ and $\mathbf{B}$ describing the elements of $aut_p(D_v)$ into a complex vector F. In order that both vectors will have the same units, we will use $c\mathbf{B}$ instead of $\mathbf{B}$. An element of $aut_p(D_v)$ will now be described by $F = \mathbf{E} + ic\mathbf{B}$ in system K and by $F' = \mathbf{E}' + ic\mathbf{B}'$ in system K'. From the formulas (1.121) and (1.126), we get

$$F = \begin{pmatrix} \mathbf{E}_1 + ic\mathbf{B}_1 \\ \mathbf{E}_2 + ic\mathbf{B}_2 \\ \mathbf{E}_3 + ic\mathbf{B}_3 \end{pmatrix}$$

$$= \begin{pmatrix} \mathbf{E}'_1 + ic\mathbf{B}'_1 \\ \alpha^{-1}(\mathbf{E}'_2 + ic\mathbf{B}'_2 - c\mathbf{B}'_3 a/c + i\mathbf{E}'_3 a/c) \\ \alpha^{-1}(\mathbf{E}'_3 + ic\mathbf{B}'_3 + c\mathbf{B}'_2 a/c - i\mathbf{E}'_2 a/c) \end{pmatrix} = \begin{pmatrix} F'_1 \\ \alpha^{-1}(F'_2 + iF'_3 a/c) \\ \alpha^{-1}(F'_3 - iF'_2 a/c) \end{pmatrix}. \tag{1.128}$$

Define the complex quantity F^2 by $F^2 = F_1^2 + F_2^2 + F_3^2$. By (1.128), we get

$$F^2 = (F'_1)^2 + \alpha^{-2}((F'_2 + iF'_3 a/c)^2 + (F'_3 - iF'_2 a/c)^2) \tag{1.129}$$

$$= (F'_1)^2 + (F'_2)^2 + (F'_3)^2 = (F')^2, \tag{1.130}$$

implying that F^2 is preserved by the transformation between inertial systems and is a Lorentz invariant for the electromagnetic field.

1.5.5 Relativistic evolution equation

Evolution described by a relativistic dynamic equation must preserve the ball D_v of all relativistically admissible velocities. As mentioned above, we consider the force as an element of $aut_p(D_v)$. The equation of evolution of a charged particle with charge q and *rest-mass* m_0 using the generator $\delta_{\mathbf{E},\mathbf{B}} \in aut_p(D_v)$ defined by (1.95) is

$$\frac{d\mathbf{v}(\tau)}{d\tau} = \frac{q}{m_0}\delta_{\mathbf{E},\mathbf{B}}(\mathbf{v}(\tau))$$

or

$$\frac{d\mathbf{v}(\tau)}{d\tau} = \frac{q}{m_0}(\mathbf{E} + \mathbf{v}(\tau) \times \mathbf{B} - c^{-2}\langle \mathbf{v}(\tau)|\mathbf{E}\rangle \mathbf{v}(\tau)), \tag{1.131}$$

where τ is a real parameter related to time, which turns out to be the proper time of the particle. We will show that this equation coincides with the known relativistic equation for the *evolution of a charged particle under the Lorentz force* of an electromagnetic field. Let us introduce a new variable $F(\tau)$ (which may depend on position and time), representing the Lorentz force acting on the object, by

$$F(\tau) = q(\mathbf{E} + \mathbf{v}(\tau) \times \mathbf{B}), \quad \mathbf{v}(\tau) \in D_v. \tag{1.132}$$

Using the fact that $\mathbf{v}(\tau) \times \mathbf{B}$ is perpendicular to $\mathbf{v}(\tau)$, we can rewrite (1.131) as

$$m_0 \frac{d\mathbf{v}}{d\tau} = F - c^{-2}\langle \mathbf{v}(\tau)|F\rangle \mathbf{v}(\tau). \tag{1.133}$$

Consider an inertial system K_0 moving with the same velocity as our object at time $t = 0$. We may assume that Newton's Second Law holds in K_0. In other words,

$$m_0 \frac{d\mathbf{v}}{dt_0} = F, \tag{1.134}$$

where t_0 denotes the time in K_0. This implies that F is the force acting on our object if the object was at rest in K_0 at time $t = 0$ and $\tau = t_0$ is the *proper time* of the object. If our object has velocity $\mathbf{v}$ at time $t = 0$, by (1.17) and (1.55), we have

$$d\tau = \sqrt{1 - |\mathbf{v}|^2/c^2}\, dt. \tag{1.135}$$

Thus, we can rewrite (1.133) as

$$m_0(1 - |\mathbf{v}|^2/c^2)^{-1/2} \frac{d\mathbf{v}}{dt} = F - c^{-2}\langle \mathbf{v}(t)|F\rangle \mathbf{v}(t). \tag{1.136}$$

Taking the scalar product of this equation with $\mathbf{v}$, we get

$$m_0(1 - |\mathbf{v}|^2/c^2)^{-1/2}\langle \mathbf{v}|\frac{d\mathbf{v}}{dt}\rangle = \langle \mathbf{v}|F\rangle(1 - |\mathbf{v}|^2/c^2), \tag{1.137}$$

or

$$\langle \mathbf{v}|F\rangle = m_0(1 - |\mathbf{v}|^2/c^2)^{-3/2}\langle \mathbf{v}|\frac{d\mathbf{v}}{dt}\rangle. \tag{1.138}$$

Finally, from (1.136) we have

$$F = m_0(1 - |\mathbf{v}|^2/c^2)^{-1/2} \frac{d\mathbf{v}}{dt} + c^{-2}\langle \mathbf{v}|F\rangle \mathbf{v}$$

$$= m_0(1 - |\mathbf{v}|^2/c^2)^{-1/2} \frac{d\mathbf{v}}{dt} + c^{-2} m_0(1 - |\mathbf{v}|^2/c^2)^{-3/2}\langle \mathbf{v}|\frac{d\mathbf{v}}{dt}\rangle \mathbf{v}$$

$$= \frac{d}{dt}(m_0(1 - |\mathbf{v}|^2/c^2)^{-1/2}\mathbf{v}) = \frac{d(m\mathbf{v})}{dt}, \tag{1.139}$$

where $m = \gamma(\mathbf{v})m_0 = m_0(1 - |\mathbf{v}|^2/c^2)^{-1/2}$. This is the usual *relativistic dynamics formula.*

We have shown that the equation of evolution (1.131) with the generator $\delta_{\mathbf{E},\mathbf{B}}$ defined by (1.95) for a charged particle of rest-mass m_0 and charge q coincides with the well-known formula

$$\frac{d(m\mathbf{v})}{dt} = q(\mathbf{E} + \mathbf{v} \times \mathbf{B}).$$

Thus, the flow generated by an electromagnetic field is defined by elements of the Lie algebra $aut_p(D_v)$, which are, in turn, vector field polynomials in $\mathbf{v}$ of degree 2. The linear term of this field comes from the magnetic force, while the constant and the quadratic terms come from the electric field. The dynamic equation of evolution in relativistic *mechanics* is also given by elements of $aut_p(D_v)$. This follows from the above discussion if we set $\mathbf{B} = 0$.

For a constant electromagnetic field, the equation of evolution (1.131) with the generator $\delta_{\mathbf{E},\mathbf{B}}$ from the Lie algebra $aut_p(D_v)$ generates a one-parameter (commutative) subgroup $g(\tau)$ of the Lie group $Aut_p(D_v)$ of projective automorphisms. From (1.77), it follows that

$$g(\tau) = \varphi_{\mathbf{a}(\tau),U(\tau)}, \tag{1.140}$$

where $\mathbf{a}(\tau)$ is the solution of (1.131) with the initial condition $\mathbf{a}(0) = 0$. From (1.91), we have

$$\frac{d}{d\tau}g(\tau)|_{\tau=0}(\mathbf{v}) = \frac{q}{m_0}(\mathbf{E} + \mathbf{v} \times \mathbf{B} - c^{-2}\langle\mathbf{v}|\mathbf{E}\rangle\mathbf{v}) \tag{1.141}$$

for any $\mathbf{v} \in D_v$. We will show now that

$$\mathbf{v}(\tau) = g(\tau)(\mathbf{v}^0) = \varphi_{\mathbf{a}(\tau)}U(\tau)(\mathbf{v}^0), \tag{1.142}$$

is a solution of the initial-value problem consisting of the differential equation (1.131) and the initial condition

$$\mathbf{v}(0) = \mathbf{v}^0. \tag{1.143}$$

Since $\mathbf{a}(0) = 0$ and $U(0) = I$, the initial condition is satisfied. To show that $\mathbf{v}(\tau)$ satisfies (1.131), note that from (1.79), we have

$$g(s + \tau) = g(s)g(\tau) = g(\tau)g(s). \tag{1.144}$$

Thus, from (1.141) and (1.142), we get

$$\frac{d\mathbf{v}(\tau)}{d\tau} = g(s)\Big|_{s=0}\frac{d}{d\tau}g(\tau)(\mathbf{v}^0) = \frac{d}{d\tau}g(s+\tau)\Big|_{s=0}(\mathbf{v}^0) = \frac{d}{ds}g(s+\tau)\Big|_{s=0}(\mathbf{v}^0)$$

$$= \frac{d}{ds} g(s)\Big|_{s=0} \mathbf{v}(\tau) = \frac{q}{m_0} (\mathbf{E} + \mathbf{v}(\tau) \times \mathbf{B} - c^{-2} \langle \mathbf{v}(\tau) | \mathbf{E} \rangle \mathbf{v}(\tau)). \quad (1.145)$$

Thus, $\mathbf{v}(\tau)$ solves the initial-value problem. Thus, we have produced a method for solving the initial-value problem for the relativistic equation with a constant electromagnetic field, given *any* initial condition.

There is an alternative way to extend Newton's Second Law to relativity. Instead of using the hybrid connection, which led us to the bounded domain D_v of relativistically admissible velocities, we could have used the cascade connection, obtaining the domain of relativistically admissible *proper* velocities. Proper velocity was defined in (1.4) as $\mathbf{u} = d\mathbf{r}/d\tau$, where $d\tau = \sqrt{1 - |\mathbf{v}|^2/c^2} dt$, so $\mathbf{u} = \mathbf{v}/\sqrt{1 - |\mathbf{v}|^2/c^2}$. The set of all proper velocities is, therefore, not bounded, and, indeed, it is R^3. As a result, a constant vector field (force) on this set is possible as a generator of evolution. Thus, the equation

$$m_0 \frac{d\mathbf{u}}{dt} = F \quad (1.146)$$

makes sense, and, in fact, it coincides with the dynamic equation of relativistic mechanics (1.139) given by elements of $aut_p(D_v)$.

1.5.6 Charged particle in a constant uniform electromagnetic field

In this section, we will obtain an explicit description of the motion of a charged particle of rest-mass m_0 and charge q in three different constant electromagnetic fields $\mathbf{E}, \mathbf{B}$. In all three cases, we will solve the initial-value problem (1.131) with initial condition $\mathbf{v}(0) = \mathbf{v}^0 = (v_1^0, v_2^0, v_3^0)$. The first case is that of a constant electric field E $(B = 0)$. The second case is that of a constant magnetic field B $(E = 0)$. The third case is that of a constant electromagnetic field E, B in which the vectors E and B are parallel.

Constant electric field E

If the charge is in a constant electric field $\mathbf{E}$, then its motion will be described by integrating its velocity $\mathbf{v}(t)$ with respect to t. We will use the evolution equation (1.131) to find the velocity of the particle $\mathbf{v}(\tau)$ as a function of proper time. To do this, we have to solve the equation

$$m_0 \frac{d\mathbf{v}(\tau)}{d\tau} = q(\mathbf{E} - c^{-2} \langle \mathbf{v}(\tau) | \mathbf{E} \rangle \mathbf{v}(\tau)), \quad (1.147)$$

with the initial condition

$$\mathbf{v}(0) = \mathbf{v}^0 = (v_1^0, v_2^0, v_3^0). \quad (1.148)$$

Let us denote by $\mathbf{a}(\tau)$ the solution of the problem for $\mathbf{a}(0) = \mathbf{v}^0 = 0$. This implies that

$$\left.\frac{d\mathbf{a}(\tau)}{d\tau}\right|_{\tau=0} = \frac{q\mathbf{E}}{m_0}. \tag{1.149}$$

Without loss of generality, we may choose the axes so that the vector $\mathbf{E}$ points in the direction of the positive x-axis. Set $\mathbf{E} = (|\mathbf{E}|, 0, 0)$ and $\mathbf{a} = (a_1, a_2, a_3)$. Then, examining equation (1.147) in each coordinate, we see that

$$a_2(\tau) = a_3(\tau) = 0, \tag{1.150}$$

is a solution of (1.147) with $\mathbf{a}(0) = 0$. Thus, it remains only to find $a_1(\tau)$.

For $a_1(\tau)$, equation (1.147) becomes

$$\frac{da_1(\tau)}{d\tau} = \frac{q|\mathbf{E}|}{c^2 m_0}(c^2 - a_1^2(\tau)), \tag{1.151}$$

with $a_1(0) = 0$. Separating variables, we obtain

$$\frac{da_1(\tau)}{c^2 - a_1^2(\tau)} = \frac{q|\mathbf{E}|}{c^2 m_0} d\tau, \tag{1.152}$$

implying that

$$\ln \frac{c + a_1(\tau)}{c - a_1(\tau)} = 2\frac{q|\mathbf{E}|}{c m_0}\tau + 2c_0. \tag{1.153}$$

Define

$$\Omega = \frac{q|\mathbf{E}|}{c m_0}. \tag{1.154}$$

Taking the exponent of both sides, we get

$$a_1(\tau) = c\tanh(\Omega\tau + c_0). \tag{1.155}$$

From the initial condition $a_1(0) = 0$, it follows that $c_0 = 0$ and

$$\mathbf{a}(\tau) = (c\tanh(\Omega\tau), 0, 0), \tag{1.156}$$

with Ω defined by (1.154).

We can define now a one-parameter subgroup

$$g(\tau) = \varphi_{\mathbf{a}(\tau)}, \tag{1.157}$$

where $\mathbf{a}(\tau)$ is given by (1.156). Then, from (1.91) and (1.149), we have

$$\frac{d}{d\tau} g(\tau)|_{\tau=0}(\mathbf{v}) = \frac{d}{d\tau}\varphi_{\mathbf{a}(\tau)}|_{\tau=0}(\mathbf{v}) = \frac{q}{m_0}(\mathbf{E} - c^{-2}\langle\mathbf{v}|\mathbf{E}\rangle\mathbf{v}) \tag{1.158}$$

for any $\mathbf{v} \in D_v$. As shown in the previous section,

$$\mathbf{v}(\tau) = g(\tau)(\mathbf{v}^0) = \varphi_{\mathbf{a}(\tau)}(\mathbf{v}^0) \tag{1.159}$$

is the solution of the initial-value problem (1.147) and (1.148). By use of (1.66), we get

$$\mathbf{v}(\tau) = \mathbf{a}(\tau) \oplus_E \mathbf{v}^0 = (\mathbf{a}(\tau) \oplus_E \mathbf{v}_1) + \delta \mathbf{v}_2,$$

where $\mathbf{v}_1 = (v_1^0, 0, 0)$, $\mathbf{v}_2 = (0, v_2^0, v_3^0)$ and

$$\delta = \frac{\alpha(\mathbf{a}(\tau))}{1 + \langle \mathbf{a}(\tau) | \mathbf{v}_1 \rangle / c^2}.$$

Next, define τ_0 so that

$$c \tanh(\Omega \tau_0) = v_1^0, \tag{1.160}$$

which implies that $\gamma(\mathbf{v}_1) = \cosh(\Omega \tau_0)$. Then, from (1.64), we get

$$\mathbf{a}(\tau) \oplus_E \mathbf{v}_1 = (c \tanh(\Omega(\tau + \tau_0)), 0, 0), \tag{1.161}$$

and, by use of hyperbolic function identities, we get

$$\delta = \frac{\cosh(\Omega \tau_0)}{\cosh(\Omega(\tau + \tau_0))}. \tag{1.162}$$

Thus,

$$\mathbf{v}(\tau) = \frac{c \cosh(\Omega \tau_0)}{\cosh(\Omega(\tau + \tau_0))} \left(\frac{\sinh(\Omega(\tau + \tau_0))}{\cosh(\Omega \tau_0)}, v_2^0/c, v_3^0/c \right), \tag{1.163}$$

with Ω defined by (1.154) and τ_0 by (1.160). From this, it follows that

$$\gamma(\mathbf{v}(\tau)) = \gamma(\mathbf{v}^0) \frac{\cosh(\Omega(\tau + \tau_0))}{\cosh(\Omega \tau_0)}. \tag{1.164}$$

The space trajectory $\mathbf{r}(\tau)$ of the particle is obtained by adding the integral of $\mathbf{v}dt$ to its position $\mathbf{r}(0)$ at $t = 0$. Thus

$$\mathbf{r}(\tau) - \mathbf{r}(0) = \int_0^\tau \mathbf{v}(\tau) \gamma(\mathbf{v}(\tau)) d\tau$$

$$= \gamma(\mathbf{v}^0) \left(\frac{c}{\Omega} \left(\frac{\cosh(\Omega(\tau + \tau_0))}{\cosh(\Omega \tau_0)} - 1 \right), v_2^0 \tau, v_3^0 \tau \right), \tag{1.165}$$

which is called hyperbolic motion. The connection between time and proper time on the trajectory can be found via

$$t(\tau) = \int_0^\tau \gamma(\mathbf{v}(\tau)) = \frac{\gamma(\mathbf{v}^0)}{\Omega \cosh(\Omega \tau_0)} \sinh(\Omega \tau),$$

implying that

$$\tau = \frac{1}{\Omega} \sinh^{-1}(\frac{\Omega \cosh(\Omega \tau_0) t}{\gamma(\mathbf{v}^0)}). \tag{1.166}$$

Substituting this into (1.163), we obtain the solution $\mathbf{v}(t)$ of the initial-value problem. To find the trajectory $\mathbf{r}(t)$ of the particle, substitute (1.166) into (1.165).

Constant magnetic field B

Consider now the motion of a charged particle of rest-mass m_0 and charge q in a constant magnetic field $\mathbf{B}$. The equation of motion for such a particle is described by the evolution equation (1.131), with $\mathbf{E} = 0$. Hence, the initial-value problem to be solved is

$$m_0 \frac{d\mathbf{v}(\tau)}{d\tau} = q(\mathbf{v}(\tau) \times \mathbf{B}), \tag{1.167}$$

with the initial condition

$$\mathbf{v}(0) = \mathbf{v}^0 = (v_1^0, v_2^0, v_3^0). \tag{1.168}$$

Without loss of generality, we may choose the axes so that the vector $\mathbf{B}$ points in the direction of the positive x-axis. Set $\mathbf{B} = (|\mathbf{B}|, 0, 0)$ and $\mathbf{v}(\tau) = (v_1(\tau), v_2(\tau), v_3(\tau))$. Complexify the y-z plane by defining $z(\tau) = v_2(\tau) + iv_3(\tau)$. Define v_r^0 and α by

$$v_2^0 + iv_3^0 = e^{-i\alpha} v_r^0. \tag{1.169}$$

Then the initial-value problem (1.167) becomes

$$m_0 \frac{dz(\tau)}{d\tau} = -iq|\mathbf{B}|z(\tau), \quad \frac{dv_1(\tau)}{d\tau} = 0. \tag{1.170}$$

The solution of these equations is

$$z(\tau) = e^{-i(\omega\tau + \alpha)} v_r^0, \quad v_1(\tau) = v_1^0, \tag{1.171}$$

where

$$\omega = q|\mathbf{B}|/m_0. \tag{1.172}$$

This solution can be written as

$$\mathbf{v}(\tau) = (v_1^0, v_r^0 \cos(\omega\tau + \alpha), -v_r^0 \sin(\omega\tau + \alpha)), \tag{1.173}$$

or, equivalently,

$$\mathbf{v}(\tau) = (v_1^0, v_2^0 \cos\omega\tau + v_3^0 \sin\omega\tau, -v_2^0 \sin\omega\tau + v_3^0 \cos\omega\tau). \tag{1.174}$$

Note that $|\mathbf{v}(\tau)| = \sqrt{(v_1^0)^2 + (v_r^0)^2}$ is constant and $\gamma(\mathbf{v}(\tau)) = \gamma(\mathbf{v}^0)$, implying that $t = \gamma(\mathbf{v}^0)\tau$.

Let $\mathcal{R}_\varphi^{\mathbf{b}}$ denote the operator of rotation around the axis through the origin in the direction $\mathbf{b}$ by an angle φ. This operator can be expressed by an exponent of the vector field $\mathbf{v} \times \mathbf{b}/|\mathbf{b}|$, which generates the rotation, as

$$\mathcal{R}_\varphi^{\mathbf{b}} = \exp(\varphi J_{\mathbf{b}}), \;\; J_{\mathbf{b}}(\mathbf{v}) = \mathbf{v} \times \mathbf{b}/|\mathbf{b}|. \tag{1.175}$$

If $\mathbf{b} = \mathbf{i}$ (the direction of the positive x-axis), then the matrix representing $\mathcal{R}_\varphi^{\mathbf{i}}$ is

$$\mathcal{R}_\varphi^{\mathbf{i}} = \begin{pmatrix} 1 & 0 & 0 \\ 0 & \cos\varphi & -\sin\varphi \\ 0 & \sin\varphi & \cos\varphi \end{pmatrix}. \tag{1.176}$$

With this notation, we can express the solution of the initial-value problem (1.167) and (1.168) as

$$\mathbf{v}(\tau) = \mathcal{R}_{\omega\tau}^{\mathbf{B}} \mathbf{v}^0 = \varphi_{0,U(\tau)}(\mathbf{v}^0) = U(\tau)(\mathbf{v}^0), \tag{1.177}$$

where $U(\tau) = \mathcal{R}_{\omega\tau}^{\mathbf{B}}$ denotes the one-parameter subgroup generated by the magnetic field.

The space trajectory $\mathbf{r}(\tau)$ of the particle is obtained by adding the integral of $\mathbf{v}dt$ to its position $\mathbf{r}(0)$ at $t = 0$. Thus, from (1.173), we get

$$\mathbf{r}(\tau) - \mathbf{r}(0) = \int_0^\tau \mathbf{v}(\tau)\gamma(\mathbf{v}(\tau))d\tau$$

$$= \gamma(\mathbf{v}^0)(v_1^0\tau, \frac{v_r^0}{\omega}\sin(\omega\tau + \alpha), \frac{v_r^0}{\omega}\cos(\omega\tau + \alpha)). \tag{1.178}$$

Switching to t, we obtain

$$\mathbf{r}(t) - \mathbf{r}(0) = (v_1^0 t, \frac{v_r^0}{\omega_0}\sin(\omega_0 t + \alpha), \frac{v_r^0}{\omega_0}\cos(\omega_0 t + \alpha)), \tag{1.179}$$

where $\omega_0 = \omega\gamma^{-1}(\mathbf{v}^0)$ and ω is defined by (1.172). Thus, a particle in a constant magnetic field moves with angular velocity ω_0 along a helix whose axis is in the direction of the magnetic field.

Constant and parallel electric field E and magnetic field B

We consider now the motion of a charged particle of rest-mass m_0 and charge q in a constant electric field $\mathbf{E}$ and magnetic field $\mathbf{B}$ in which the vectors $\mathbf{E}$

and $\mathbf{B}$ are parallel. The equation of motion for such a particle is described by the evolution equation (1.131). Hence, the initial-value problem to be solved is

$$m_0 \frac{d\mathbf{v}(\tau)}{d\tau} = q(\mathbf{E} + \mathbf{v}(\tau) \times \mathbf{B} - c^{-2}\langle \mathbf{v}(\tau)|\mathbf{E}\rangle \mathbf{v}(\tau)), \tag{1.180}$$

with the initial condition

$$\mathbf{v}(0) = \mathbf{v}^0 = (v_1^0, v_2^0, v_3^0). \tag{1.181}$$

As mentioned at the end of section 1.5.2, page 39, in this case the generator $\delta_{\mathbf{E},0}$ of the electric field and the generator $\delta_{0,\mathbf{B}}$ of the magnetic field commute. This implies that the flows on the velocity ball D_v generated by each field individually also commute. Thus, we can solve the problem separately for each field and compose the results to obtain the flow generated by the combined field.

More precisely, without loss of generality, we may choose the axes so that the vector $\mathbf{B}$ points in the direction of the positive x-axis. Then $\mathbf{E}$ is also parallel to the x-axis. The flow generated by $\mathbf{E}$, by (1.159), is given by $\mathbf{v}^0 \to \varphi_{\mathbf{a}(\tau)}\mathbf{v}^0$, where $\mathbf{a}(\tau)$ is defined by (1.156), with Ω defined by (1.154). Similarly, the flow generated by $\mathbf{B}$ is given by (1.177) and is $\mathbf{v}^0 \to \mathcal{R}^{\mathbf{B}}_{\omega\tau}\mathbf{v}^0$, where $\mathcal{R}^{\mathbf{B}}_{\omega\tau}$ is defined by (1.176) and ω is defined by (1.172). We will show now that

$$\mathbf{v}(\tau) = \varphi_{\mathbf{a}(\tau)}(\mathcal{R}^{\mathbf{B}}_{\omega\tau}\mathbf{v}^0) = \mathbf{a}(\tau) \oplus_E (\mathcal{R}^{\mathbf{B}}_{\omega\tau}\mathbf{v}^0) \tag{1.182}$$

solves the initial-value problem (1.180)–(1.181).

Obviously, $\mathbf{v}(\tau)$ satisfies the initial condition $\mathbf{v}(0) = \mathbf{v}^0$. From (1.173), (1.177) and (1.163), we get

$$\mathbf{v}(\tau) = \frac{\cosh(\Omega\tau_0)}{\cosh(\Omega(\tau+\tau_0))}\left(\frac{c\sinh(\Omega(\tau+\tau_0))}{\cosh(\Omega\tau_0)} v_r^0 \cos(\omega\tau+\alpha), -v_r^0 \sin(\omega\tau+\alpha)\right), \tag{1.183}$$

where Ω is defined by (1.154), ω is defined by (1.172), τ_0 by (1.160) and v_r^0, α by (1.169). From this, it follows that

$$\gamma(\mathbf{v}(\tau)) = \gamma(\mathbf{v}^0)\frac{\cosh(\Omega(\tau+\tau_0))}{\cosh(\Omega\tau_0)}. \tag{1.184}$$

To find the trajectory $\mathbf{r}(\tau)$ of the particle, we need to add to its position at $t = 0$ the integral of $\mathbf{v}dt$. By use of (1.165) and (1.178), we get

$$\mathbf{r}(\tau) - \mathbf{r}(0) = \int_0^t \mathbf{v}(\tau)\gamma(\mathbf{v}(\tau))d\tau$$

$$= \gamma(\mathbf{v}^0)\left(\frac{c}{\Omega}(\frac{\cosh(\Omega(\tau+\tau_0))}{\cosh(\Omega\tau_0)} - 1), \frac{v_r^0\tau}{\omega}\sin(\omega\tau+\alpha), \frac{v_r^0\tau}{\omega}\cos(\omega\tau+\alpha)\right). \tag{1.185}$$

1.6 Notes

Einstein's original axiomatic derivation [20] of the Lorentz transformations is based on two assumptions:

- the principle of special relativity,
- the constancy of the speed of light in all inertial frames.

From 1910 until the present, much research has been done to show that the Lorentz transformations can be derived from weaker assumptions. See, for example, [63]. A derivation of the Lorentz transformations from the principle of special relativity and a symmetry based on space-time invariance was obtained by J. H. Field in [24]. In [66], Y. Terletskii derived the Lorentz transformations from the principle of relativity, isotropy of space and homogeneity of space and time. He also reversed the space axes to preserve the symmetry. As was shown by C. Marchal in [54], the Lorentz transformations, up to a constant, are a direct consequence of the principle of special relativity and the symmetry of the transformations between two inertial systems.

In [50], A. Lee and T.M. Kalotas showed that the Lorentz transformations up to an unknown constant are a manifestation of the properties of the space-time of inertial systems, such as homogeneity of time and isotropy of space. They derived a relativistic velocity-addition formula and showed that this constant is non-negative. We have used here their argument to show the non-negativity of our constant e. In [55], D. Mermin showed that the relativistic addition law of parallel velocities with some universal constant can be derived directly from the principle of relativity and assumptions of homogeneity, smoothness and symmetry, *without* making use of the constancy of the speed of light. From these assumptions, he showed that there is an invariant velocity depending on this universal constant. This chapter is based mainly on ideas which first appeared in [30] and [25] and were developed further in [28] for special relativity and in [29] for accelerated systems.

The main deviations of the approach in this chapter from the standard approaches to relativity are

- the formulation of the principle of special relativity as a symmetry,
- the choice of axes to preserve the symmetry,
- the consistency of inputs and outputs for the transformations and the description of the systems,
- the choice of parameters to simplify the transformations,
- the introduction of a weight on time which makes the eigenvectors of the symmetry orthogonal (as in the Sturm–Liouville theory), thus leading to conservation of intervals,
- the use of the algebraic structure of the conserved bounded symmetric domain to describe the evolution of systems.

The evolution equation (1.131) and the one-parameter group associated with it was used in the last section to derive an explicit description of the motion of particles in a constant, uniform (a) electric, (b) magnetic and (c) parallel electric and magnetic fields for any initial conditions. Similar solutions may be found in [51], pp. 52–57 and [42], pp. 579–592. In the next chapter, we will obtain an *explicit* solution for the motion of a charged particle in a constant, uniform electromagnetic field in which E and B are perpendicular.

2 The real spin domain

In the previous chapter, we used the principle of special relativity to obtain the real bounded symmetric domain D_v. This domain is symmetric with respect to the projective automorphisms and is a domain of type I in the Cartan classification of bounded symmetric domains. In this chapter, we will discuss another real domain, called the *real spin factor*, which is a domain of type IV in the Cartan classification. The *complex* spin factor will be studied in Chapter 3.

We introduce the real spin factor as the ball D_s of symmetric velocities, defined in Chapter 1, section 1.2.1. We derive a formula for the addition of symmetric velocities and define an automorphism group based on this addition. We show that this group is exactly the group $Aut_c(D_s)$ of *conformal* automorphisms of D_s and that D_s is symmetric with respect to $Aut_c(D_s)$. We then show that the elements of the Lie algebra of $Aut_c(D_s)$ are expressible in terms of a triple product, which we call the *spin triple product*.

Next, we show that the relativistic evolution equations of mechanics and electromagnetism can be written by use of the above Lie algebra. This produces a new method of solving relativistic dynamic equations. If the motion has an invariant plane, the equation of evolution for the symmetric velocity becomes a first-order *analytic* equation in one complex variable. We apply this method to the description of the motion of a charged particle in uniform, constant and mutually perpendicular electric and magnetic fields. We find explicit analytic solutions for this problem. We also obtain a conformal group representation of the Lorentz group.

2.1 Symmetric velocity addition

2.1.1 The meaning of s-velocity and s-velocity addition

In the previous chapter, we defined the symmetric velocity (1.38) and (1.67) as the velocity of the eigenspace corresponding to the 1-eigenvalue, or the fixed points, of the relativistic transformations between two inertial systems. From this point on, we will assume the conservation of the speed of light. Equation (1.71) shows that the new dynamic variable, *symmetric velocity*, is the relativistic half of the corresponding velocity. The symmetric velocity $\mathbf{w}_1$ and its corresponding velocity $\mathbf{v}$ are related by

$$\mathbf{v} = \mathbf{w}_1 \oplus_E \mathbf{w}_1 = \frac{\mathbf{w}_1 + \mathbf{w}_1}{1 + \frac{|\mathbf{w}_1|}{c}\frac{|\mathbf{w}_1|}{c}} = \frac{2\mathbf{w}_1}{1 + |\mathbf{w}_1|^2/c^2},$$

where $\oplus_E$ denotes Einstein velocity addition. Instead of $\mathbf{w}_1$, we prefer to use a unit-free vector $\mathbf{w} = \mathbf{w}_1/c$ and call it *s-velocity*. Thus, the relationship between an s-velocity $\mathbf{w}$ and its corresponding velocity $\mathbf{v}$ is given by

$$\mathbf{v} = \Phi(\mathbf{w}) = \frac{2c\mathbf{w}}{1 + |\mathbf{w}|^2}, \tag{2.1}$$

where Φ denotes the function mapping the s-velocity $\mathbf{w}$ to its velocity $\mathbf{v}$.

Conversely, the s-velocity $\mathbf{w}$ can be expressed in terms of $\mathbf{v}$ by

$$\mathbf{w} = \Phi^{-1}(\mathbf{v}) = \frac{\mathbf{v}/c}{1 + \sqrt{1 - |\mathbf{v}|^2/c^2}} = \frac{\gamma\boldsymbol{\beta}}{1+\gamma}, \tag{2.2}$$

where $\gamma = 1/\sqrt{1 - |\mathbf{v}|^2/c^2}$ and $\boldsymbol{\beta} = \mathbf{v}/c$. From this we see that $\mathbf{w} \to \boldsymbol{\beta}/2$ as $\boldsymbol{\beta} \to 0$, and $\mathbf{w} \to \boldsymbol{\beta}$ as $\boldsymbol{\beta} \to 1$. The set of all relativistically admissible s-velocities form a unit ball

$$D_s = \{\mathbf{w} \in R^3 : \ |\mathbf{w}| < 1\}. \tag{2.3}$$

Recall the physical interpretation of the symmetric velocity. Consider two inertial systems with relative velocity $\mathbf{v}$ between them. Place two objects of equal mass (test masses) at the origin of each inertial system. The center of mass of the two objects will be called the *center of the two inertial systems.* The symmetric velocity is the velocity of each system with respect to the center of the systems, and the s-velocity is the unit-free velocity of the systems with respect to their center (see Figure 2.1).

We will show that the ball of all relativistically admissible s-velocities D_s is a bounded symmetric domain with respect to the automorphisms generated by s-velocity addition. To define this addition, we shall consider three inertial systems K_1, K_2 and K_3. We choose the space axes of K_2 to be parallel to the axes of K_1 and the axes of K_3 to be parallel to those of K_2. Denote their origins by O_1, O_2 and O_3, respectively. Denote by $\mathbf{a}$ the s-velocity of system K_2 with respect to K_1 and by $\mathbf{w}$ the s-velocity of system K_3 with respect to K_2. Then the s-velocity $\mathbf{w}_3$ of system K_3 with respect to K_1 (*i.e.*, the velocity of K_3 with respect to the center of systems K_1 and K_3) will be called the *sum of the s-velocities* $\mathbf{a}$ and $\mathbf{w}$ and will be denoted by $\mathbf{a} \oplus_s \mathbf{w}$ (see Figure 2.2).

2.1.2 Derivation of the s-velocity addition formula

We now calculate the s-velocity sum $\mathbf{a} \oplus_s \mathbf{w}$. Let us denote by $\mathbf{v}$ the relative velocity of system K_2 with respect to K_1 and by $\mathbf{u}$ the relative velocity of system K_3 with respect to K_2. Then

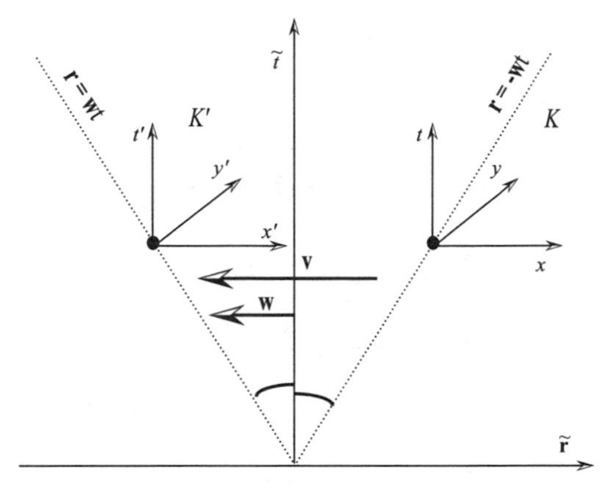

Fig. 2.1. The physical meaning of symmetric velocity. Two inertial systems K and K' with relative velocity $\mathbf{v}$ between them are viewed from the system connected to their center. In this system, K and K' are each moving with velocity $\pm\mathbf{w}$.

$$\mathbf{v} = \frac{2c\mathbf{a}}{1+|\mathbf{a}|^2}, \quad \mathbf{u} = \frac{2c\mathbf{w}}{1+|\mathbf{w}|^2}. \tag{2.4}$$

From the definition of Einstein velocity addition, the relative velocity of system K_3 with respect to K_1 is $\mathbf{v} \oplus_E \mathbf{u}$, which, using (2.2), gives

$$\mathbf{a} \oplus_s \mathbf{w} = \frac{(\mathbf{v} \oplus_E \mathbf{u})/c}{1+\alpha(\mathbf{v} \oplus_E \mathbf{u})}, \tag{2.5}$$

where $\alpha(\mathbf{v}) = \sqrt{1-|\mathbf{v}|^2/c^2}$ for any velocity $\mathbf{v}$. By (1.57) from page 25, we have

$$\mathbf{v} \oplus_E \mathbf{u} = \frac{1}{1+\langle \mathbf{v}|\mathbf{u}\rangle/c^2}(\mathbf{v} + \alpha(\mathbf{v})\mathbf{u} + \frac{1}{(1+\alpha(\mathbf{v}))c^2}\langle \mathbf{v}|\mathbf{u}\rangle \mathbf{v}), \tag{2.6}$$

and by (1.58), we have

$$\alpha(\mathbf{v} \oplus_E \mathbf{u}) = \frac{\alpha(\mathbf{v})\alpha(\mathbf{u})}{1+\langle \mathbf{v} \mid \mathbf{u}\rangle/c^2}, \tag{2.7}$$

implying that

$$\mathbf{a} \oplus_s \mathbf{w} = \frac{(\mathbf{v} + \alpha(\mathbf{v})\mathbf{u} + \frac{1}{(1+\alpha(\mathbf{v}))c^2}\langle \mathbf{v}|\mathbf{u}\rangle \mathbf{v})/c}{1+\langle \mathbf{v}|\mathbf{u}\rangle/c^2 + \alpha(\mathbf{v})\alpha(\mathbf{u})} \tag{2.8}$$

From the definition of α, we have

$$\alpha(\mathbf{v}) = \frac{1-|\mathbf{a}|^2}{1+|\mathbf{a}|^2}, \quad \alpha(\mathbf{u}) = \frac{1-|\mathbf{w}|^2}{1+|\mathbf{w}|^2}, \tag{2.9}$$

and so

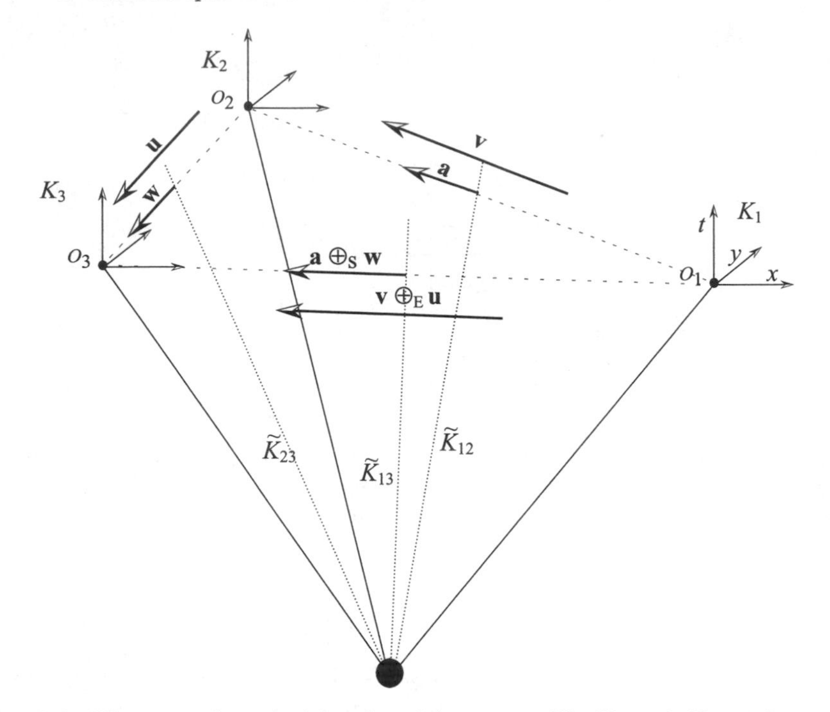

Fig. 2.2. The sum of s-velocities. Inertial systems K_1, K_2 and K_3, with origins O_1, O_2 and O_3, respectively, had a common origin at time $t = 0$. The line $\widetilde{K}_{12}$ is the world-line of the center of the two inertial systems K_1 and K_2. Similarly, the lines $\widetilde{K}_{23}$ and $\widetilde{K}_{13}$ represent the world-lines of the centers of the systems K_2, K_3 and K_1, K_3, respectively. The velocity of system K_2 with respect to system K_1 is $\mathbf{v}$, and its s-velocity $\mathbf{a}$ is the velocity of K_2 with respect to $\widetilde{K}_{12}$. Similarly, the velocity of system K_3 with respect to system K_2 is $\mathbf{u}$, and its s-velocity $\mathbf{w}$ is the velocity of K_3 with respect to $\widetilde{K}_{23}$. The velocity of system K_3 with respect to system K_1 is, by definition of Einstein velocity addition, equal to $\mathbf{v} \oplus_E \mathbf{u}$. The s-velocity of K_3 with respect to K_1, meaning the unit-free velocity of K_3 with respect to $\widetilde{K}_{13}$, is called the sum of symmetric velocities $\mathbf{a}$ and $\mathbf{w}$ and is denoted by $\mathbf{a} \oplus_s \mathbf{w}$.

$$1 + \alpha(\mathbf{v}) = \frac{2}{1 + |\mathbf{a}|^2}, \quad \frac{1}{1 + \alpha(\mathbf{v})} = \frac{1 + |\mathbf{a}|^2}{2}. \tag{2.10}$$

Substituting these expressions into (2.8), we obtain

$$\mathbf{a} \oplus_s \mathbf{w} = \frac{\left(\frac{2c\mathbf{a}}{1+|\mathbf{a}|^2} + \frac{1-|\mathbf{a}|^2}{1+|\mathbf{a}|^2}\frac{2c\mathbf{w}}{1+|\mathbf{w}|^2} + \frac{1+|\mathbf{a}|^2}{2}\frac{4\langle \mathbf{a}|\mathbf{w}\rangle}{(1+|\mathbf{a}|^2)(1+|\mathbf{w}|^2)}\frac{2c\mathbf{a}}{1+|\mathbf{a}|^2}\right)/c}{1 + \frac{4\langle \mathbf{a}|\mathbf{w}\rangle}{(1+|\mathbf{a}|^2)(1+|\mathbf{w}|^2)} + \frac{1-|\mathbf{a}|^2}{1+|\mathbf{a}|^2}\frac{1-|\mathbf{w}|^2}{1+|\mathbf{w}|^2}}$$

$$= \frac{(1 + |\mathbf{w}|^2 + 2\langle \mathbf{a} \mid \mathbf{w}\rangle)\mathbf{a} + (1 - |\mathbf{a}|^2)\mathbf{w}}{1 + |\mathbf{a}|^2|\mathbf{w}|^2 + 2\langle \mathbf{a} \mid \mathbf{w}\rangle}. \tag{2.11}$$

Thus, we obtain the s-velocity-addition formula

$$\mathbf{a} \oplus_s \mathbf{w} = \frac{(1 + |\mathbf{w}|^2 + 2\langle \mathbf{a} \mid \mathbf{w} \rangle)\mathbf{a} + (1 - |\mathbf{a}|^2)\mathbf{w}}{1 + |\mathbf{a}|^2|\mathbf{w}|^2 + 2\langle \mathbf{a} \mid \mathbf{w} \rangle}. \tag{2.12}$$

It is sometimes useful to express the Einstein velocity addition $\mathbf{v} \oplus_E \mathbf{u}$ in terms of the addition of their corresponding s-velocities. From (2.2) and the definition of s-velocity addition, it follows that for any two velocities $\mathbf{v}$ and $\mathbf{u}$, we have

$$\mathbf{v} \oplus_E \mathbf{u} = \Phi(\Phi^{-1}(\mathbf{v}) \oplus_s \Phi^{-1}(\mathbf{u})). \tag{2.13}$$

2.1.3 *S*-velocity addition on the complex plane

To understand (2.12), note that $\mathbf{a} \oplus_s \mathbf{w}$ is a linear combination of $\mathbf{a}$ and $\mathbf{w}$ and therefore belongs to the plane Π generated by $\mathbf{a}$ and $\mathbf{w}$. We introduce a complex structure on Π in such a way that the disk $\Delta = D_s \cap \Pi$ is homeomorphic to the unit disc $|z| < 1$. Denote by a the complex number corresponding to the vector $\mathbf{a}$ and by w the complex number corresponding to the vector $\mathbf{w}$. It is known that

$$Re\,\langle a \mid w \rangle = \frac{\overline{a}w + a\overline{w}}{2},\ |w|^2 = w\overline{w}, \tag{2.14}$$

where the bar denotes complex conjugation. Substituting this into (2.12), we get

$$a \oplus_s w = \frac{(1 + w\overline{w} + \overline{a}w + a\overline{w})a + (1 - a\overline{a})w}{1 + a\overline{a}w\overline{w} + \overline{a}w + a\overline{w}} = \frac{(a+w)(1+a\overline{w})}{(1+\overline{a}w)(1+a\overline{w})} = \frac{a+w}{1+\overline{a}w}, \tag{2.15}$$

which is the well-known Möbius transformation of the complex unit disk. Thus, s-velocity addition is a generalization of the Möbius addition of complex numbers (see Figure 2.3). Note that formula (2.15) does not depend on the choice of the complexification of the disk $\Delta = D_s \cap \Pi$. For if we map $\mathbf{a}$ to $e^{i\theta}a$ (instead of to a) and $\mathbf{w}$ to $e^{i\theta}w$ (instead of to w), then

$$e^{i\theta}a \oplus_s e^{i\theta}w = \frac{e^{i\theta}a + e^{i\theta}w}{1 + \overline{e^{i\theta}a}e^{i\theta}w} = \frac{e^{i\theta}a + e^{i\theta}w}{1 + \overline{a}w} = e^{i\theta}(a \oplus_s w). \tag{2.16}$$

Note that (2.12) has meaning not only for vectors in R^3, but also for vectors in R^n, for arbitrary n. If we define

$$D_s^n = \{\mathbf{w} \in R^n :\ \ |\mathbf{w}| < 1\}, \tag{2.17}$$

then from the connection of s-velocity addition and the Möbius transformation, it follows that if $\mathbf{a}$ and $\mathbf{w}$ belong to D_s^n, then the sum $\mathbf{a} \oplus_s \mathbf{w}$, defined by (2.12), also belongs to D_s^n.

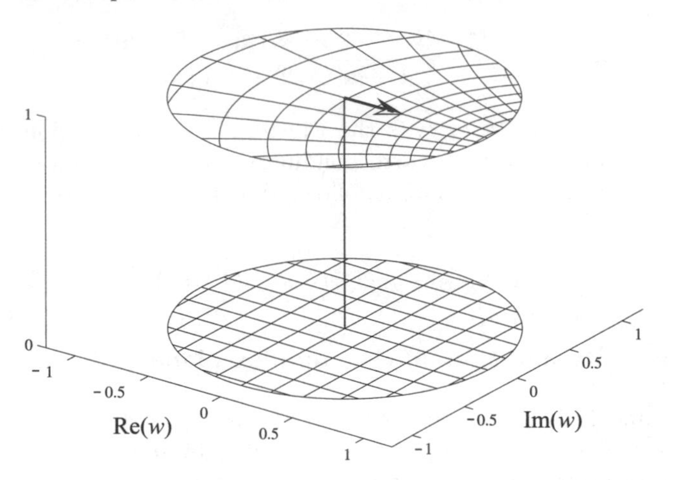

Fig. 2.3. Symmetric velocity addition $a \oplus_s w$ for $a = 0.4$. The lower circle in the figure is the unit disc of the complex plane, representing a two-dimensional section of the s-velocity ball D_s. The upper circle is the image of the lower circle under the transformation $w \to \frac{w+a}{1+\overline{a}w}$. Each circle is enhanced with a grid to highlight the effect of this transformation. Notice how a typical square of the lower grid is deformed and changes in size under the transformation.

2.2 Projective and conformal commutativity and associativity

2.2.1 Non-commutativity of s-velocity addition

Let a and w be two arbitrary complex numbers in the unit disc of the complex plane. If we switch the roles of a and w in $a \oplus_s w = \frac{a+w}{1+\overline{a}w}$, the numerator will remain the same, but the denominator will transform to its conjugate and hence will not be the same unless $\overline{a}w$ is real. But $\overline{a}w$ is real if and only if a and w are parallel. Thus $\mathbf{a} \oplus_s \mathbf{w}$ is equal to $\mathbf{w} \oplus_s \mathbf{a}$ if and only if the vectors $\mathbf{a}$ and $\mathbf{w}$ are parallel. In this case, $\mathbf{a} \oplus_s \mathbf{w} = (c\mathbf{a} \oplus_E c\mathbf{w})/c$, implying that for parallel vectors, Einstein and symmetric velocity addition coincide up to scaling.

Observe that, in general,

$$a \oplus_s w = \frac{a+w}{1+\overline{a}w} = \frac{1+a\overline{w}}{1+\overline{a}w} \cdot \frac{w+a}{1+a\overline{w}} = \lambda(w \oplus_s a), \tag{2.18}$$

where

$$\lambda = \frac{1+a\overline{w}}{1+\overline{a}w}. \tag{2.19}$$

Since $|\lambda| = 1$, we have $\lambda = e^{i\beta}$ for some angle β. Hence,

$$a \oplus_s w = e^{i\beta}(w \oplus_s a). \tag{2.20}$$

From the definition of λ, we have

$$\beta = \arg\lambda = \arg(\frac{1+a\overline{w}}{1+\overline{a}w}) = -2\arg(1+\overline{a}w). \tag{2.21}$$

So $\arg(1+\overline{a}w) = -\beta/2$, or

$$\tan(\beta/2) = -\frac{Im(1+\overline{a}w)}{Re(1+\overline{a}w)}.$$

For the next step, notice that the complex number $1+\overline{a}w$ is independent of the complexification, for if $\mathbf{a} \to e^{i\theta}a$ and $\mathbf{w} \to e^{i\theta}w$, then

$$e^{-i\theta}\overline{a}e^{i\theta}w = \overline{a}w.$$

So we choose a complexification in which a is real and positive. Let θ be the angle from a to w, or, equivalently, the angle between the corresponding symmetric velocities $\mathbf{a}$ and $\mathbf{w}$. Then

$$\tan(\beta/2) = -\frac{Im(1+\overline{a}w)}{Re(1+\overline{a}w)} = -\frac{aIm(w)}{1+aRe(w)}$$

$$= -\frac{a\,|w|\sin\theta}{1+a\,|w|\cos\theta} = -\frac{|\mathbf{a}|\,|\mathbf{w}|\sin\theta}{1+|\mathbf{a}|\,|\mathbf{w}|\cos\theta}. \tag{2.22}$$

Thus, the non-commutativity of the addition of two symmetric velocities $\mathbf{a}$ and $\mathbf{w}$ is given by an operator of rotation by an angle β, defined by (2.22), in the plane Π (generated by $\mathbf{a}$ and $\mathbf{w}$) with respect to the axis through the origin in the direction $\mathbf{a}\times\mathbf{w}$, which is perpendicular to Π (see Figure 2.4). This operator was called the *gyration operator* by A.A. Ungar [67] and denoted $\text{gyr}[\mathbf{a},\mathbf{w}]$. For the Möbius addition in the complex plane, the gyration operator is multiplication by the number $e^{i\beta}$. Thus, (2.20) can be written as

$$a \oplus_s w = e^{i\beta}(w \oplus_s a) = \text{gyr}_c[a,w](w \oplus_s a), \tag{2.23}$$

where

$$\text{gyr}_c[a,w] = \frac{1+a\overline{w}}{1+\overline{a}w}. \tag{2.24}$$

Recall that in Chapter 1, page 50, we introduced a rotation operator $\mathcal{R}_{\varphi}^{\mathbf{b}}$ denoting rotation around the axis through the origin in the direction $\mathbf{b}$ by an angle φ. By use of this operator, we can express the gyration operator as

$$\text{gyr}_c[\mathbf{a},\mathbf{w}] = \mathcal{R}_{\beta}^{\mathbf{a}\times\mathbf{w}}, \tag{2.25}$$

where β is defined by (2.22). The gyration operator expresses the non-commutativity of addition in D_s:

$$\mathbf{a} \oplus_s \mathbf{w} = \text{gyr}_c[\mathbf{a},\mathbf{w}](\mathbf{w} \oplus_s \mathbf{a}). \tag{2.26}$$

See Figure 2.4**(a)**. We will follow [67] and call equation (2.26) the *commutative law for conformal geometry.*

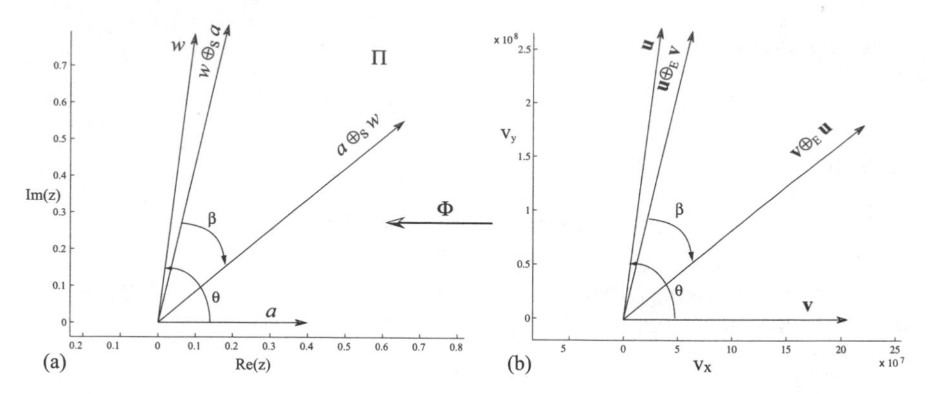

Fig. 2.4. The non-commutativity of s-velocity and Einstein velocity additions. **(a)** Two s-velocities $a = 0.4$ and $w = 0.1 + i0.85$. The sum $a \oplus_s w = 0.676 + i0.596$, while the sum $w \oplus_s a = 0.193 + i0.88$. The angle θ is the angle between a and w, while β is the angle between $a \oplus_s w$ and $w \oplus_s a$. The two angles are related by (2.22). The two sums $a \oplus_s w$ and $w \oplus_s a$ have the same length, and $a \oplus_s w = e^{i\beta}(w \oplus_s a) = \text{gyr}[a, w](w \oplus_s a)$. **(b)** The two velocities $\mathbf{v} = (2.07, 0, 0)10^8 m/s$, corresponding to s-velocity a, and $\mathbf{u} = (0.35, 2.94, 0)10^8 m/s$, corresponding to w. The sum $\mathbf{v} \oplus_E \mathbf{u} = (2.24, 1.98, 0)10^8 m/s$, while the sum $\mathbf{u} \oplus_E \mathbf{v} = (0.64, 2.91, 0)10^8 m/s$. The angle θ is the angle between $\mathbf{v}$ and $\mathbf{u}$, while β is the angle between $\mathbf{v} \oplus_E \mathbf{u}$ and $\mathbf{u} \oplus_E \mathbf{v}$. The two angles are related by (2.30). The two sums have the same length, and $\mathbf{v} \oplus_E \mathbf{u} = \text{gyr}_p[\mathbf{v}, \mathbf{u}](\mathbf{u} \oplus_E \mathbf{v})$.

2.2.2 Non-commutativity of Einstein velocity addition

Using the map Φ defined by (2.2), connecting a velocity and its s-velocity, and formula (2.13), connecting Einstein velocity addition and s-velocity addition, we can apply (2.26) to express the non-commutativity of Einstein velocity addition. For any two velocities $\mathbf{v}$ and $\mathbf{u}$, we have

$$\Phi(\mathbf{v} \oplus_E \mathbf{u}) = \Phi(\mathbf{v}) \oplus_s \Phi(\mathbf{u}) = \text{gyr}_c[\Phi(\mathbf{v}), \Phi(\mathbf{u})](\Phi(\mathbf{u}) \oplus_s \Phi(\mathbf{v}))$$

$$= \text{gyr}_c[\Phi(\mathbf{v}), \Phi(\mathbf{u})]\Phi(\mathbf{u} \oplus_E \mathbf{v}). \tag{2.27}$$

The map Φ and the operator $\text{gyr}_c[\Phi(\mathbf{v}), \Phi(\mathbf{u})]$ commute because Φ is a radial function and $\text{gyr}_c[\Phi(\mathbf{v}), \Phi(\mathbf{u})]$ is a rotation. Thus,

$$\Phi(\mathbf{v} \oplus_E \mathbf{u}) = \Phi(\text{gyr}_c[\Phi(\mathbf{v}), \Phi(\mathbf{u})](\mathbf{u} \oplus_E \mathbf{v})). \tag{2.28}$$

We define the *projective gyration operator* by

$$\text{gyr}_p[\mathbf{v}, \mathbf{u}] = \text{gyr}_c[\Phi(\mathbf{v}), \Phi(\mathbf{u})]. \tag{2.29}$$

This operator is a rotation in the plane Π generated by $\mathbf{v}$ and $\mathbf{u}$ by an angle β defined by

$$\tan(\beta/2) = -\frac{|\Phi(\mathbf{v})|\,|\Phi(\mathbf{u})|\sin\theta}{1+|\Phi(\mathbf{v})|\,|\Phi(\mathbf{u})|\cos\theta}, \tag{2.30}$$

where θ is the angle between $\mathbf{v}$ and $\mathbf{u}$. Thus,

$$\mathrm{gyr}_p[\mathbf{v},\mathbf{u}] = \mathcal{R}_\beta^{\mathbf{v}\times\mathbf{u}}. \tag{2.31}$$

From (2.28), we get the commutative law for Einstein velocity addition, which is also called the commutative law of projective geometry:

$$\mathbf{v} \oplus_E \mathbf{u} = \mathrm{gyr}_p[\mathbf{v},\mathbf{u}](\mathbf{u} \oplus_E \mathbf{v}), \tag{2.32}$$

(see Figure 2.4 **(b)**).

2.2.3 Non-associativity of s-velocity and Einstein velocity addition

Now we want to derive the analogs of (2.26) and (2.32) for the associative law. For s-velocity addition, this means finding the connection between $\mathbf{a} \oplus_s (\mathbf{b} \oplus_s \mathbf{w})$ and $(\mathbf{a} \oplus_s \mathbf{b}) \oplus_s \mathbf{w}$ for any s-velocities $\mathbf{a}, \mathbf{b}$ and $\mathbf{w}$. Let Π be the plane generated by $\mathbf{a}$ and $\mathbf{b}$, and assume first that $\mathbf{w} \in \Pi$. Complexify Π and observe that

$$a \oplus_s (b \oplus_s w) = a \oplus_s \left(\frac{b+w}{1+\bar{b}w}\right) = \frac{a+\frac{b+w}{1+\bar{b}w}}{1+\bar{a}\frac{b+w}{1+\bar{b}w}} = \frac{a+a\bar{b}w+b+w}{1+\bar{b}w+\bar{a}b+\bar{a}w}$$

$$= \frac{\frac{a+b}{1+\bar{a}b}+\frac{1+a\bar{b}}{1+\bar{a}b}w}{1+\frac{\bar{a}+\bar{b}}{1+\bar{a}b}w} = \frac{\frac{a+b}{1+\bar{a}b}+\frac{1+a\bar{b}}{1+\bar{a}b}w}{1+\frac{\bar{a}+\bar{b}}{1+a\bar{b}}\frac{1+a\bar{b}}{1+\bar{a}b}w} = (a \oplus_s b) \oplus_s \frac{1+a\bar{b}}{1+\bar{a}b}w. \tag{2.33}$$

Using the definition of the gyration operator (2.23), we can rewrite this as

$$a \oplus_s (b \oplus_s w) = (a \oplus_s b) \oplus_s \mathrm{gyr}_c[a,b]w. \tag{2.34}$$

Returning to the s-velocities and using (2.25), we get

$$\mathbf{a} \oplus_s (\mathbf{b} \oplus_s \mathbf{w}) = (\mathbf{a} \oplus_s \mathbf{b}) \oplus_s \mathrm{gyr}_c[\mathbf{a},\mathbf{b}]\mathbf{w}, \tag{2.35}$$

for any $\mathbf{w} \in \Pi$. As we will see in the next section, the map $\mathbf{w} \to \mathbf{a} \oplus_s \mathbf{w}$, for any fixed $\mathbf{a}$, is a conformal map on D_s. This implies that both sides of (2.35) are conformal maps in $\mathbf{w}$ of the unit ball $D_s \in R^3$, which coincide on the intersection of the ball with the plane Π. By a uniqueness theorem for conformal maps, they must agree for *any* $\mathbf{w} \in D_s$. Equation (2.35) is called the *associative law for s-velocities.*

Turning to Einstein velocity addition, let $\mathbf{d}, \mathbf{v}$ and $\mathbf{u}$ be any three velocities. Using (2.35), (2.13) and (2.29), we get

$$\mathbf{d} \oplus_E (\mathbf{v} \oplus_E \mathbf{u}) = (\mathbf{d} \oplus_E \mathbf{v}) \oplus_E \mathrm{gyr}_p[\mathbf{d},\mathbf{v}]\mathbf{u}. \tag{2.36}$$

This is the *associativity law for Einstein velocity addition.*

2.2.4 Expression for the non-transitivity of parallel translation

Formula (2.36) can be interpreted as the correction of non-transitivity of parallel translation between inertial systems. It is well known that if an inertial system K_2 moves parallel to system K_1 with relative velocity $\mathbf{d}$, and system K_3 moves parallel to system K_2 with relative velocity $\mathbf{v}$, then, if $\mathbf{d}$ is not collinear to $\mathbf{v}$, the system K_3 does not move parallel to system K_1. How can we measure the non-parallelism between K_1 and K_3? Note that if an object is moving with uniform velocity $\mathbf{u}$ in system K_3, its velocity in K_2 will be $\mathbf{v} \oplus_E \mathbf{u}$, and in system K_1, its velocity will be $\mathbf{d} \oplus_E (\mathbf{v} \oplus_E \mathbf{u})$. Define a space frame $\overline{K}_3$ moving together with system K_3 but parallel to K_1. In this frame, let the velocity of our object be $\overline{\mathbf{u}}$. Since the system $\overline{K}_3$ moves parallel to K_1 with relative velocity $\mathbf{d} \oplus_E \mathbf{v}$, the object's velocity in system K_1 is $(\mathbf{d} \oplus_E \mathbf{v}) \oplus_E \overline{\mathbf{u}}$. Now from (2.36), it follows that

$$\overline{\mathbf{u}} = \mathrm{gyr}_p[\mathbf{d}, \mathbf{v}]\mathbf{u}, \tag{2.37}$$

implying that the operator $\mathrm{gyr}_p[\mathbf{d}, \mathbf{v}]$, which is a rotation operator, corrects for the non-parallelism of systems K_1 and K_3.

2.3 The Lie group $Aut_c(D_s)$

In this section, we will show that the group generated by s-velocity addition is the conformal group on D_s.

2.3.1 The automorphisms of D_s generated by s-velocity addition

Given an s-velocity $\mathbf{a} \in D_s$, we define a map $\psi_{\mathbf{a}}$ by

$$\psi_{\mathbf{a}}(\mathbf{w}) = \mathbf{a} \oplus_s \mathbf{w} = \frac{(1 + |\mathbf{w}|^2 + 2\langle \mathbf{a} \mid \mathbf{w} \rangle)\mathbf{a} + (1 - |\mathbf{a}|^2)\mathbf{w}}{1 + |\mathbf{a}|^2|\mathbf{w}|^2 + 2\langle \mathbf{a} \mid \mathbf{w} \rangle}. \tag{2.38}$$

The transformation $\psi_{\mathbf{a}}(\mathbf{w})$ is shown in Figure 2.3. This formula is somewhat simpler in spherical coordinates (r, θ, φ). We choose the orientation so that $\mathbf{a}$ is on the positive part of the z-axis and thus has coordinates $(|\mathbf{a}|, 0, 0) = (a, 0, 0)$. Let the coordinates of $\mathbf{w}$ be (r, θ, φ). Then, in the complexified plane Π generated by $\mathbf{a}$ and $\mathbf{w}$, a represents $\mathbf{a}$ and $re^{i\theta}$ represents $\mathbf{w}$. By (2.15), $\psi_{\mathbf{a}}(\mathbf{w})$ is represented by

$$\psi_a(w) = \frac{a + re^{i\theta}}{1 + are^{i\theta}}. \tag{2.39}$$

If we denote the spherical coordinates of $\psi_{\mathbf{a}}(\mathbf{w})$ by (r', θ', φ'), then

$$\psi_{\mathbf{a}} \begin{pmatrix} r \\ \theta \\ \varphi \end{pmatrix} = \begin{pmatrix} r' \\ \theta' \\ \varphi' \end{pmatrix} = \begin{pmatrix} \left| \frac{re^{i\theta} + a}{1 + a\, re^{i\theta}} \right| \\ arg(\frac{re^{i\theta} + a}{1 + are^{i\theta}}) \\ \varphi \end{pmatrix}. \tag{2.40}$$

Since the transformation $\psi_{\mathbf{a}}$ acts on the disc $\Delta = D_s \cap \Pi$ like the Möbius transformation (2.15) on Δ, $\psi_{\mathbf{a}}$ is one-to-one and onto D_s. The inverse $(\psi_{\mathbf{a}})^{-1}$ is the map $\psi_{-\mathbf{a}}$, where $\psi_{-\mathbf{a}}(w) = (-\mathbf{a}) \oplus_s \mathbf{w}$. To show that $\psi_{\mathbf{a}}$ is conformal, we have to show that its derivative $\frac{d\psi_{\mathbf{a}}}{d\mathbf{w}}(\mathbf{w})$ is a multiple of an isometry.

2.3.2 The derivative of $\psi_{\mathbf{a}}$

To calculate the derivative $\frac{d\psi_{\mathbf{a}}}{d\mathbf{w}}(\mathbf{w})d\mathbf{w}$ of the map $\psi_{\mathbf{a}}$ at $\mathbf{w}$ in the direction $d\mathbf{w}$, we decompose $d\mathbf{w} = d\mathbf{w}_1 + d\mathbf{w}_2$, where $d\mathbf{w}_1$ belongs to the plane Π generated by $\mathbf{a}$ and $\mathbf{w}$ and $d\mathbf{w}_2$ is perpendicular to Π (see Figure 2.5). We

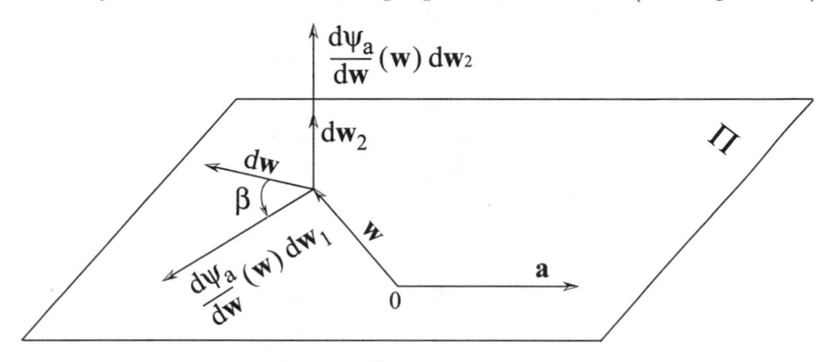

Fig. 2.5. The action of the map $\frac{d\psi_{\mathbf{a}}}{d\mathbf{w}}(\mathbf{w})d\mathbf{w}$. The vector $d\mathbf{w}$ is decomposed as $d\mathbf{w}_1 + d\mathbf{w}_2$, where $d\mathbf{w}_1$ belongs to the plane Π generated by $\mathbf{a}$ and $\mathbf{w}$ and $d\mathbf{w}_2$ is perpendicular to Π. The map rotates $d\mathbf{w}_1$ in the plane Π by an angle β and multiplies it by a constant $\delta = |\frac{1-|a|^2}{(1+\bar{a}w)^2}|$, while $d\mathbf{w}_2$ is only multiplied by δ.

complexify the plane Π and replace the vectors $\mathbf{a}, \mathbf{w}$ and $d\mathbf{w}_1$ with their corresponding complex numbers a, w and dw_1. Then, from (2.15), we get

$$\frac{d\psi_{\mathbf{a}}}{dw}(w)dw_1 = \frac{dw_1(1+\bar{a}w) - (a+w)\bar{a}dw_1}{(1+\bar{a}w)^2} = \frac{1-|a|^2}{(1+\bar{a}w)^2}dw_1, \tag{2.41}$$

which shows that $\frac{d\psi_{\mathbf{a}}}{dw}(w)dw_1$ is a rotation in the plane Π by an angle β, which is the argument of the complex number $1/(1+\bar{a}w)^2$, and multiplies its length by

$$\delta = |\frac{1-|a|^2}{(1+\bar{a}w)^2}| = \frac{1-|a|^2}{1+|a|^2|w|^2+2\langle a \mid w\rangle}. \tag{2.42}$$

Since

$$\beta = \arg\frac{1}{(1+\bar{a}w)^2} = -2\arg(1+\bar{a}w), \tag{2.43}$$

equation (2.21) and (2.24) imply that $e^{i\beta} = \mathrm{gyr}_c[a,w]$. Thus,

$$\frac{d\psi_{\mathbf{a}}}{dw}(w)dw_1 = \delta\, \mathrm{gyr}_c[a,w]dw_1. \tag{2.44}$$

Figure 2.6 shows the value of $\frac{d\psi_{0.4}}{dw}(w)0.2$ for any $w \in D_s$. Note how the angle of rotation β and the multiplication factor δ change for different values of w.

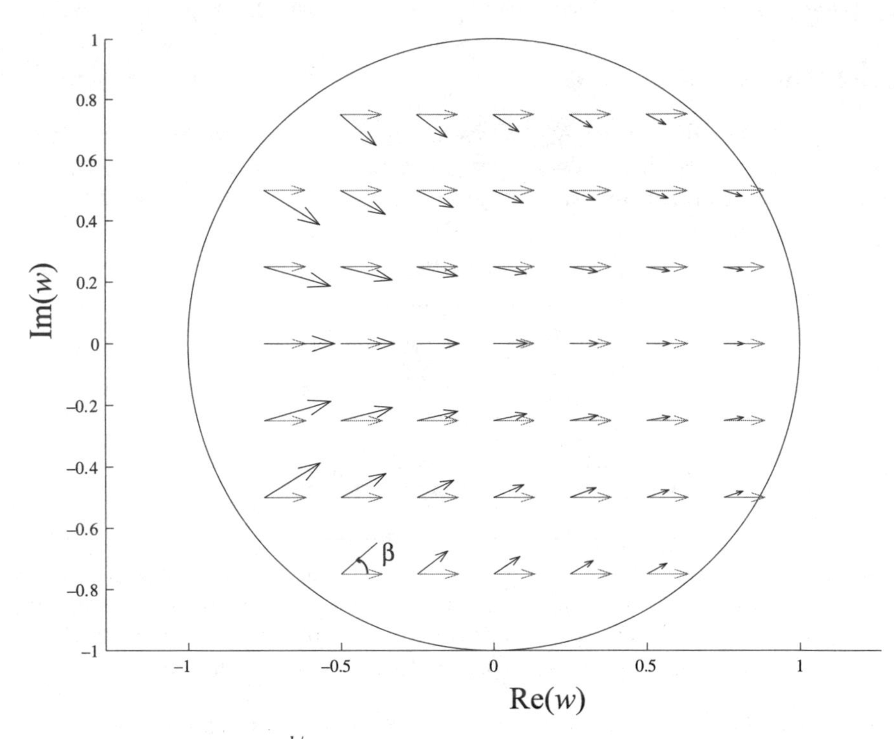

Fig. 2.6. The value of $\frac{d\psi_{0.4}}{dw}(w)0.2$ for different w. The bright arrows represent $u = 0.2$ before application of the derivative $\frac{d\psi_{0.4}}{dw}(w)$, while the dark arrow is its image $\frac{d\psi_{0.4}}{dw}(w)u$. The length-stretching coefficient δ decreases in the direction of $a = 0.4$ and depends mainly on $Re(w)$. The angle of rotation β is zero for real w and depends mainly on $Im(w)$. Compare these results with the stretching and rotation of the squares of the grid under $\psi_{0.4}$ in Figure 2.3.

For the action of the derivative on $d\mathbf{w}_2$, note that $\langle d\mathbf{w}_2|\mathbf{w}\rangle = 0$ and $\langle d\mathbf{w}_2|\mathbf{a}\rangle = 0$. Then, from (2.38), we get

$$\frac{d\psi_{\mathbf{a}}}{d\mathbf{w}}(\mathbf{w})d\mathbf{w}_2 = \frac{(1-|\mathbf{a}|^2)d\mathbf{w}_2}{1+|\mathbf{a}|^2|\mathbf{w}^2|+2\langle\mathbf{a}\mid\mathbf{w}\rangle} = \delta d\mathbf{w}_2, \tag{2.45}$$

which is multiplication by the same constant as in (2.42). Thus,

$$\frac{d\psi_{\mathbf{a}}}{d\mathbf{w}}(\mathbf{w}) = \delta\,\mathrm{gyr}_c[\mathbf{a},\mathbf{w}]$$

is a rotation with respect to the line perpendicular to the plane Π generated by $\mathbf{a}$ and $\mathbf{w}$ by an angle β defined by (2.43) followed by multiplication by the constant δ of (2.42). This implies that the map $\psi_{\mathbf{a}}$ is conformal.

We can define the map $\psi_{\mathbf{a}}$ on D_s^n by use of (2.38). The argument above shows that here also $\frac{d\psi_{\mathbf{a}}}{d\mathbf{w}}(\mathbf{w})$ is a rotation and multiplication by δ in Π and multiplication by δ in the subspace perpendicular to Π. Thus, $\psi_{\mathbf{a}}$ is conformal in this case as well.

2.3.3 The description of $Aut_c(D_s)$

We denote by $Aut_c(D_s)$ the set of all conformal automorphisms of the domain D_s. This set is a group, since composition of two conformal automorphisms is a conformal automorphism, and the inverse (which always exists) is a conformal automorphism. As we have shown in the previous section, for any $\mathbf{a} \in D_s$, the map $\psi_{\mathbf{a}}$ defined by (2.38) is conformal and, thus, an element of $Aut_c(D_s)$.

Next, we characterize the elements of $Aut_c(D_s)$. Let ψ be any conformal automorphism of D_s. Set $\mathbf{a} = \psi(0)$ and $U = \psi_{\mathbf{a}}^{-1}\psi$. Then U is a conformal map that maps $0 \to 0$ and is thus a linear map which can be represented by a 3×3 matrix. Since U maps D_s onto itself, it is an isometry and is represented by an orthogonal matrix. Since $\psi = \psi_{\mathbf{a}}U$, the group of all conformal automorphisms $Aut_c(D_s)$ is defined by

$$Aut_c(D_s) = \{\psi_{\mathbf{a}}U : \mathbf{a} \in D_s,\ U \in O(3)\}. \tag{2.46}$$

We write $\psi_{\mathbf{a},U}$ instead of $\psi_{\mathbf{a}}U$. From (2.38), we have

$$\psi_{\mathbf{a},U}(\mathbf{w}) = \frac{(1 + |\mathbf{w}|^2 + 2\langle \mathbf{a} \mid U\mathbf{w}\rangle)\mathbf{a} + (1 - |\mathbf{a}|^2)U\mathbf{w}}{1 + |\mathbf{a}|^2|\mathbf{w}|^2 + 2\langle \mathbf{a} \mid U\mathbf{w}\rangle}, \tag{2.47}$$

for $\mathbf{w} \in D_s$.

The group $Aut_c(D_s)$ is a real Lie group of dimension 6, since any element of the group is determined by an element $\mathbf{a}$ of the 3-dimensional open unit ball in R^3 and an element U of the 3-dimensional orthogonal group $O(3)$. It is easy to see that D_s is a bounded symmetric domain with respect to the conformal group $Aut_c(D_s)$. The element $S:\ S(\mathbf{w}) = -\mathbf{w}$ of $Aut_c(D_s)$ is a symmetry about the origin of the ball, and for any $\mathbf{a} \in D_s$, the map $\psi_{\mathbf{a}} \in Aut_c(D_s)$ satisfies $\psi_{\mathbf{a}}(0) = \mathbf{a}$.

For the domain D_s^n with arbitrary n, we can define $Aut_c(D_s^n)$ by use of (2.46), taking $\mathbf{a} \in D_s^n$ and $U \in O(n)$. The dimension of this Lie group is $\frac{n(n+1)}{2}$.

2.4 The Lie Algebra $aut_c(D_s)$ and the spin triple product

2.4.1 The generators of $Aut_c(D_s)$

The elements of a Lie algebra are, by definition, the tangent space to the identity of the group. To define the elements of $aut_c(D_s)$, consider differentiable curves $g(s)$ from a neighborhood I_0 of zero into $Aut_c(D_s)$, with $g(0) = \psi_{\mathbf{0},I}$, the identity of $Aut_c(D_s)$. Any such $g(s)$ has the form

$$g(s) = \psi_{\mathbf{a}(s),U(s)}, \tag{2.48}$$

where $\mathbf{a} : I_0 \to D_s$ is a differentiable function satisfying $\mathbf{a}(0) = \mathbf{0}$ and $U(s) : I_0 \to O(3)$ is differentiable and satisfies $U(0) = I$. We denote by ξ the element of $aut_c(D_s)$ generated by $g(s)$. For any fixed $\mathbf{w} \in D_s$, $g(s)(\mathbf{w})$ is a smooth curve in D_s, with $g(0) = \mathbf{w}$, and $\xi(\mathbf{w})$ is a tangent vector to this line. Thus, the elements of $aut_c(D_s)$ are vector fields $\xi(\mathbf{w})$ on D_s defined by

$$\xi(\mathbf{w}) = \frac{d}{ds} g(s)(\mathbf{w})\Big|_{s=0}. \tag{2.49}$$

We now obtain the explicit form of $\xi(\mathbf{w})$. First, define $\mathbf{b} = \mathbf{a}'(0)$, which is a vector in R^3, and $A = U'(0)$, which is a 3×3 skew-symmetric matrix (*i.e.*, $A^T = -A$). Then

$$\xi(\mathbf{w}) = \frac{d}{ds} \psi_{\mathbf{a}(s),U(s)}(\mathbf{w})\Big|_{s=0} =$$

$$\frac{d}{ds} \frac{(1 + |\mathbf{w}|^2 + 2\langle \mathbf{a}(s) \mid U(s)\mathbf{w}\rangle)\mathbf{a}(s) + (1 - |\mathbf{a}(s)|^2)U(s)\mathbf{w}}{1 + |\mathbf{a}(s)|^2|\mathbf{w}|^2 + 2\langle \mathbf{a}(s) \mid U(s)\mathbf{w}\rangle}\Big|_{s=0}. \tag{2.50}$$

Since $\frac{d}{ds}|\mathbf{a}(s)|^2|_{s=0} = 0$, we get

$$\xi(\mathbf{w}) = (1 + |\mathbf{w}|^2)\mathbf{a}'(0) + U'(0)\mathbf{w} - 2\langle \mathbf{a}'(0) \mid \mathbf{w}\rangle\mathbf{w}$$

$$= (1 + |\mathbf{w}|^2)\mathbf{b} + A\mathbf{w} - 2\langle \mathbf{b} \mid \mathbf{w}\rangle\mathbf{w}. \tag{2.51}$$

We can rewrite this expression as a polynomial of degree 2 in $\mathbf{w}$:

$$\xi(\mathbf{w}) = \mathbf{b} + A\mathbf{w} - 2\langle \mathbf{b} \mid \mathbf{w}\rangle\mathbf{w} + |\mathbf{w}|^2\mathbf{b}. \tag{2.52}$$

Thus,

$$aut_c(D_s) = \{\mathbf{b} + A\mathbf{w} - 2\langle \mathbf{b} \mid \mathbf{w}\rangle\mathbf{w} + |\mathbf{w}|^2\mathbf{b}\}, \tag{2.53}$$

where $\mathbf{b} \in R^3$ and A is a 3×3 matrix such that $A^T = -A$.

2.4.2 The triple product and the generators of translations

It will be shown in Chapter 5, section 5.3.4, that the generators of translations (meaning $A = 0$) in a bounded symmetric domain are of the form

$$\xi_{\mathbf{b}}(\mathbf{w}) = \mathbf{b} - \{\mathbf{w}, \mathbf{b}, \mathbf{w}\}, \tag{2.54}$$

where $\{\mathbf{w}, \mathbf{b}, \mathbf{w}\}$ is the triple product associated with the bounded symmetric domain.

Formulas (2.52) and (2.54) indicate that the triple product has to be defined in such a way that

$$\{\mathbf{w}, \mathbf{b}, \mathbf{w}\} = 2\langle \mathbf{b} \mid \mathbf{w}\rangle \mathbf{w} - |\mathbf{w}|^2 \mathbf{b}. \tag{2.55}$$

By substituting $\mathbf{w} = \mathbf{a} + \mathbf{c}$ in the previous equation and using the linearity of the triple product and its symmetry

$$\{\mathbf{a}, \mathbf{b}, \mathbf{c}\} = \{\mathbf{c}, \mathbf{b}, \mathbf{a}\}, \tag{2.56}$$

we obtain the following definition for the triple product:

$$\{\mathbf{a}, \mathbf{b}, \mathbf{c}\} = \langle \mathbf{a}|\mathbf{b}\rangle \mathbf{c} + \langle \mathbf{c}|\mathbf{b}\rangle \mathbf{a} - \langle \mathbf{a}|\mathbf{c}\rangle \mathbf{b}, \tag{2.57}$$

where $\mathbf{a}, \mathbf{b}, \mathbf{c} \in R^3$. This product is called the *spin triple product.* The bounded symmetric domain D_s endowed with the spin triple product is called the *spin factor* and is a domain of type IV in Cartan's classification.

We now derive the *complex* form of the spin triple product. Complexify the plane Π generated by the vectors $\mathbf{w}$ and $\mathbf{b}$. Let the complex numbers w and b represent $\mathbf{w}$ and $\mathbf{b}$, respectively. Using (2.14), the triple product $\{\mathbf{w}, \mathbf{b}, \mathbf{w}\}$ defined by equation (2.55) becomes

$$\{w, b, w\} = (b\overline{w} + \overline{b}w)w - w\overline{w}b = w^2\overline{b}. \tag{2.58}$$

Note that this product is complex analytic in w and conjugate linear in b. As above, by substituting $w = a + c$, we get a complex triple product

$$\{z, b, w\} = z\overline{b}w, \quad \text{where } z, b, w \in C, \tag{2.59}$$

called the *complex spin triple product.* In the next chapter, we will study the domain associated with this spin triple product on C^n, for arbitrary n.

2.4.3 The triple product and the generators of rotations

The Lie algebra $aut_c(D_s)$ consists of generators of boosts, described by (2.54) and (2.55) in terms of the triple product, and generators of rotations. To describe the generators of rotations on D_s, we first choose an orthonormal basis $\mathbf{e}_1, \mathbf{e}_2, \mathbf{e}_3$ in R^3, the tangent space of D_s. For any $\mathbf{a}, \mathbf{b} \in R^3$, define an operator $D(\mathbf{a}, \mathbf{b}) : R^3 \to R^3$ by

$$D(\mathbf{a}, \mathbf{b})\mathbf{c} = \{\mathbf{a}, \mathbf{b}, \mathbf{c}\}. \tag{2.60}$$

Using the definition (2.57) of the spin triple product, the operator $D(\mathbf{e}_2, \mathbf{e}_3)$ acts on the basis vectors by

$$D(\mathbf{e}_2,\mathbf{e}_3)(\mathbf{u}) = \begin{cases} 0, & \text{if } \mathbf{u}=\mathbf{e}_1, \\ -\mathbf{e}_3, & \text{if } \mathbf{u}=\mathbf{e}_2, \\ \mathbf{e}_2, & \text{if } \mathbf{u}=\mathbf{e}_3, \end{cases} \tag{2.61}$$

and its matrix in this basis is

$$\pi_c(J_1) = \begin{pmatrix} 0 & 0 & 0 \\ 0 & 0 & 1 \\ 0 & -1 & 0 \end{pmatrix}, \tag{2.62}$$

which represents the momentum J_1 of rotation about the $\mathbf{e}_1$-axis. We use the notation π_c to indicate that π_c is an element of $aut_c(D_s)$, the Lie algebra of the conformal group. Similarly, the operators $\pi_c(J_2) = D(\mathbf{e}_3,\mathbf{e}_1)$ and $\pi_c(J_3) = D(\mathbf{e}_1,\mathbf{e}_2)$ represent the momentum of rotation about the $\mathbf{e}_2$- and $\mathbf{e}_3$-axes, respectively.

A general generator of rotation, represented by a 3×3 antisymmetric matrix A, is a linear combination $A = B_1\pi_c(J_1) + B_2\pi_c(J_2) + B_3\pi_c(J_3)$. By introducing the notation

$$\pi_c(J) = \pi_c(J_1,J_2,J_3) = (D(\mathbf{e}_2,\mathbf{e}_3), D(\mathbf{e}_3,\mathbf{e}_1), D(\mathbf{e}_1,\mathbf{e}_2)), \tag{2.63}$$

the generator of rotation A may be expressed as

$$A\mathbf{w} = (\mathbf{B}\cdot\pi_c(J))(\mathbf{w}) = \mathbf{w}\times\mathbf{B},$$

where $\mathbf{B} = (B_1,B_2,B_3) \in R^3$. We can now express the elements of the Lie algebra $aut_c(D_s)$ in terms of the spin triple product. From (2.52) and (2.55), it follows that any element ξ of $aut_c(D_s)$ has the form

$$\xi = \xi_{\mathbf{b},\mathbf{B}}(\mathbf{w}) = \mathbf{b} + (\mathbf{B}\cdot\pi_c(J))(\mathbf{w}) - \{\mathbf{w},\mathbf{b},\mathbf{w}\}$$

$$= \mathbf{b} + \mathbf{w}\times\mathbf{B} - \{\mathbf{w},\mathbf{b},\mathbf{w}\}, \tag{2.64}$$

where $\mathbf{b},\mathbf{B} \in R^3$. See Figures 2.7 and 2.8 for two examples of these vector fields.

2.4.4 The Lie bracket on $aut_c(D_s)$.

To show that the set $aut_c(D_s)$ defined by (2.64) is a Lie algebra, it remains to check that this set is closed under the Lie bracket. Let $\xi_{\mathbf{b},\mathbf{B}}$ and $\xi_{\widetilde{\mathbf{b}},\widetilde{\mathbf{B}}}$ be any two elements of $aut_c(D_s)$. Since these elements are vector fields, the Lie bracket is defined by

$$[\xi_{\mathbf{b},\mathbf{B}},\xi_{\widetilde{\mathbf{b}},\widetilde{\mathbf{B}}}](\mathbf{w}) = \frac{d\xi_{\mathbf{b},\mathbf{B}}}{d\mathbf{w}}(\mathbf{w})\xi_{\widetilde{\mathbf{b}},\widetilde{\mathbf{B}}}(\mathbf{w}) - \frac{d\xi_{\widetilde{\mathbf{b}},\widetilde{\mathbf{B}}}}{d\mathbf{w}}(\mathbf{w})\xi_{\mathbf{b},\mathbf{B}}(\mathbf{w}) \tag{2.65}$$

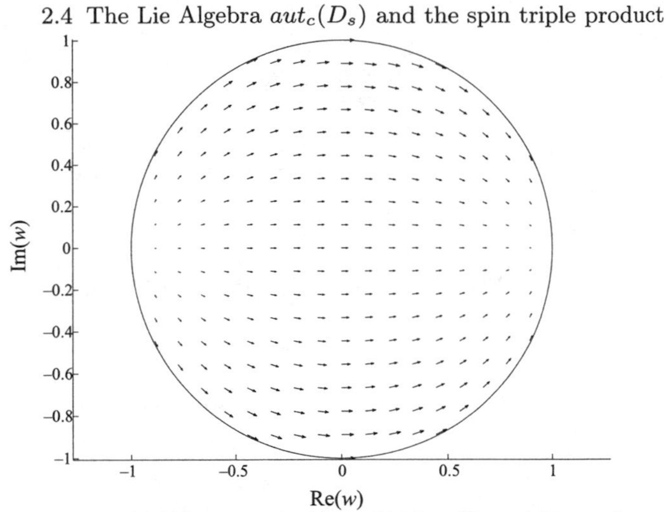

Fig. 2.7. The vector field $\xi_{\mathbf{b},\mathbf{B}}$, with $\mathbf{b} = (0.07, 0, 0)$ and $\mathbf{B} = 0$, on a two-dimensional section of the s-velocity ball D_s. Note that this vector field is similar to the corresponding one for the Lie algebra $aut_p(D_v)$ of the velocity ball (see Figure 1.30 of Chapter 1).

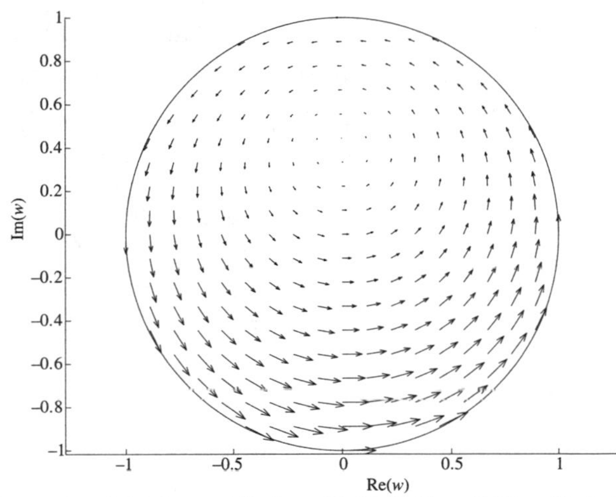

Fig. 2.8. The vector field $\xi_{\mathbf{b},\mathbf{B}}$ with $\mathbf{b} = (0.07, 0, 0)$ and $\mathbf{B} = (0, 0, 0.1)$, on a two-dimensional section of the s-velocity ball D_s. Note that this vector field is similar to the corresponding one for the Lie algebra $aut_p(D_v)$ of the velocity ball (see Figure 1.31) of Chapter 1.

for $\mathbf{w} \in D_s$, where $\frac{d\xi_{\mathbf{b},\mathbf{B}}}{d\mathbf{w}}(\mathbf{w})\xi_{\widetilde{\mathbf{b}},\widetilde{\mathbf{B}}}(\mathbf{w})$ denotes the derivative of $\xi_{\mathbf{b},\mathbf{B}}$ at the point $\mathbf{w}$ in the direction of the vector $\xi_{\widetilde{\mathbf{b}},\widetilde{\mathbf{B}}}(\mathbf{w})$. To show that $aut_c(D_s)$ is closed under the Lie bracket, we will calculate $[\xi_{\mathbf{b},\mathbf{B}}, \xi_{\widetilde{\mathbf{b}},\widetilde{\mathbf{B}}}]$ and show that it has the form (2.64).

From (2.55) and (2.64), we have

$$\frac{d\xi_{\mathbf{b},\mathbf{B}}}{d\mathbf{w}}(\mathbf{w})d\mathbf{w} =$$

$$d\mathbf{w} \times \mathbf{B} - 2\langle d\mathbf{w}|\mathbf{b}\rangle\mathbf{w} - 2\langle\mathbf{w}|\mathbf{b}\rangle d\mathbf{w} + 2\langle d\mathbf{w}|\mathbf{w}\rangle\mathbf{b}. \tag{2.66}$$

Using the identity

$$\langle d\mathbf{w}|\mathbf{w}\rangle\mathbf{b} - \langle d\mathbf{w}|\mathbf{b}\rangle\mathbf{w} = d\mathbf{w} \times (\mathbf{b} \times \mathbf{w}), \tag{2.67}$$

we have

$$\frac{d\xi_{\mathbf{b},\mathbf{B}}}{d\mathbf{w}}(\mathbf{w})d\mathbf{w} = d\mathbf{w} \times (\mathbf{B} + \mathbf{b} \times 2\mathbf{w}) - 2\langle\mathbf{w}|\mathbf{b}\rangle d\mathbf{w}. \tag{2.68}$$

Thus,

$$[\xi_{\mathbf{b},\mathbf{B}}, \xi_{\widetilde{\mathbf{b}},\widetilde{\mathbf{B}}}](\mathbf{w}) = \xi_{\widetilde{\mathbf{b}},\widetilde{\mathbf{B}}}(\mathbf{w}) \times (\mathbf{B} + \mathbf{b} \times 2\mathbf{w}) - 2\langle\mathbf{w}|\mathbf{b}\rangle\xi_{\widetilde{\mathbf{b}},\widetilde{\mathbf{B}}}(\mathbf{w})$$

$$-\xi_{\mathbf{b},\mathbf{B}}(\mathbf{w}) \times (\widetilde{B} + \widetilde{\mathbf{b}} \times 2\mathbf{w}) + 2\langle\mathbf{w}|\widetilde{\mathbf{b}}\rangle\xi_{\mathbf{b},\mathbf{B}}(\mathbf{w}). \tag{2.69}$$

Using (2.55) and (2.64), the previous expression becomes a second-degree polynomial in $\mathbf{w}$, with constant term $\widetilde{\mathbf{b}} \times \mathbf{B} - \mathbf{b} \times \widetilde{B}$ and linear term

$$(\mathbf{w} \times \widetilde{B}) \times \mathbf{B} - (\mathbf{w} \times \mathbf{B}) \times \widetilde{B}$$

$$+2\widetilde{\mathbf{b}} \times (\mathbf{b} \times \mathbf{w}) - 2\mathbf{b} \times (\widetilde{\mathbf{b}} \times \mathbf{w}) - 2\langle\mathbf{w}|\mathbf{b}\rangle\widetilde{\mathbf{b}} + 2\langle\mathbf{w}|\widetilde{\mathbf{b}}\rangle\mathbf{b}. \tag{2.70}$$

By using (2.67) and the identity

$$(\mathbf{w} \times \widetilde{B}) \times \mathbf{B} - (\mathbf{w} \times \mathbf{B}) \times \widetilde{B} = \mathbf{w} \times (\widetilde{B} \times \mathbf{B}), \tag{2.71}$$

the linear term can be written as

$$\mathbf{w} \times (\mathbf{B} \times \widetilde{B} + 4\mathbf{b} \times \widetilde{\mathbf{b}}). \tag{2.72}$$

The quadratic term can be simplified to

$$-\{\mathbf{w}, \widetilde{\mathbf{b}} \times \mathbf{B} - \mathbf{b} \times \widetilde{B}, \mathbf{w}\}. \tag{2.73}$$

Thus, from (2.69), we have

$$[\xi_{\mathbf{b},\mathbf{B}}, \xi_{\widetilde{\mathbf{b}},\widetilde{\mathbf{B}}}] = \xi_{\widetilde{\mathbf{b}}\times\mathbf{B}-\mathbf{b}\times\widetilde{B},\mathbf{B}\times\widetilde{B}+4\mathbf{b}\times\widetilde{\mathbf{b}}}, \tag{2.74}$$

an element of $aut_c(D_s)$.

For arbitrary n, the Lie algebra $aut_c(D_s^n)$ of the Lie group $Aut_c(D_s^n)$ is defined by (2.53), where $\mathbf{b} \in R^n$ and A is an $n \times n$ skew-symmetric matrix. The associated spin triple product is given by (2.57). Also here it can be shown that $aut_c(D_s^n)$ is closed under the Lie bracket.

2.4.5 The representation of the Lorentz group in $aut_c(D_s)$

In section 2.4.3, we defined a representation π_c of the group of rotations into the Lie algebra $aut_c(D_s)$ by (2.63), which defines the representation on the generators J_1, J_2 and J_3 of rotation about the basis vectors $\mathbf{e}_1, \mathbf{e}_2$ and $\mathbf{e}_3$, respectively. We want to extend this representation to a representation of the Lorentz group. To achieve this, we need to find a representation for the generators K_1, K_2 and K_3 of boosts in the direction of the basis vectors $\mathbf{e}_1, \mathbf{e}_2$ and $\mathbf{e}_3$, respectively, and to show that they satisfy the commutation relations of the Lorentz group.

Recall that the Lie algebra of the Lorentz group is the real span of J_k, K_k, for $k = 1, 2, 3$. As we have shown in section 1.5.3, page 39, the generators of the rotation group satisfy

$$[J_1, J_2] = -J_3, \ [J_2, J_3] = -J_1, \ [J_3, J_1] = -J_2, \tag{2.75}$$

and the remaining commutation relations for the generators of the group are

$$[J_1, K_1] = 0, \ [J_1, K_2] = -K_3, \ [J_1, K_3] = K_2, \tag{2.76}$$

$$[J_2, K_1] = K_3, \ [J_2, K_2] = 0, \ [J_2, K_3] = -K_1, \tag{2.77}$$

$$[J_3, K_1] = -K_2, \ [J_3, K_2] = K_1, \ [J_3, K_3] = 0, \tag{2.78}$$

$$[K_1, K_2] = J_3, \ [K_2, K_3] = J_1, \ [K_3, K_1] = J_2. \tag{2.79}$$

As defined earlier,

$$\pi_c(J_1) = D(\mathbf{e}_2, \mathbf{e}_3) = \xi_{\mathbf{0},\mathbf{e}_1}, \ \pi_c(J_2) = D(\mathbf{e}_3, \mathbf{e}_1) = \xi_{\mathbf{0},\mathbf{e}_2},$$

$$\pi_c(J_3) = D(\mathbf{e}_1, \mathbf{e}_2) = \xi_{\mathbf{0},\mathbf{e}_3}. \tag{2.80}$$

Using (2.74), it is easily verified that (2.75) holds. Motivated by the results of the previous subsection, we define

$$\pi_c(K_1) = \xi_{\mathbf{e}_1/2,0}, \ \pi_c(K_2) = \xi_{\mathbf{e}_2/2,0}, \ \pi_c(K_3) = \xi_{\mathbf{e}_3/2,0}. \tag{2.81}$$

Again using (2.74), one can check that (2.76),(2.77), (2.78) and (2.79) hold. Thus π_c, defined by (2.80) and (2.81), is a representation of the Lorentz group into $aut_c(D_s)$.

2.5 Relativistic dynamic equations on D_s

We now derive the relativistic dynamic equation for symmetric velocity as a new dynamic variable. Suppose $\mathbf{w}$ is the symmetric velocity corresponding to the velocity $\mathbf{v}$. Using the identity $\gamma^2\beta^2 = (\gamma - 1)(\gamma + 1)$ and (2.2), we get

$$|\mathbf{w}|^2 = \frac{\gamma - 1}{\gamma + 1} \quad \text{and} \quad \gamma = \frac{1 + |\mathbf{w}|^2}{1 - |\mathbf{w}|^2}, \tag{2.82}$$

and, thus,

$$m\mathbf{v} = m_0\gamma\mathbf{v} = m_0\frac{1 + |\mathbf{w}|^2}{1 - |\mathbf{w}|^2}\frac{2c\mathbf{w}}{1 + |\mathbf{w}|^2} = m_0\frac{2c\mathbf{w}}{1 - |\mathbf{w}|^2}, \tag{2.83}$$

where m_0 is the rest-mass of the object.

Substituting this into the relativistic dynamic equation

$$\mathbf{F} = \frac{d}{dt}(m\mathbf{v}),$$

we have

$$\mathbf{F} = \frac{d}{dt}m_0\frac{2c\mathbf{w}}{1 - |\mathbf{w}|^2}$$

$$= 2m_0c\left(\frac{1}{1 - |\mathbf{w}|^2}\frac{d\mathbf{w}}{dt} + \frac{2\mathbf{w}}{(1 - |\mathbf{w}|^2)^2}\langle\frac{d\mathbf{w}}{dt}|\mathbf{w}\rangle\right). \tag{2.84}$$

By taking the inner product with $\mathbf{w}$, we obtain

$$\langle\mathbf{F}|\mathbf{w}\rangle = 2m_0c\langle\frac{d\mathbf{w}}{dt}|\mathbf{w}\rangle\frac{1 + |\mathbf{w}|^2}{(1 - |\mathbf{w}|^2)^2}. \tag{2.85}$$

By substituting $\langle\frac{d\mathbf{w}}{dt}|\mathbf{w}\rangle$ from (2.85) into (2.84), we obtain

$$\frac{2m_0c}{1 - |\mathbf{w}|^2}\frac{d\mathbf{w}}{dt} = \mathbf{F} - \frac{2\mathbf{w}}{1 + |\mathbf{w}|^2}\langle\mathbf{F}|\mathbf{w}\rangle. \tag{2.86}$$

Multiplying both sides of (2.86) by $1 + |\mathbf{w}|^2$, we get

$$2m_0c\frac{1 + |\mathbf{w}|^2}{1 - |\mathbf{w}|^2}\frac{d\mathbf{w}}{dt} = \mathbf{F}(1 + |\mathbf{w}|^2) - 2\langle\mathbf{F}|\mathbf{w}\rangle\mathbf{w}. \tag{2.87}$$

Using the relation

$$d\tau = \sqrt{1 - |\mathbf{v}|^2/c^2}dt = \frac{1 - |\mathbf{w}|^2}{1 + |\mathbf{w}|^2}dt, \tag{2.88}$$

we obtain *the relativistic dynamic equation for symmetric velocities*

$$m_0 \frac{d\mathbf{w}}{d\tau} = \mathbf{F}/(2c) - \{\mathbf{w}, \mathbf{F}/(2c), \mathbf{w}\} = \xi_{\mathbf{F}/(2c),0}(\mathbf{w}), \tag{2.89}$$

where τ denotes proper time, the triple product is the spin triple product defined by (2.57), and $\xi_{\mathbf{F}/(2c),0}$ is given by (2.64). Thus, the action of a constant force $\mathbf{F}$ on D_s is described by an element $\xi_{\mathbf{b},\mathbf{B}}$ of $aut_c(D_s)$, with $\mathbf{b} = \mathbf{F}/(2c)$ and $\mathbf{B} = 0$. Since for small velocities $\mathbf{w} = \mathbf{v}/(2c)$, we have a factor $1/(2c)$ for the generator. Note that the flow generated by a constant force on D_s will be conformal.

Next, we will derive the relativistic dynamic equation for the electromagnetic field for symmetric velocities. Let $\mathbf{E}$ denote the strength of the electric field, and let $\mathbf{B}$ denote the strength of the magnetic field. Then, from the formula for the Lorentz force for the electromagnetic field, the dynamic equation becomes

$$\frac{d}{dt}(m\mathbf{v}) = q(\mathbf{E} + \mathbf{v} \times \mathbf{B}).$$

By using equation (2.83), we obtain

$$q(\mathbf{E} + \frac{2c\mathbf{w}}{1+|\mathbf{w}|^2} \times \mathbf{B}) = \frac{d}{dt} m_0 \frac{2c\mathbf{w}}{1-|\mathbf{w}|^2} \tag{2.90}$$

$$= 2m_0 c(\frac{1}{1-|\mathbf{w}|^2}\frac{d\mathbf{w}}{dt} + \frac{2\mathbf{w}}{(1-|\mathbf{w}|^2)^2}\langle \frac{d\mathbf{w}}{dt}|\mathbf{w}\rangle).$$

By taking the inner product with $\mathbf{w}$, we get

$$q\langle \mathbf{E}|\mathbf{w}\rangle = 2m_0 c\langle \frac{d\mathbf{w}}{dt}|\mathbf{w}\rangle \frac{1+|\mathbf{w}|^2}{(1-|\mathbf{w}|^2)^2}. \tag{2.91}$$

By substituting $\langle \frac{d\mathbf{w}}{dt}|\mathbf{w}\rangle$ from (2.91) into (2.90), we obtain

$$\frac{2m_0 c}{1-|\mathbf{w}|^2}\frac{d\mathbf{w}}{dt} = q(\mathbf{E} + \frac{2c\mathbf{w}}{1+|\mathbf{w}|^2} \times \mathbf{B} - \frac{2\mathbf{w}}{1+|\mathbf{w}|^2}\langle \mathbf{E}|\mathbf{w}\rangle). \tag{2.92}$$

Multiplying both sides of the previous equation by $1 + |\mathbf{w}|^2$ and switching from dt to $d\tau$, the dynamic equation becomes

$$m_0 c\, d\mathbf{w}/d\tau = q(\mathbf{E}/2 + \mathbf{w} \times cB - \mathbf{w}\langle \mathbf{w}|\mathbf{E}\rangle + |\mathbf{w}|^2\mathbf{E}/2), \tag{2.93}$$

the relativistic dynamic equation for the electromagnetic field.

Using (2.55), this equation becomes

$$m_0 c\, d\mathbf{w}/d\tau = q(\mathbf{E}/2 + \mathbf{w} \times c\mathbf{B} - \{\mathbf{w}, \mathbf{E}/2, \mathbf{w}\}) = q\xi_{\mathbf{E}/2, c\mathbf{B}}(\mathbf{w}), \tag{2.94}$$

showing that this dynamic equation is given by an element (2.64) of $aut_c(D_s)$ if we take $\mathbf{b} = \mathbf{E}/2$. Thus, the flow on D_s generated by a constant electromagnetic field is a one-parameter *conformal* flow in $Aut(D_s)$. By (2.46), this flow is of the form $\psi_{\mathbf{a}(\tau),U(\tau)}$. Note that $\mathbf{a}(\tau)$ is the trajectory of the s-velocity of a particle with zero initial velocity (or s-velocity) under the field. In case, there is a plane Π which is preserved under the evolution, $U(\tau)$ is a rotation with respect to the line perpendicular to Π and is uniquely defined by its action on Π.

To obtain the space trajectory $\mathbf{r}(\tau)$ of the particle, we have to add to its initial position $\mathbf{r}(0)$ the integral $\int_0^\tau \mathbf{v}dt = \int_0^\tau \mathbf{v}(\tau)\gamma(\mathbf{v}(\tau))d\tau$. Using (2.1) and (2.88), we get

$$\mathbf{r}(\tau) = \mathbf{r}(0) + 2c\int_0^\tau \frac{\mathbf{w}(\tau)}{1-|\mathbf{w}(\tau)|^2}d\tau, \tag{2.95}$$

and the proper velocity of the particle, defined by (1.4), is

$$\mathbf{u}(\tau) = \frac{d\mathbf{r}(\tau)}{d\tau} = \frac{2c\mathbf{w}(\tau)}{1-|\mathbf{w}(\tau)|^2} = \gamma\Phi(\mathbf{w}(\tau)), \tag{2.96}$$

with Φ defined by (2.1). If we want to use time t as a parameter on the space trajectory, we have to replace τ by a function of t, which can be defined from the equation

$$t = \int_0^\tau \gamma(\mathbf{v}(\tau))d\tau = \int_0^\tau \frac{1+|\mathbf{w}(\tau)|^2}{1-|\mathbf{w}(\tau)|^2}d\tau. \tag{2.97}$$

2.6 Perpendicular electric and magnetic fields

2.6.1 General setup of the problem

We will now use equation (2.94) to find an analytic solution for the motion of an electric charge q in a uniform, constant electromagnetic field $\mathbf{E}, \mathbf{B}$ in which the vector $\mathbf{B}$ is perpendicular to $\mathbf{E}$. We will assume first that the initial velocity of the charge is perpendicular to $\mathbf{B}$. In this case, the charge will stay in the plane Π which is perpendicular to $\mathbf{B}$ and passes through its initial position. This follows from the fact that the right side of (2.94) is in Π at $\tau = 0$ and $d\mathbf{w}/d\tau$ belongs to this plane.

We will complexify the plane Π so that the vector $\mathbf{E} \in \Pi$ lies on the positive part of the imaginary axis. We associate to any s-velocity $\mathbf{w}$ a complex vector $w = w_1 + iw_2$, with real w_1, w_2. Note that w is unit-free. The vector $\mathbf{E}$ will be represented by the complex number $i|\mathbf{E}|$. In this representation, the vector $\mathbf{w} \times c\mathbf{B}$, which is in Π, is equal to $c|\mathbf{B}|(w_2 - iw_1) = -ic|\mathbf{B}|w$. By use of (2.58), the vector $\{\mathbf{w}, \mathbf{E}/2, \mathbf{w}\}$ is represented by the complex number

$$\{w, E/2, w\} = -i(|\mathbf{E}|/2)w^2. \tag{2.98}$$

The equation (2.94) of evolution of $w(\tau)$ now becomes

$$dw/d\tau = \frac{q}{m_0 c}(i|\mathbf{E}|/2 - ic|\mathbf{B}|w + i(|\mathbf{E}|/2)w^2) = \frac{iq|\mathbf{E}|}{2m_0 c}(1 - 2c\frac{|\mathbf{B}|}{|\mathbf{E}|}w + w^2). \tag{2.99}$$

Rewrite this differential equation as

$$dw(\tau)/d\tau = i\Omega(w(\tau)^2 - 2\widetilde{B}w(\tau) + 1), \tag{2.100}$$

where the constants are

$$\Omega = \frac{q|\mathbf{E}|}{2m_0 c}, \quad \widetilde{B} = \frac{c|\mathbf{B}|}{|\mathbf{E}|}. \tag{2.101}$$

Note that we get a first-order complex analytic differential equation, which by a well-known theorem from differential equations has an analytic solution. The solution is unique for a given initial condition

$$w(0) = w_0, \tag{2.102}$$

where the complex number w_0 represents the initial s-velocity w_0 of the charge. Of course, w_0 could be calculated from the initial velocity $\mathbf{v}_0$ by

$$w_0 = \frac{\mathbf{v}_0}{c(1 + \sqrt{1 + |\mathbf{v}_0|/c^2})}. \tag{2.103}$$

The differential equation (2.100) can be solved by separation of variables. We have

$$\frac{dw(\tau)}{w(\tau)^2 - 2\widetilde{B}w(\tau) + 1} = i\Omega d\tau, \tag{2.104}$$

and, integrating both sides, we get

$$\int \frac{dw(\tau)}{w(\tau)^2 - 2\widetilde{B}w(\tau) + 1} = i\Omega\tau + C, \tag{2.105}$$

where the constant C depends on the initial condition (2.102). The explicit form of this integral depends on the sign of the discriminant $4\widetilde{B}^2 - 4$ of the denominator in the integral. Let

$$\Delta = \widetilde{B}^2 - 1 = \frac{(c|\mathbf{B}|)^2 - |\mathbf{E}|^2}{|\mathbf{E}|^2}. \tag{2.106}$$

We will consider three cases: **case 1** $|\mathbf{E}| < c|\mathbf{B}|$, **case 2** $|\mathbf{E}| = c|\mathbf{B}|$ and **case 3** $|\mathbf{E}| > c|\mathbf{B}|$. To simplify the notation, we introduce two new constants

$$\delta = \sqrt{|\Delta|} = \frac{\sqrt{|(c|\mathbf{B}|)^2 - |\mathbf{E}|^2|}}{|\mathbf{E}|}, \quad \beta = \delta\Omega = \frac{q\sqrt{|(c|\mathbf{B}|)^2 - |\mathbf{E}|^2|}}{2m_0 c}. \tag{2.107}$$

2.6.2 The solution for the case $|\mathbf{E}| < c|\mathbf{B}|$

Case 1 Consider first the case $\Delta = ((c|\mathbf{B}|)^2 - |\mathbf{E}|^2)/|\mathbf{E}|^2 > 0$, meaning that $|\mathbf{E}| < c|\mathbf{B}|$. Rewrite the denominator of the integral of (2.105) as

$$w(\tau)^2 - 2\widetilde{B}w(\tau) + 1 = (w(\tau) - w_1)(w(\tau) - w_2),$$

where the roots of this quadratic polynomial are

$$w_1 = \widetilde{B} - \sqrt{\widetilde{B}^2 - 1} = \widetilde{B} - \delta, \quad w_2 = \widetilde{B} + \delta. \tag{2.108}$$

Then, by decomposing into partial fractions, the integral is

$$\int \frac{dw(\tau)}{w(\tau)^2 - 2\widetilde{B}w(\tau) + 1} = \frac{1}{w_2 - w_1} \ln \frac{w(\tau) - w_2}{w(\tau) - w_1} + C.$$

Substituting this into (2.105), we get

$$\frac{1}{w_2 - w_1} \ln \frac{w(\tau) - w_2}{w(\tau) - w_1} = i\Omega\tau + C, \tag{2.109}$$

or by (2.108),

$$\ln \frac{w(\tau) - w_2}{w(\tau) - w_1} = i2\Omega\delta\tau + C = i2\beta\tau + C.$$

By exponentiating both sides, we get

$$\frac{w(\tau) - w_2}{w(\tau) - w_1} = Ce^{i2\beta\tau}, \tag{2.110}$$

which is a periodic solution with period $T = 2\pi/(2\beta) = \pi/\beta$. From the initial condition $w(0) = w_0$, we get

$$C = \frac{w_0 - w_2}{w_0 - w_1}, \tag{2.111}$$

which, in general, is a complex number.

Equation (2.110) implies that

$$w(\tau) = \frac{w_2 - w_1 Ce^{i2\beta\tau}}{1 - Ce^{i2\beta\tau}}, \tag{2.112}$$

where C is defined by (2.111), w_1, w_2 by (2.108) and β by (2.101) and (2.107).

Note that the linear fractional transformation in (2.112) has real coefficients. Such a transformation maps the circle $Ce^{i2\beta\tau}$, which is symmetric with respect to the reals, into a circle which is also symmetric with respect to the reals. Thus, the center of the circle described by $w(\tau)$ is on the real axis. To find the intersection of this circle with the real line, we chose τ_1 and τ_2 such that

$$Ce^{i2\beta\tau_1} = |C|, \quad Ce^{i2\beta\tau_2} = -|C|. \tag{2.113}$$

Then

$$w(\tau_1) = \frac{w_2 - w_1|C|}{1-|C|}, \quad w(\tau_2) = \frac{w_2 + w_1|C|}{1+|C|} \tag{2.114}$$

are the two intersections of the circle with the real axis. This implies that the center of the circle is

$$d = \frac{w_2 - w_1|C|^2}{1-|C|^2}.$$

Substituting the value C from (2.111) and w_1, w_2 from (2.108), we get

$$d = \frac{1-|w_0|^2}{2(\widetilde{B} - Re(w_0))}. \tag{2.115}$$

The radius of the circle is

$$R = |w_0 - d|. \tag{2.116}$$

Let $a(\tau)$ be the solution which corresponds to $w_0 = 0$. From (2.111) we get $C = w_2/w_1$. In this case, the center of the circle is $d_0 = 1/(2\widetilde{B}) = |\mathbf{E}|/(2c|\mathbf{B}|)$, and $R_0 = |\mathbf{E}|/(2c|\mathbf{B}|)$. See Figure 2.9 for $w(\tau)$ with different initial conditions.

We calculate now the flow on D_s generated by the electromagnetic field. Substituting the initial condition (2.111) into the solution (2.112) and using the fact that $w_1 w_2 = 1$, we get

$$w(\tau) = \frac{w_1(e^{i2\beta\tau} - 1) + (1 - w_1^2 e^{i2\beta\tau})w_0}{e^{i2\beta\tau} - w_1^2 + w_1(1 - e^{i2\beta\tau})w_0}. \tag{2.117}$$

Dividing the numerator and the denominator by $e^{i2\beta\tau} - w_1^2$, we get that $w(\tau)$ has the form

$$w(\tau) = \psi_{a(\tau)}(U(\tau)w_0), \tag{2.118}$$

where $\psi_a(w)$ is defined by (2.15) and

$$a(\tau) = \frac{w_1(e^{i2\beta\tau} - 1)}{e^{i2\beta\tau} - w_1^2}, \quad U(\tau) = \frac{1 - w_1^2 e^{i2\beta\tau}}{e^{i2\beta\tau} - w_1^2}. \tag{2.119}$$

We can express the connection between the rotation, given by $U(\tau)$, and the translation, given by $a(\tau)$, as

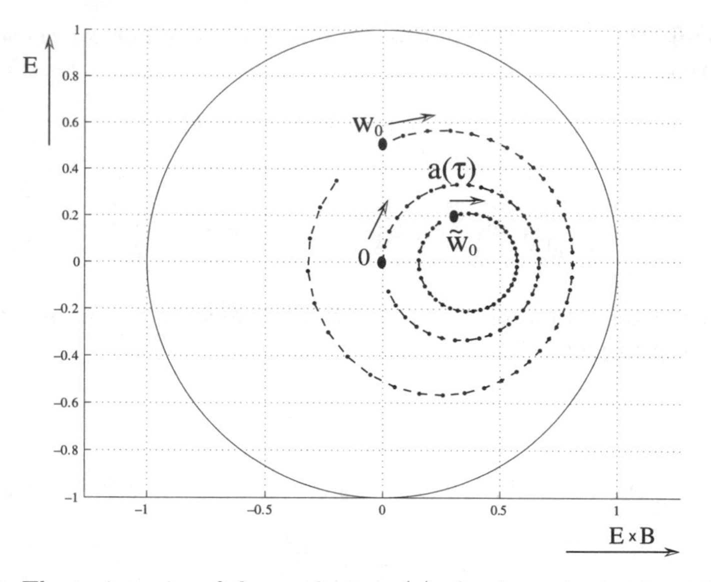

Fig. 2.9. The trajectories of the s-velocity $w(\tau)$ of a charged particle with $q/m = 10^7 C/kg$ in a constant, uniform field, with $\mathbf{E} = 1V/m$ and $cB = 1.5V/m$. The initial conditions are $w_0 = -0.02 + i0.5$ and $\widetilde{w}_0 = 0.3 + i0.2$. Also shown is $a(\tau)$, corresponding to $w_0 = 0$. Note that the trajectories are circles.

$$U(\tau) = \frac{1 - \widetilde{B}a(\tau)}{1 - \widetilde{B}\overline{a}(\tau)}. \tag{2.120}$$

We use (2.95) to calculate the position of the charged particle in each case and under different initial conditions. In particular, we will derive the explicit solution $\mathbf{r}(t)$ for the initial condition $w_0 = 0$. In our case, from (2.119) we have

$$\frac{a(\tau)}{1 - |a(\tau)|^2} = \frac{\widetilde{B} - \delta}{2\widetilde{B}\delta}((1 + \frac{1}{\widetilde{B}\delta})(1 - \cos(2\beta\tau)) + i\sin(2\beta\tau)). \tag{2.121}$$

Substituting this into (2.95), we get

$$\mathbf{r}(t) = \frac{\widetilde{B} - \delta}{4\beta\widetilde{B}\delta}((1 + \frac{1}{\widetilde{B}\delta})(2\beta t - \sin(2\beta t)), -\cos(2\beta t), 0) \tag{2.122}$$

and

$$\mathbf{v}(t) = \frac{\widetilde{B} - \delta}{2\widetilde{B}\delta}((1 + \frac{1}{\widetilde{B}\delta})(1 - \cos(2\beta t)), \sin(2\beta t), 0). \tag{2.123}$$

This shows that the particle moves along a cycloid path with $\mathbf{E} \times \mathbf{B}$ drift given by the constant velocity $\frac{\widetilde{B}-\delta}{2\widetilde{B}\delta}(1 + \frac{1}{\widetilde{B}\delta})$.

Figures 2.10 and 2.11 illustrate the evolution of velocity and position, respectively, of the particles with initial conditions as in Figure 2.9.

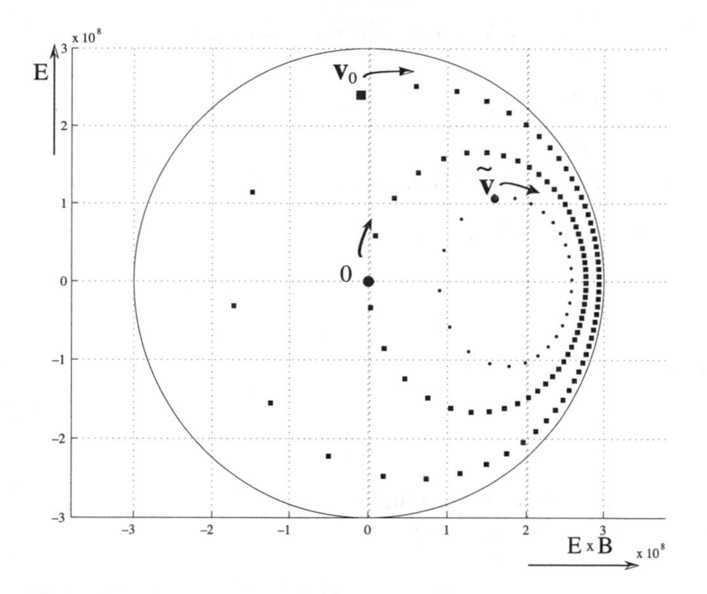

Fig. 2.10. The velocity trajectories $\mathbf{v}(t)$ on D_v of the test particle of Figure 2.9 in the same electromagnetic field. The initial velocities are $\mathbf{v}_0 = (-0.1, 2.4, 0)10^8 m/s$, $\widetilde{\mathbf{v}}_0 = (1.59, 1.06, 0)10^8 m/s$ and 0. The velocity of the particle is shown at time intervals $dt = 10s$.

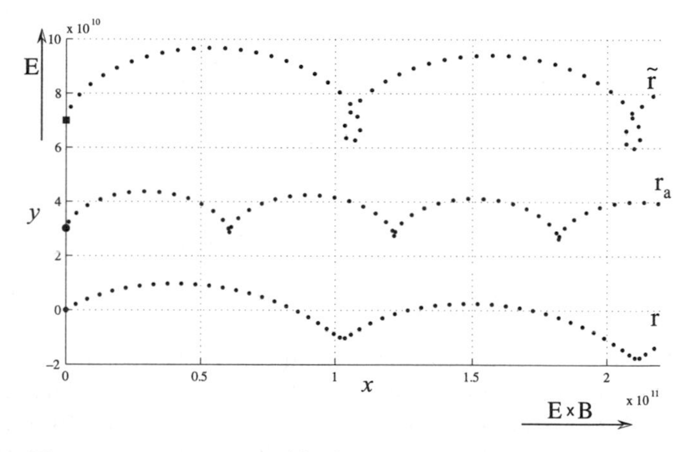

Fig. 2.11. The space trajectories $\mathbf{r}(t)$ of the test particle of Figure 2.9 in the same electromagnetic field during 1200 seconds. The position of the particle is shown at fixed time intervals $dt = 10s$.

2.6.3 The solution for the case $|\mathbf{E}| = c|\mathbf{B}|$

Case 2 Consider now the case $\Delta = ((c|\mathbf{B}|)^2 - |\mathbf{E}|^2)/|\mathbf{E}|^2 = 0$, meaning that $|\mathbf{E}| = c|\mathbf{B}|$, or $\widetilde{B} = 1$. Rewrite the denominator of the integral of (2.105) as

$$w(\tau)^2 - 2w(\tau) + 1 = (w(\tau) - 1)^2.$$

Then the integral is

$$\int \frac{dw(\tau)}{w(\tau)^2 - 2\widetilde{B}w(\tau) + 1} = -\frac{1}{w(\tau) - 1} + C.$$

Substituting this into (2.105), we get

$$-\frac{1}{w(\tau) - 1} = i\Omega\tau + C. \tag{2.124}$$

From the initial condition, we get

$$C = -\frac{1}{w_0 - 1}, \tag{2.125}$$

which, in general, is a complex number.

Equation (2.124) implies that

$$w(\tau) = 1 - \frac{1}{i\Omega\tau + C} = \frac{w_0 - i\Omega\tau w_0 + i\Omega\tau}{1 - i\Omega\tau w_0 + i\Omega\tau}, \tag{2.126}$$

where Ω is defined by (2.101).

Dividing the numerator and the denominator of (2.126) by $1 + i\Omega\tau$, we get (2.118), where

$$a(\tau) = \frac{i\Omega\tau}{1 + i\Omega\tau}, \quad U(\tau) = \frac{1 - i\Omega\tau}{1 + i\Omega\tau}. \tag{2.127}$$

Here, also, the connection between $a(\tau)$ and $U(\tau)$ is

$$U(\tau) = \frac{1 - a(\tau)}{1 - \overline{a}(\tau)} = \frac{1 - \widetilde{B}a(\tau)}{1 - \widetilde{B}\overline{a}(\tau)}. \tag{2.128}$$

This defines the conformal flow in $Aut_c(D_s)$ generated by this electromagnetic field.

Note that $w(\tau)$ is an arc of a circle, since it is the image under a linear fractional transformation of a half line $\{i\Omega\tau : \tau \in [0, \infty)\}$. This arc starts at w_0 at $\tau = 0$ and approaches the point $w_\infty = 1$ as τ goes to infinity (see Figure 2.12). For large τ, the charge moves with speed approaching the speed of light in the direction $\mathbf{E} \times \mathbf{B}$. Since also $w_{-\infty} = 1$, the center of this arc is on the real axis, at

$$d = \frac{1 - |w_0|^2}{2(1 - Re(w_0))},$$

which is similar to the result in the previous case. Since $a(\tau)$ corresponds to $w(\tau)$ with $w_0 = 0$, we get that $a(\tau)$ belongs to a circle with center $d_0 = 1/2$ and radius $R_0 = 1/2$ (see Figure 2.12).

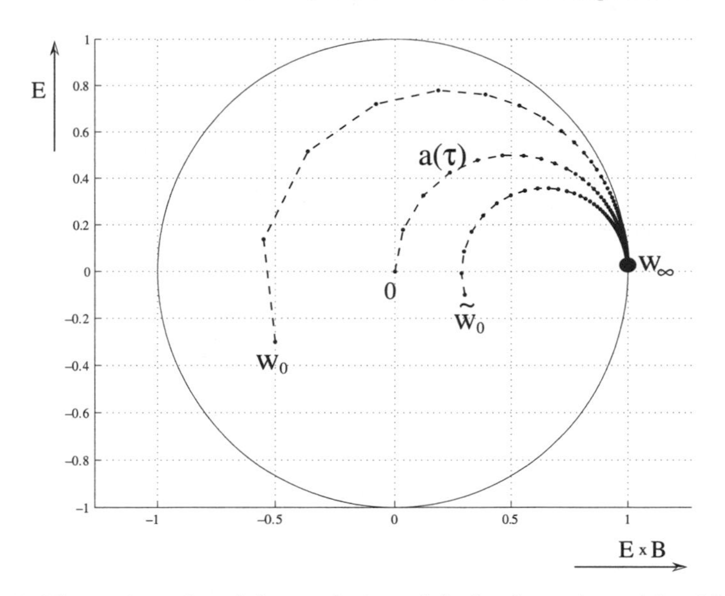

Fig. 2.12. The trajectories of the s-velocity $w(\tau)$ of a charged particle with $q/m = 10^7 C/kg$ in a constant, uniform field, with $\mathbf{E} = 1V/m$ and $c\mathbf{B} = 1V/m$. The initial conditions are $w_0 = -0.5 - i0.3$ and $\widetilde{w}_0 = 0.3 - i0.1$. Also shown is $a(\tau)$, corresponding to $w_0 = 0$. Note that the trajectories all end at $w_\infty = 1$.

We use (2.1) and (2.88) to derive the evolution of velocity, and (2.95) to calculate the position of the charged particle, under different initial conditions. For the the initial condition $w_0 = 0$, we get

$$\mathbf{r}(t) = (\frac{2}{3}c\Omega^2 t^3, c\Omega t^2, 0) \tag{2.129}$$

and

$$\mathbf{v}(t) = (2c\Omega^2 t^2, 2c\Omega t, 0). \tag{2.130}$$

Figures 2.13 and 2.14 illustrate the results of these calculations.

2.6.4 The solution for the case $|\mathbf{E}| > c|\mathbf{B}|$

Case 3 Finally, consider the case $\Delta = ((c|\mathbf{B}|)^2 - |\mathbf{E}|^2)/|\mathbf{E}|^2 < 0$, meaning that $|\mathbf{E}| > c|\mathbf{B}|$, or $\widetilde{B} < 1$. Rewrite the denominator of the integral of (2.105) as

$$w(\tau)^2 - 2\widetilde{B}w(\tau) + 1 = (w(\tau) - \widetilde{B})^2 + 1 - \widetilde{B}^2 + C.$$

Then, from the table of integrals, we get

$$\int \frac{dw(\tau)}{w(\tau)^2 - 2\widetilde{B}w(\tau) + 1} = \frac{1}{\delta}\tan^{-1}\frac{w(\tau) - \widetilde{B}}{\delta}.$$

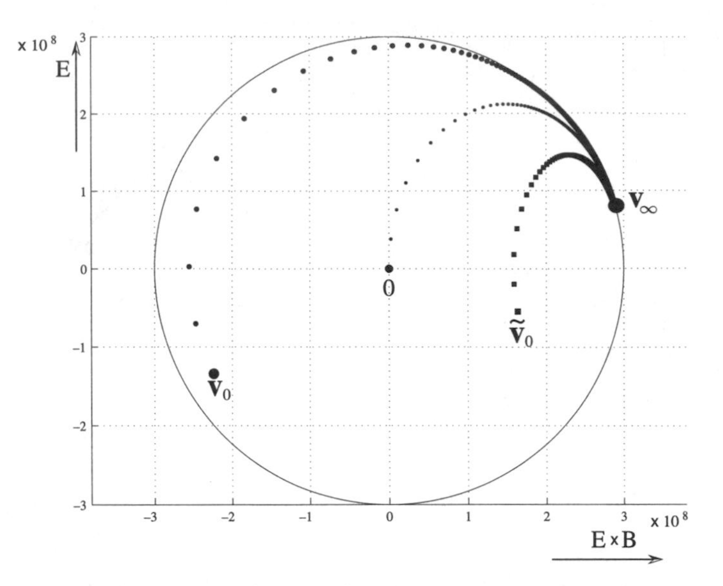

Fig. 2.13. The velocity trajectories $\mathbf{v}(t)$ on D_v of the test particle of Figure 2.12 in the same electromagnetic field. The initial velocities are $\mathbf{v}_0 = (-2.24, -1.34, 0)10^8 m/s$, $\widetilde{\mathbf{v}}_0 = (1.64, -0.55, 0)10^8 m/s$ and 0. The velocity of the particle is shown at time intervals $dt = 10s$.

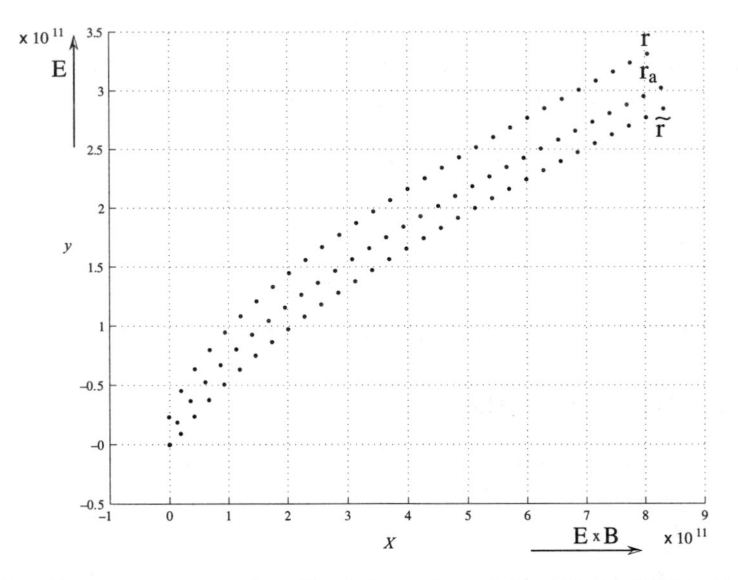

Fig. 2.14. The space trajectories $\mathbf{r}(t)$ of the test particle of Figure 2.12 in the same electromagnetic field during 3000 seconds. The position of the particle is shown at fixed time intervals $dt = 100s$. Note that in this case, it takes the particle a much longer time to get close to its limiting velocity.

Substituting this into (2.105), we get

$$\tan^{-1}\frac{w(\tau)-\widetilde{B}}{\delta}=i\beta\tau+C. \tag{2.131}$$

From the initial condition $w(0)=w_0$, we get

$$C=\tan^{-1}\frac{w_0-\widetilde{B}}{\delta}, \tag{2.132}$$

which, in general, is a complex number.

Equation (2.131) implies that

$$\frac{w(\tau)-\widetilde{B}}{\delta}=\tan(i\beta\tau+C)=\frac{i\tanh(\beta\tau)+(w_0-\widetilde{B})/\delta}{1-i\tanh(\beta\tau)(w_0-\widetilde{B})/\delta},$$

and, therefore,

$$w(\tau)=\widetilde{B}+\delta\frac{i\delta\tanh(\beta\tau)+w_0-\widetilde{B}}{\delta-i\tanh(\beta\tau)(w_0-\widetilde{B})}, \tag{2.133}$$

where the constants are defined by (2.101) and (2.107). Since in this case, $\delta^2=1-\widetilde{B}^2$, we can rewrite equation (2.133) as

$$w(\tau)=\frac{i\tanh(\beta\tau)+w_0(\delta-i\widetilde{B}\tanh(\beta\tau))}{\delta+i\widetilde{B}\tanh(\beta\tau)-i\tanh(\beta\tau)w_0}. \tag{2.134}$$

Dividing the numerator and the denominator by $\delta+i\widetilde{B}\tanh(\beta\tau)$, we get (2.118), where

$$a(\tau)=\frac{i\tanh(\beta\tau)}{\delta+i\widetilde{B}\tanh(\beta\tau)},\quad U(\tau)=\frac{\delta-i\widetilde{B}\tanh(\beta\tau)}{\delta+i\widetilde{B}\tanh(\beta\tau)}. \tag{2.135}$$

By use of the gyration operator, defined by (2.24), we can express the connection between the rotation, given by $U(\tau)$, and the translation, given by $a(\tau)$, as

$$U(\tau)=\frac{1-\widetilde{B}a(\tau)}{1-\widetilde{B}\overline{a}(\tau)}=\operatorname{gyr}[a(\tau),-\widetilde{B}]. \tag{2.136}$$

Equation (2.118) defines the one-parameter subgroup of the conformal group $Aut_c(D_s)$ generated by our electromagnetic field.

Observe that $w(\tau)$ (for a fixed w_0), as defined by (2.133), is the image of a line segment $i\tanh\beta\tau$ under a linear fractional transformation. Thus, $w(\tau)$ is an arc of a circle. To identify this circle, we calculate the limit $w_{\pm\infty}=\lim_{\tau\to\pm\infty}w(\tau)$. Since $\lim_{\tau\to\pm\infty}\tanh(\beta\tau)=\pm1$, we get

$$w_{\pm\infty} = \lim_{\tau \to \pm\infty} w(\tau) = \widetilde{B} + \delta \frac{\pm i\delta + w_0 - \widetilde{B}}{\delta \mp i(w_0 - \widetilde{B})} = \widetilde{B} \pm i\delta. \tag{2.137}$$

For large τ, the charge moves with speed approaching the speed of light in the direction $w_\infty = \widetilde{B} + i\delta$, which is independent of the initial condition. This direction does not depend on the magnitude of the field, but only on the ratio between the magnitudes of the electric and magnetic components of the field (see Figure 2.15).

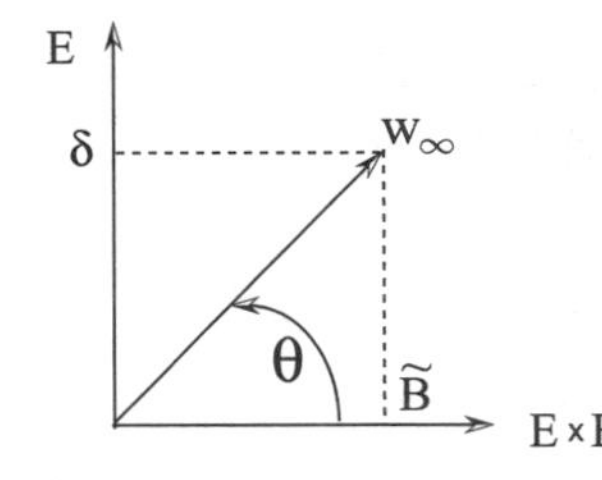

Fig. 2.15. The direction of the limiting velocity $w_\infty = \widetilde{B} + i\delta$. The angle θ is defined by $\tan\theta = \sqrt{|\mathbf{E}|^2 - c|\mathbf{B}|^2}/(c|\mathbf{B}|)$.

Since both points $\widetilde{B} + i\delta$ and $\widetilde{B} - i\delta$ belong to the circle, the center of the circle is represented by a positive real number d, which satisfies the equation

$$|w_0 - d|^2 = |d - \widetilde{B} - i\delta|^2. \tag{2.138}$$

This implies that

$$d = \frac{1 - |w_0|^2}{2(\widetilde{B} - Re(w_0))}. \tag{2.139}$$

The radius of the circle is

$$R = |w_0 - d|. \tag{2.140}$$

Since $a(\tau)$, defined above, corresponds to $w(\tau)$ when $w_0 = 0$, we get that $a(\tau)$ belongs to a circle with center $d_0 = 1/(2\widetilde{B}) = |\mathbf{E}|/(2c|\mathbf{B}|)$ and radius $R_0 = |\mathbf{E}|/(2c|\mathbf{B}|)$ (see Figure 2.16).

We derive now the explicit solution $\mathbf{r}(t)$ for the initial condition $w_0 = 0$. From (2.135), we have

$$\frac{a(\tau)}{1 - |a(\tau)|^2} = \frac{i}{\delta}\sinh(\beta\tau)\cosh(\beta\tau) + \frac{\widetilde{B}}{\delta^2}\sinh^2(\beta\tau). \tag{2.141}$$

Substituting this in (2.95), we get

$$\mathbf{r}(t) = \mathbf{r}(0) + (\frac{c\widetilde{B}}{2\beta\delta^2}(\sinh(2\beta t) - 2\beta t), \frac{c}{2\beta\delta}\cosh(2\beta t), 0) \tag{2.142}$$

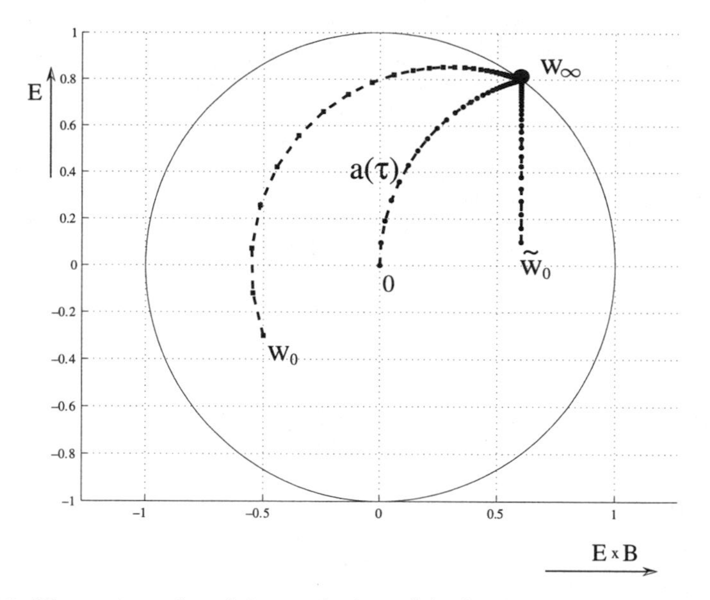

Fig. 2.16. The trajectories of the s-velocity $w(\tau)$ of a charged particle with $q/m = 10^7 C/kg$ in a constant, uniform field, with $\mathbf{E} = 1V/m$ and $c\mathbf{B} = 0.6V/m$. The initial conditions are $w_0 = -0.5 - i0.3$ and $\widetilde{w}_0 = 0.6 + i0.1$. Also shown is $a(\tau)$, corresponding to $w_0 = 0$. Note that the trajectories all end at $w_\infty = 0.6 + i0.8$.

and

$$\mathbf{v}(t) = \left(\frac{c\widetilde{B}}{2\delta^2}(\cosh(2\beta t) - 1), \frac{c}{2\delta}\sinh(2\beta t), 0 \right). \tag{2.143}$$

Figures 2.17 and 2.18 illustrate the results of these calculations.

In all cases, the s-velocity trajectory with zero initial condition $a(\tau)$ is on a circle with center $d_\cap = 1/(2\widetilde{B}) = |\mathbf{E}|/(2c|\mathbf{B}|)$ and radius $R_\cap = |\mathbf{E}|/(2c|\mathbf{B}|)$. The solution of the initial-value problem (2.100) and (2.102), for any initial condition $w_0 \in D_s$, is given by

$$w(\tau) = \psi_{a(\tau)}(U(\tau)w_0), \tag{2.144}$$

where $a(\tau)$ differs from case to case, and

$$U(\tau) = \frac{1 - \widetilde{B}a(\tau)}{1 - \widetilde{B}\overline{a}(\tau)} \tag{2.145}$$

in all three cases.

2.7 Notes

The connection between symmetric velocity and conformal geometry was observed in [30] and further explored in [27], in which the relativistic equation

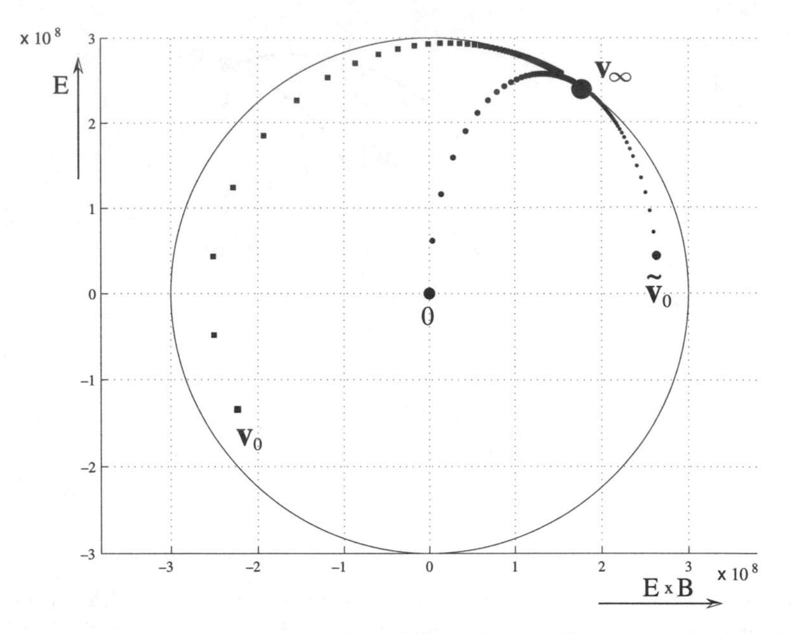

Fig. 2.17. The velocity trajectories $\mathbf{v}(t)$ on D_v of the test particle of Figure 2.16 in the same electromagnetic field. The initial velocities are $\mathbf{v}_0 = (-2.24, -1.34, 0)10^8 m/s$, $\widetilde{\mathbf{v}}_0 = (2.63, 0.44, 0)10^8 m/s$ and 0. The velocity of the particle is shown at time intervals $dt = 10s$.

for symmetric velocity was also derived. The formula for addition in the conformal group on the unit disc in R^n can be found in [1]. The formula (2.55) for the generator of the conformal group appeared already in [45]. The gyration operator for the conformal and projective geometries and its properties is well described in [67].

The relativistic motion of charged particles in a constant, uniform electromagnetic field $\mathbf{E}, \mathbf{B}$ is studied in [51] for an electric field $\mathbf{E}$ alone, a magnetic field $\mathbf{B}$ alone, parallel electric $\mathbf{E}$ and magnetic $\mathbf{B}$ fields, and mutually perpendicular $\mathbf{E}, \mathbf{B}$ (meaning $\mathbf{E} \cdot \mathbf{B} = 0$) of equal strength (meaning $|\mathbf{E}| = c|\mathbf{B}|$). Recently, Takeuchi [65] obtained an exact solution of the relativistic equation of motion of a charged particle in electric and magnetic fields that are constant, uniform and mutually perpendicular. Our solution of the problem in section 1.5.6 of Chapter 1 and in section 2.6 of Chapter 2 is new and will appear in [26].

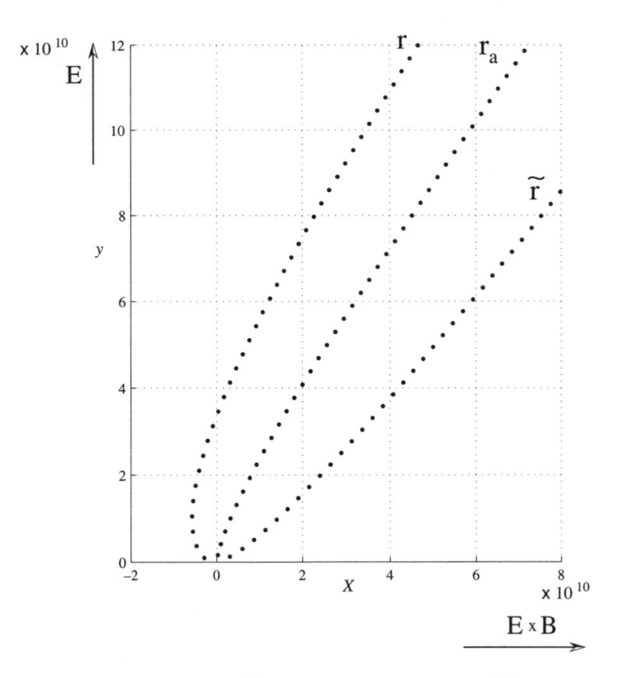

Fig. 2.18. The space trajectories $\mathbf{r}(t)$ of the test particle of Figure 2.16 in the same electromagnetic field during 500 seconds. The position of the particle is shown at fixed time intervals $dt = 10s$.

3 The complex spin factor and applications

In this chapter, we will discuss the complex spin factor, a domain of type IV in the Cartan classification. This domain is symmetric with respect to the analytic automorphisms. In fact, in the previous chapter, we used the analyticity of the spin factor on a two-dimensional plane to solve equations of evolution.

We start by extending the real spin triple product to the complex case. Then we study the algebraic properties and the geometry of the unit ball of the spin factor and its dual. Since this is our first example of a non-trivial bounded symmetric domain, the concepts of the general theory of bounded symmetric domains will be illustrated here, in order to help the reader become familiar with BSDs.

In this chapter, we will study the duality between minimal and maximal tripotents. This duality plays an important role in the study of the geometry of the unit ball of the spin factor and its dual. We describe different representations of the Lorentz group as linear transformations of the spin factor which preserve the determinant. The duality between maximal and minimal tripotents allows us to construct both spin 1 *and* spin 1/2 representations on the same spin factor. Thus, we can incorporate particles of integer and half-integer spin in one model. As a result, the complex spin factor with its triple product is a new model for supersymmetry.

The spin factor plays an important role in physics. It was shown in [35] that the state space of any two-state quantum system is the dual of a complex spin factor. The spin factor can be used to represent efficiently the electromagnetic field strength. Its basis satisfies the Canonical Anticommutation Relations (CAR). The basic operators of the complex spin triple product are closely related to the geometric product of Clifford algebras. Recently, it was shown [10] that Clifford algebras provide a model for different physical phenomena.

3.1 The algebraic structure of the complex spin factor

In the previous chapter, we saw that s-velocity addition generates the group $Aut_c(D_s)$ of conformal automorphisms of the unit ball of R^3. For any n, the elements of the Lie algebra $aut_c(D_s^n)$ of the conformal group $Aut_c(D_s^n)$ can be described by use of the triple product, defined as

$$\{\mathbf{a}, \mathbf{b}, \mathbf{c}\} = \langle \mathbf{a}|\mathbf{b}\rangle \mathbf{c} + \langle \mathbf{c}|\mathbf{b}\rangle \mathbf{a} - \langle \mathbf{a}|\mathbf{c}\rangle \mathbf{b}, \tag{3.1}$$

where $\mathbf{a}, \mathbf{b}, \mathbf{c} \in R^n$. In case the evolution has an invariant plane Π, we may introduce a complex structure on Π, and the above triple product becomes

$$\{z, b, w\} = z\bar{b}w, \quad \text{for } z, b, w \in C, \tag{3.2}$$

which is complex linear in the first and third variables (z and w) and conjugate linear in the second variable (b). We have also seen in section 2.6 of Chapter 2 that the relativistic evolution equation in this structure has an explicit solution and is given by analytic linear fractional transformations. These considerations suggest extending the triple product to C^n in such a way that on the real part it will coincide with (3.1) and on the complex plane with (3.2).

3.1.1 The triple product structure on C^n and its advantage over the geometric product

Let C^n denote n-dimensional (finite or infinite) complex Euclidean space with the natural basis

$$\mathbf{e}_1 = (1, 0, \ldots, 0), \mathbf{e}_2 = (0, 1, \ldots, 0), \cdots, \mathbf{e}_n = (0, \ldots, 0, 1)$$

and the usual inner product

$$\langle \mathbf{a}|\mathbf{b}\rangle = a_1\bar{b}_1 + a_2\bar{b}_2 + \cdots + a_n\bar{b}_n, \tag{3.3}$$

where $\mathbf{a} = (a_1, \ldots, a_n)$, $\mathbf{b} = (b_1, \ldots, b_n)$. The Euclidean norm of $\mathbf{a}$ is defined by $|\mathbf{a}| = \langle \mathbf{a}|\mathbf{a}\rangle^{1/2}$. For any $\mathbf{a}, \mathbf{b}, \mathbf{c} \in C^n$, we define a triple product by

$$\{\mathbf{a}, \mathbf{b}, \mathbf{c}\} = \langle \mathbf{a}|\mathbf{b}\rangle \mathbf{c} + \langle \mathbf{c}|\mathbf{b}\rangle \mathbf{a} - \langle \mathbf{a}|\overline{\mathbf{c}}\rangle \overline{\mathbf{b}}, \tag{3.4}$$

where $\overline{\mathbf{b}} = (\bar{b}_1, \ldots, \bar{b}_n)$ denotes the complex conjugate of $\mathbf{b}$. This product is called the *spin triple product* and is an extension of the real spin triple product (3.1) and the one-dimensional complex spin product (3.2).

Note that this triple product is linear in the first and third variables ($\mathbf{a}$ and $\mathbf{c}$) and conjugate linear in the second variable ($\mathbf{b}$). Since, by the definition of the inner product, we have $\langle \mathbf{a}|\overline{\mathbf{c}}\rangle = \langle \mathbf{c}|\overline{\mathbf{a}}\rangle$, the triple product is symmetric in the outer variables, *i.e.*,

$$\{\mathbf{a}, \mathbf{b}, \mathbf{c}\} = \{\mathbf{c}, \mathbf{b}, \mathbf{a}\}. \tag{3.5}$$

The space C^n with the above triple product is called the *complex spin triple factor* and will be denoted by S^n. We use this name because if we define a norm based on this triple product, then the unit ball of $\mathcal{S}^n$ is a domain of Cartan type IV known as the spin domain. The *real part of the spin factor*, denoted $\mathcal{S}^n_R$, is the subspace of $\mathcal{S}^n$ defined by

$$\mathcal{S}^n_R = \{\mathbf{a} \in \mathcal{S}^n : \quad \overline{\mathbf{a}} = \mathbf{a}\},$$

or, equivalently,

$$\mathcal{S}^n_R = \text{span}_R\{\mathbf{e}_j\}. \tag{3.6}$$

This subspace is identical to R^n with the triple product defined by (3.1). For instance, the ball D_s of s-velocities considered in the previous chapter, endowed with the triple product derived from the Lie algebra $aut_c(D_s)$, is the real spin factor $\mathcal{S}^3_R$.

For any $\mathbf{a}, \mathbf{b} \in \mathcal{S}^n$, we define a complex linear map $D(\mathbf{a}, \mathbf{b}) : \mathcal{S}^n \to \mathcal{S}^n$ by

$$D(\mathbf{a}, \mathbf{b})\mathbf{z} = \{\mathbf{a}, \mathbf{b}, \mathbf{z}\} = \langle \mathbf{a}|\mathbf{b}\rangle \mathbf{z} + \langle \mathbf{z}|\mathbf{b}\rangle \mathbf{a} - \langle \mathbf{a}|\overline{\mathbf{z}}\rangle \overline{\mathbf{b}}. \tag{3.7}$$

The linear map $D(\mathbf{a}, \mathbf{b})$ is equal to

$$\langle \mathbf{a}|\mathbf{b}\rangle\, I + \mathbf{a} \wedge \mathbf{b}, \tag{3.8}$$

where I denotes the identity operator and

$$(\mathbf{a} \wedge \mathbf{b})(\mathbf{z}) = \langle \mathbf{z}|\mathbf{b}\rangle \mathbf{a} - \langle \mathbf{a}|\overline{\mathbf{z}}\rangle \overline{\mathbf{b}}.$$

Thus, the map $D(\mathbf{a}, \mathbf{b})$ resembles the geometric product of $\mathbf{a}$ and $\mathbf{b}$, defined by

$$\mathbf{ab} = \langle \mathbf{a}|\mathbf{b}\rangle + \mathbf{a} \wedge \mathbf{b}, \tag{3.9}$$

where the sum of a scalar $\langle \mathbf{a}|\mathbf{b}\rangle$ and bivector $\mathbf{a} \wedge \mathbf{b}$ belongs to the Clifford algebra. Hence, the operator $D(\mathbf{a}, \mathbf{b})$ is a natural operator on the spin factor and plays a role similar to that of the geometric product.

It is worth comparing the representations of the geometric product as the product in the Clifford algebra and as operators on the complex spin triple product. In the first case, in order to represent n canonical anticommutation relations, we need an algebra of dimension 2^n, while in the second case, it is enough to consider the space $\mathcal{S}^n$ of complex dimension n, along with the operators defined by the spin triple product on it.

The complex spin triple factor arises naturally in physics. In Chapter 2, the spin triple product was constructed, in the real case, directly from the conformal group. The complex spin triple product was effectively used in section 2.6 to describe the relativistic evolution of a charged particle in mutually perpendicular electric and magnetic fields. In the complex case, the spin triple product is built solely on the geometry of a Cartan domain of type IV which represents two-state systems in quantum mechanics, as it was shown in [35]. The advantage of the Clifford algebra approach, on the other hand, is the ability to express the equations of physics in a more compact form.

The Lorentz group is represented in both cases by a spin-half representation. As we will show later, the spin factor has a spin 1 representation as well. Like any bounded symmetric domain, the spin factor possesses a well-developed harmonic analysis, has an explicitly defined invariant measure, and supports a spectral theorem as well as quantization and representation as operators on a Hilbert space. However, since the spin factor representation is more compact, we are currently missing several techniques that play an important role in the Clifford algebra approach. For instance, here we do not have multivectors of order 3 or higher, nor do we have the analog of the I operator. But we believe that it is possible to overcome these difficulties.

3.1.2 The triple product representation of the Canonical Anticommutation Relations

The canonical anticommutation relations (CAR) are the basic relations used in the description of fermion fields.

We will show now that the natural basis of $\mathcal{S}^n$ satisfies a triple analog of the CAR. Recall that the classical definition of CAR involves a sequence p_k of elements of an associative algebra which satisfy the relations

$$p_l p_k + p_k p_l = 2\delta_{kl}, \tag{3.10}$$

where

$$\delta_{kl} = \begin{cases} 1, & \text{if } k = l, \\ 0, & \text{otherwise.} \end{cases} \tag{3.11}$$

This implies that $p_k^2 = 1$, and, therefore,

$$p_k p_k p_l = p_l \text{ for any } 1 \leq k, l \leq n. \tag{3.12}$$

Multiplying (3.10) on the left by p_l, we get

$$p_l p_k p_l = -p_k \text{ for } k \neq l. \tag{3.13}$$

We call the relations (3.12) and (3.13) the *triple canonical anticommutation relations* (TCAR).

Using definition (3.4) of the spin triple product, it is easy to verify that the elements $\mathbf{e}_1, \mathbf{e}_2, \ldots, \mathbf{e}_n$ of the natural basis of the spin triple factor satisfy the following relations:

$$\{\mathbf{e}_l, \mathbf{e}_k, \mathbf{e}_l\} = -\mathbf{e}_k, \quad \text{for } k \neq l, \tag{3.14}$$

$$\{\mathbf{e}_l, \mathbf{e}_k, \mathbf{e}_k\} = \{\mathbf{e}_k, \mathbf{e}_k, \mathbf{e}_l\} = \mathbf{e}_l, \quad \text{for any } \ k, l, \tag{3.15}$$

$$\{\mathbf{e}_k, \mathbf{e}_l, \mathbf{e}_m\} = 0 \ \text{ for } \ k, l, m \text{ distinct.} \tag{3.16}$$

Thus, the natural basis of the spin triple factor $\mathcal{S}^n$ or $\mathcal{S}^n_R$ satisfies the TCAR. Conversely, if we define a ternary operation on $\{\mathbf{e}_l, \mathbf{e}_2, \dots, \mathbf{e}_n\}$ which satisfies (3.14)–(3.16), then the resulting triple product on $\mathcal{S}^n$ will be exactly the spin triple product.

We will say that a basis $\{\mathbf{u}_1, \mathbf{u}_2, \dots, \mathbf{u}_n\}$ of $\mathcal{S}^n$ is a *TCAR basis* if it satisfies (3.14)–(3.16). We show now that a TCAR basis is *orthonormal.* The converse is false. For example, $\{\mathbf{e}_1, i\mathbf{e}_2, \mathbf{e}_3, \dots, \mathbf{e}_n\}$ is an orthonormal basis of $\mathcal{S}^n$ but not a TCAR basis because if $l \neq 2$, then $\{\mathbf{e}_l, i\mathbf{e}_2, \mathbf{e}_l\} = i\mathbf{e}_2$, in violation of (3.14).

By the definition of the triple product, for any $1 \leq k \leq n$, we have $\{\mathbf{u}_k, \mathbf{u}_k, \mathbf{u}_k\} = 2|\mathbf{u}_k|^2\mathbf{u}_k - \langle \mathbf{u}_k | \overline{\mathbf{u}}_k \rangle \overline{\mathbf{u}}_k$. But any element of a TCAR basis satisfies $\{\mathbf{u}_k, \mathbf{u}_k, \mathbf{u}_k\} = \mathbf{u}_k$. Thus,

$$2|\mathbf{u}_k|^2\mathbf{u}_k - \langle \mathbf{u}_k | \overline{\mathbf{u}}_k \rangle \overline{\mathbf{u}}_k = \mathbf{u}_k. \tag{3.17}$$

Hence, there is a complex number λ_k such that $\overline{\mathbf{u}}_k = \lambda_k \mathbf{u}_k$. Since $|\overline{\mathbf{u}}_k| = |\mathbf{u}_k|$, λ_k has absolute value 1. Thus,

$$\overline{\mathbf{u}}_k = \lambda_k \mathbf{u}_k : \ \ |\lambda_k| = 1. \tag{3.18}$$

This implies that

$$\langle \mathbf{u}_k | \overline{\mathbf{u}}_k \rangle \overline{\mathbf{u}}_k = \langle \mathbf{u}_k | \lambda_k \mathbf{u}_k \rangle \lambda_k \mathbf{u}_k = |\mathbf{u}_k|^2 \mathbf{u}_k.$$

Substituting this last expression into (3.17), it follows that $|\mathbf{u}_k| = 1$. Moreover, for any $1 \leq j \leq k \leq n$, we have

$$\{\mathbf{u}_k, \mathbf{u}_j, \mathbf{u}_k\} = 2\langle \mathbf{u}_k | \mathbf{u}_j \rangle \mathbf{u}_k - \langle \mathbf{u}_k | \overline{\mathbf{u}}_k \rangle \overline{\mathbf{u}}_j,$$

and for such elements in a TCAR basis, we have $\{\mathbf{u}_k, \mathbf{u}_j, \mathbf{u}_k\} = -\mathbf{u}_j$. Since the vector $\mathbf{u}_k$ is linearly independent of both $\mathbf{u}_j$ and $\overline{\mathbf{u}}_j$ (which is proportional to $\mathbf{u}_j$), the vectors $\mathbf{u}_k$ and $\mathbf{u}_j$ are orthogonal. Thus, any TCAR basis is an *orthonormal basis* of C^n.

3.1.3 The automorphism group Taut ($\mathcal{S}^n$) and its Lie algebra

The natural morphisms of the complex spin triple factor $\mathcal{S}^n$ are the linear, invertible maps (bijections) $T : \mathcal{S}^n \to \mathcal{S}^n$ which preserve the triple product. This means that

$$T\{\mathbf{a}, \mathbf{b}, \mathbf{c}\} = \{T\mathbf{a}, T\mathbf{b}, T\mathbf{c}\}. \tag{3.19}$$

Such a linear map is called a *triple automorphism* of $\mathcal{S}^n$. We denote by Taut $(\mathcal{S}^n)$ the group of all triple automorphisms of $\mathcal{S}^n$.

Since the definition of a TCAR basis involves only the triple product, it is obvious that a triple automorphism T maps a TCAR basis into a TCAR basis. In particular, the image of the natural basis $\{\mathbf{e}_1, \mathbf{e}_2, \dots, \mathbf{e}_n\}$ is a TCAR basis. For any $1 \le k \le n$, let $\mathbf{u}_k = T\mathbf{e}_k$. Since $\{\mathbf{u}_1, \mathbf{u}_2, \dots, \mathbf{u}_n\}$ is a TCAR basis of $\mathcal{S}^n$, it is also an orthonormal basis of C^n and from (3.18) it follows that for any k there is a number λ_k such that $|\lambda_k| = 1$ and $\overline{\mathbf{u}}_k = \lambda_k \mathbf{u}_k$. Moreover, for any $1 \le j \ne k \le n$, from (3.14) we have

$$\mathbf{u}_j = -\{\mathbf{u}_k, \mathbf{u}_j, \mathbf{u}_k\} = \langle \mathbf{u}_k | \overline{\mathbf{u}}_k \rangle \overline{\mathbf{u}_j} = \langle \mathbf{u}_k | \lambda_k \mathbf{u}_k \rangle \lambda_j \mathbf{u}_j = \langle \mathbf{u}_k | \mathbf{u}_k \rangle \overline{\lambda_k} \lambda_j \mathbf{u}_j,$$

implying that $\overline{\lambda}_k \lambda_j = 1$. Hence, $\lambda_k = \lambda_j$. Call this common constant μ. So $|\mu| = 1$, and for any $1 \le k \le n$, we have $\overline{\mathbf{u}}_k = \mu \mathbf{u}_k$. Define $\lambda = \overline{\mu}^{1/2}$ and $U = \overline{\lambda} T$. Then

$$\overline{U\mathbf{e}_k} = \overline{\overline{\lambda} T\mathbf{e}_k} = \lambda \overline{T\mathbf{e}_k} = \lambda \overline{\mathbf{u}}_k = \lambda \mu \mathbf{u}_k = \overline{\lambda} \mathbf{u}_k = \overline{\lambda} T\mathbf{e}_k = U\mathbf{e}_k.$$

This implies that the matrix of U in the natural basis has real entries, and, since it maps an orthonormal basis to an orthonormal basis, U is orthogonal. Thus, we have shown that any map T of the spin triple factor $\mathcal{S}^n$ which preserves the triple product has the form $T = \lambda U$, where λ is a complex number of absolute value 1 and U is orthogonal.

Conversely, suppose a linear map T of the complex spin triple factor $\mathcal{S}^n$ has the form λU. Using the fact that an orthogonal map preserves the triple product, we have, for any k, l and m,

$$\{T\mathbf{u}_k, T\mathbf{u}_l, T\mathbf{u}_m\} = \{\lambda U\mathbf{u}_k, \lambda U\mathbf{u}_l, \lambda U\mathbf{u}_m\} = \lambda \{U\mathbf{u}_k, U\mathbf{u}_l, U\mathbf{u}_m\}$$

$$= \lambda U\{\mathbf{u}_k, \mathbf{u}_l, \mathbf{u}_m\} = T\{\mathbf{u}_k, \mathbf{u}_l, \mathbf{u}_m\},$$

showing that T preserves the triple product. From our discussion, it follows that

$$\text{Taut}\,(\mathcal{S}^n) = U(1) \times O(n), \tag{3.20}$$

where $U(1)$ is the group of rotations in the complex plane and $O(n)$ is the orthogonal group of dimension n. Thus, *Taut* $(\mathcal{S}^n)$ is a Lie group with real dimension $n(n-1)/2 + 1$.

This group is a natural candidate for the description of the state space of a quantum system. The state description of a quantum system is often given by a complex-valued wave function $\psi(\mathbf{r})$, where $\mathbf{r} \in R^3$. This description is invariant under the choice of the orthogonal basis in R^3, implying that there is a natural action of the group $O(3)$ on the state space. In the presence of an electromagnetic field, the gauge of the field induces a multiple of the state by a complex number λ, $|\lambda| = 1$, which will not affect any meaningful results.

Multiplication of all $\psi(\mathbf{r})$ by such a λ corresponds to an action of the group $U(1)$ on this state space. Moreover, even without gauge invariance, all meaningful quantities in quantum mechanics are invariant under multiplication by a complex number of absolute value 1, resulting in an action of $U(1)$. Thus, Taut $(\mathcal{S}^n)$ acts naturally on the state space of quantum systems. A similar result holds for quantum fields.

We now describe the Lie algebra taut $(\mathcal{S}^n)$ of the Lie group Taut $(\mathcal{S}^n)$. This Lie algebra consists of sums of generators of the group $U(1)$ of rotations in the complex plane and generators of $O(n)$, the orthogonal group of dimension n. It is well known and easy to verify that the first type of generator is described by a pure imaginary number and that the second type of generator is described by an $n \times n$ real antisymmetric matrix. From the TCAR, it follows that the matrix of $D(\mathbf{u}_k, \mathbf{u}_l)$ with respect to the basis $\mathbf{u}_1, \dots, \mathbf{u}_n$ is a basic antisymmetric matrix. For instance,

$$D(\mathbf{u}_1, \mathbf{u}_2) = \begin{pmatrix} 0 & 1 & 0 & \cdots & 0 \\ -1 & 0 & 0 & \cdots & 0 \\ \vdots & \vdots & \vdots & \cdots & \vdots \\ 0 & 0 & 0 & \cdots & 0 \end{pmatrix}. \tag{3.21}$$

Thus, the Lie algebra taut $(\mathcal{S}^n)$ is the direct sum of iR and the algebra of real antisymmetric $n \times n$ matrices, that is,

$$\text{taut}\,(\mathcal{S}^n) = \{diI + \sum_{k<l} d_{kl} D(\mathbf{u}_k, \mathbf{u}_l) : \ d, d_{kl} \in R\}. \tag{3.22}$$

3.1.4 Tripotents in $\mathcal{S}^n$

For binary operations, the building blocks are the projections, which are the *idempotents* of the operation, that is, non-zero elements p that satisfy $p^2 = p$. For a ternary operation, the building blocks are the *tripotents*, non-zero elements $\mathbf{u}$ satisfying $\{\mathbf{u}, \mathbf{u}, \mathbf{u}\} = \mathbf{u}$.

We will describe now the tripotents $\mathbf{u} \in \mathcal{S}^n$. To do this, we define first the notion of determinant for elements of $\mathcal{S}^n$. For any $\mathbf{a} \in \mathcal{S}^n$, the *determinant* of $\mathbf{a}$, denoted $\det \mathbf{a}$, is

$$\det \mathbf{a} = \langle \mathbf{a} | \overline{\mathbf{a}} \rangle = \sum_{i=1}^{n} a_i^2. \tag{3.23}$$

In case the elements of $\mathcal{S}^n$ can be represented by matrices, this definition agrees with the ordinary determinant of a matrix. Note that elements with zero determinant are called *null-vectors* in the literature.

From (3.4), it follows that if an element $\mathbf{u}$ is a tripotent, then

$$\mathbf{u} = \{\mathbf{u}, \mathbf{u}, \mathbf{u}\} = 2\langle \mathbf{u} | \mathbf{u} \rangle \mathbf{u} - (\det \mathbf{u}) \overline{\mathbf{u}}.$$

Thus, only the following two cases can occur. In case 1,

$$\det \mathbf{u} = 0 \text{ and } \langle \mathbf{u}|\mathbf{u}\rangle = 1/2. \tag{3.24}$$

Such a $\mathbf{u}$ is called a *minimal tripotent.* If $\det \mathbf{u} \neq 0$, then there is a constant λ such that $\overline{\mathbf{u}} = \lambda\mathbf{u}$. Since $|\overline{\mathbf{u}}| = |\mathbf{u}|$, we must have $|\lambda| = 1$. Define $\overline{\mu} = \lambda^{1/2}$. Then $\overline{\mu}\mathbf{u} = \mu\overline{\mathbf{u}}$, implying that $\overline{\mu}\mathbf{u} = \mathbf{r}$ is a real vector. This leads us to case 2, in which

$$\mathbf{u} = \mu\mathbf{r} \text{ and } \langle \mathbf{r}|\mathbf{r}\rangle = 1, \tag{3.25}$$

where $\mathbf{r} \in \mathcal{S}^n_R$ and $|\mu| = 1$. In this case, $\mathbf{u}$ is called a *maximal tripotent.*

We say that an element $\mathbf{w} \in \mathcal{S}^n$ is *algebraically orthogonal* to a tripotent $\mathbf{u}$ in $\mathcal{S}^n$ (*i.e.*, orthogonal in the sense of the algebraic structure and not in the sense of the inner product) if

$$D(\mathbf{u})\mathbf{w} = 0, \tag{3.26}$$

where the operator D is defined by (3.7), and $D(\mathbf{u})$ is an abbreviation for $D(\mathbf{u},\mathbf{u})$. Suppose $\mathbf{u} = \mu\mathbf{r}$ is a maximal tripotent. Then, for any $\mathbf{a} \in \mathcal{S}^n$,

$$D(\mathbf{u})\mathbf{a} = \langle \mathbf{u}|\mathbf{u}\rangle\mathbf{a} + \langle \mathbf{a}|\mathbf{u}\rangle\mathbf{u} - \langle \mathbf{u}|\overline{\mathbf{a}}\rangle\overline{\mathbf{u}}$$

$$= \langle \mathbf{r}|\mathbf{r}\rangle\mathbf{a} + \langle \mathbf{a}|\mathbf{r}\rangle\mathbf{r} - \langle \mathbf{a}|\mathbf{r}\rangle\mathbf{r} = \mathbf{a},$$

implying that $D(\mathbf{u}) = I$. This implies that there are no tripotents algebraically orthogonal to $\mathbf{u}$, explaining why $\mathbf{u}$ is called a maximal tripotent. Note that from (3.25), each element $\mathbf{e}_j$ of the natural basis is a maximal tripotent.

Suppose $\mathbf{v}$ is a minimal tripotent. Then (3.24) implies that $\overline{\mathbf{v}}$ is also a minimal tripotent. Since $\det \mathbf{v} = \langle \mathbf{v}|\overline{\mathbf{v}}\rangle = 0$, we have $D(\mathbf{v})\overline{\mathbf{v}} = 0$, and so $\overline{\mathbf{v}}$ is algebraically orthogonal to $\mathbf{v}$. Since $\langle \mathbf{v}|\mathbf{v}\rangle = 1/2$, the orthogonal (in C^n) projections $P_{\mathbf{v}}$ and $P_{\overline{\mathbf{v}}}$ are

$$P_{\mathbf{v}}\mathbf{a} = 2\langle \mathbf{a}|\mathbf{v}\rangle\mathbf{v}, \quad P_{\overline{\mathbf{v}}}\mathbf{a} = 2\langle \mathbf{a}|\overline{\mathbf{v}}\rangle\overline{\mathbf{v}}.$$

These two projections are algebraically orthogonal projections in the sense that $P_{\mathbf{v}}P_{\overline{\mathbf{v}}} = 0$. Moreover, since

$$D(\mathbf{v})\mathbf{a} = \langle \mathbf{v}|\mathbf{v}\rangle\mathbf{a} + \langle \mathbf{a}|\mathbf{v}\rangle\mathbf{v} - \langle \mathbf{v}|\overline{\mathbf{a}}\rangle\overline{\mathbf{v}},$$

we can write $D(\mathbf{v})$ in terms of $P_{\mathbf{v}}$ and $P_{\overline{\mathbf{v}}}$ as

$$D(\mathbf{v}) = \frac{1}{2}(I + P_{\mathbf{v}} - P_{\overline{\mathbf{v}}}). \tag{3.27}$$

The spectrum of the operator $D(\mathbf{v})$ is the set $\{1, 1/2, 0\}$, where the eigenvalue 1 is obtained on multiples of $\mathbf{v}$ (*i.e.*, on the image of $P_{\mathbf{v}}$), the eigenvalue

0 is obtained on multiples of $\overline{\mathbf{v}}$ (on the image of $P_{\overline{\mathbf{v}}}$) and the eigenvalue 1/2 is obtained on the image of the projection $I - P_{\mathbf{v}} - P_{\overline{\mathbf{v}}}$. This leads to the *Peirce decomposition* of $\mathcal{S}^n$ with respect to a minimal tripotent $\mathbf{v}$ as a direct sum of the 1, 0 and 1/2 eigenspaces of the operator $D(\mathbf{v})$ (see section 3.1.6 below).

We claim that if $\mathbf{w} \in \mathcal{S}^n$ is algebraically orthogonal to a minimal tripotent $\mathbf{v}$, then

$$\mathbf{w} = \lambda \overline{\mathbf{v}}, \text{ with } \lambda \in C. \tag{3.28}$$

To see this, note first that

$$0 = D(\mathbf{v}, \mathbf{v})\mathbf{w} = \frac{1}{2}\mathbf{w} + \langle \mathbf{w}|\mathbf{v}\rangle \mathbf{v} - \langle \mathbf{v}|\overline{\mathbf{w}}\rangle \overline{\mathbf{v}}.$$

Taking the inner product of this expression with $\mathbf{v}$ and substituting $\langle \overline{\mathbf{v}}|\mathbf{v}\rangle = 0$, we get

$$\frac{1}{2}\langle \mathbf{w}|\mathbf{v}\rangle + \frac{1}{2}\langle \mathbf{w}|\mathbf{v}\rangle = 0,$$

implying that $\langle \mathbf{w}|\mathbf{v}\rangle = 0$. Thus $\mathbf{w} = 2\langle \mathbf{v}|\overline{\mathbf{w}}\rangle \overline{\mathbf{v}} = \lambda \overline{\mathbf{v}}$.

From the definition of the triple product, we have $\{\mathbf{v}, \overline{\mathbf{v}}, \mathbf{v}\} = 0$. Thus, for any $\mathbf{a} = \alpha \mathbf{v} + \beta \overline{\mathbf{v}}$, with $\alpha, \beta \in C$, we have

$$\{\mathbf{a}, \mathbf{a}, \mathbf{a}\} = \alpha |\alpha|^2 \mathbf{v} + \beta |\beta|^2 \overline{\mathbf{v}}. \tag{3.29}$$

In particular, both $\mathbf{v} + \overline{\mathbf{v}}$ and $\mathbf{v} - \overline{\mathbf{v}}$ satisfy (3.25) and thus are maximal tripotents and have determinant 1. Since all the tripotents of the complex spin triple factor $\mathcal{S}^n$ are either maximal or minimal, $\mathbf{v}$ cannot be written as a sum of two orthogonal tripotents. This explains the terminology *minimal* tripotent. Moreover, there cannot be more than *two* mutually orthogonal tripotents in $\mathcal{S}^n$. Such a triple is said to be of *rank 2*. Thus, $\mathcal{S}^n$ is a rank 2 triple.

Note also that if we decompose a minimal tripotent $\mathbf{v}$ as

$$\mathbf{v} = \mathbf{x} + i\mathbf{y}, \qquad \mathbf{x}, \mathbf{y} \in \mathcal{S}^n_R, \tag{3.30}$$

then, from the definition of the determinant, we have

$$\det \mathbf{v} = |\mathbf{x}|^2 - |\mathbf{y}|^2 + 2i\langle \mathbf{x}|\mathbf{y}\rangle, \tag{3.31}$$

and from (3.24), the condition $\det \mathbf{v} = 0$ implies

$$|\mathbf{x}| = |\mathbf{y}| \quad \text{and} \quad \langle \mathbf{x}|\mathbf{y}\rangle = 0. \tag{3.32}$$

Finally, the condition $\langle \mathbf{v}|\mathbf{v}\rangle = 1/2$ implies

$$|\mathbf{x}| = |\mathbf{y}| = 1/2. \tag{3.33}$$

Thus the real and imaginary parts of a minimal tripotent satisfy (3.32) and (3.33). Conversely, if two real vectors $\mathbf{x}$ and $\mathbf{y}$ satisfy these conditions, the vector $\mathbf{v}$, defined by (3.30), is a minimal tripotent. Table 3.1 summarizes the properties of the two types of tripotents in a complex spin triple factor.

Type	Norm	det	$D(\mathbf{u},\mathbf{u})$	Decomposition into real and imaginary parts
Maximal	$<\mathbf{u}\mid\mathbf{u}>=1$	$\lvert\det\mathbf{u}\rvert=1$	$D(\mathbf{u},\mathbf{u})=I$	$\mathbf{u}=\cos\theta\ \mathbf{r}+i\ \sin\theta\ \mathbf{r}$ $\lvert\mathbf{r}\rvert=1$
Minimal	$<\mathbf{v}\mid\mathbf{v}>=\frac{1}{2}$	$\det\mathbf{v}=0$	$D(\mathbf{v},\mathbf{v})=\frac{1}{2}(I+P_{\mathbf{v}}-P_{\bar{\mathbf{v}}})$	$\mathbf{v}=\mathbf{x}+i\mathbf{y}$ $<\mathbf{x}\mid\mathbf{y}>=0\quad\lvert\mathbf{x}\rvert=\lvert\mathbf{y}\rvert=\frac{1}{2}$

Table 3.1. The algebraic properties of tripotents in $\mathcal{S}^n$.

3.1.5 Singular decomposition in $\mathcal{S}^n$

In this subsection, we explain how to obtain the *singular decomposition* of an element of $\mathcal{S}^n$. This concept plays a major role in the investigation of spin factors.

Let $\mathbf{a}$ be any element in $\mathcal{S}^n$. If $\det\mathbf{a}=0$, then it follows from (3.24) that $\mathbf{a}$ is a positive multiple of a minimal tripotent. In fact, $\mathbf{u}:=\frac{1}{\sqrt{2}|\mathbf{a}|}\mathbf{a}$ is a minimal tripotent. If $\det\mathbf{a}\neq 0$, then, as we will show, there exist an algebraically orthogonal pair $\mathbf{v}_1,\mathbf{v}_2$ of minimal tripotents and a pair of non-negative real numbers s_1,s_2, called the *singular numbers* of $\mathbf{a}$, such that

$$s_1\geq s_2\geq 0$$

and

$$\mathbf{a}=s_1\mathbf{v}_1+s_2\mathbf{v}_2. \tag{3.34}$$

This decomposition is called the *singular decomposition of* $\mathbf{a}$. If $\mathbf{a}$ is not a multiple of a maximal tripotent, then $s_1>s_2$ and the decomposition is unique. If $\mathbf{a}$ *is* a multiple of a maximal tripotent, then $s_1=s_2$, and the decomposition is, in general, not unique.

To obtain the singular decomposition of an element $\mathbf{a}$ in $\mathcal{S}^n$ we define first the element's *polar decomposition*. Recall that the *polar form* of a complex number $z=x+iy$ is $re^{i\theta}$, where $r=|z|$ is the *modulus* of z and $e^{i\theta}$ is of modulus 1, called also unimodular, where $\theta=\arctan\left(\frac{y}{x}\right)$ is the *argument*, or *phase*, of z.

For $\mathbf{a}$ in $\mathcal{S}^n$ with $\det\mathbf{a}\neq 0$, we define the *argument* of $\det\mathbf{a}$ to be

$$\arg\det\mathbf{a}=\frac{\det\mathbf{a}}{|\det\mathbf{a}|}.$$

Note that for any element $\mathbf{a}\in\mathcal{S}^n$ with $\det\mathbf{a}\neq 0$, if μ is a complex number with $|\mu|=1$, then from (3.23), it follows that $\arg\det(\mu\mathbf{a})=\mu^2\arg\det\mathbf{a}$. Set $\lambda=(\arg\det\mathbf{a})^{1/2}$ and $\mathbf{a}_+=\overline{\lambda}\mathbf{a}$. Then $\mathbf{a}_+$ has positive determinant (in fact, $\det\mathbf{a}_+=|\det\mathbf{a}|$) and

$$\mathbf{a}=\lambda\mathbf{a}_+\,. \tag{3.35}$$

This is the *polar decomposition of* $\mathbf{a}$ as a product of a complex number λ of modulus 1 and an element $\mathbf{a}_+$ in $\mathcal{S}^n$ with $\det \mathbf{a}_+ > 0$. Since $|\lambda| = 1$, we have

$$|\mathbf{a}| = |\mathbf{a}_+| \,. \tag{3.36}$$

Note also that $\tilde{\lambda} = -\lambda$ satisfies $\tilde{\lambda} = (\arg \det \mathbf{a})^{1/2}$ and that the element $\tilde{\mathbf{a}}_+ = -\overline{\lambda}\mathbf{a} = -\mathbf{a}_+$ has non-negative determinant. Thus, any element $\mathbf{a} \in \mathcal{S}^n$ has an additional polar decomposition

$$\mathbf{a} = -\lambda(-\mathbf{a}_+)\,.$$

Decompose $\mathbf{a}_+$ into real and imaginary parts $Re(\mathbf{a}_+)$ and $Im(\mathbf{a}_+)$, respectively. Then

$$\det \mathbf{a}_+ = |Re(\mathbf{a}_+)|^2 - |Im(\mathbf{a}_+)|^2 + 2i\langle Re(\mathbf{a}_+)|Im(\mathbf{a}_+)\rangle, \tag{3.37}$$

and the condition $\det \mathbf{a}_+ > 0$ implies

$$\langle Re(\mathbf{a}_+)|Im(\mathbf{a}_+)\rangle = 0 \quad \text{and} \quad |Re(\mathbf{a}_+)| > |Im(\mathbf{a}_+)|. \tag{3.38}$$

Therefore, in the above notation, we have

$$\mathbf{a} = \lambda(Re(\mathbf{a}_+) + iIm(\mathbf{a}_+)), \;\; Re(\mathbf{a}_+), Im(\mathbf{a}_+) \in \mathcal{S}^n_R, \tag{3.39}$$

where $\lambda = (\arg \det \mathbf{a})^{1/2}$ and $Re(\mathbf{a}_+), Im(\mathbf{a}_+)$ satisfy (3.38).

Using (3.32) and (3.33), we see that the two elements

$$\mathbf{w}_1 = \frac{Re(\mathbf{a}_+)}{2|Re(\mathbf{a}_+)|} + i\frac{Im(\mathbf{a}_+)}{2|Im(\mathbf{a}_+)|} \quad \text{and} \quad \mathbf{w}_2 = \overline{\mathbf{w}}_1 \tag{3.40}$$

are orthogonal minimal tripotents. Then

$$\mathbf{a}_+ = (|Re(\mathbf{a}_+)| + |Im(\mathbf{a}_+)|)\mathbf{w}_1 + (|Re(\mathbf{a}_+)| - |Im(\mathbf{a}_+)|)\mathbf{w}_2. \tag{3.41}$$

The desired singular decomposition (3.34) of $\mathbf{a}$ is now obtained by defining the singular numbers s_1, s_2 to be

$$s_1 = |Re(\mathbf{a}_+)| + |Im(\mathbf{a}_+)|, \;\; s_2 = |Re(\mathbf{a}_+)| - |Im(\mathbf{a}_+)| \tag{3.42}$$

and taking as minimal tripotents the two elements

$$\mathbf{v}_1 = \lambda\mathbf{w}_1, \;\; \mathbf{v}_2 = \lambda\mathbf{w}_2. \tag{3.43}$$

The minimality and orthogonality of the tripotents $\mathbf{v}_1$ and $\mathbf{v}_2$ follow from the corresponding properties for $\mathbf{w}_1$ and $\mathbf{w}_2$. Note that from (3.39) and (3.42) we obtain

$$|\mathbf{a}|^2 = |Re(\mathbf{a}_+)|^2 + |Im(\mathbf{a}_+)|^2 = (\frac{s_1 + s_2}{2})^2 + (\frac{s_1 - s_2}{2})^2 = \frac{s_1^2 + s_2^2}{2}. \tag{3.44}$$

If $s_1 = s_2$, then $s_1^{-1}\mathbf{a} = \mathbf{v}_1 + \mathbf{v}_2$, a sum of orthogonal minimal tripotents, and, hence, a maximal tripotent. Thus, if $\mathbf{a}$ is not a multiple of a maximal tripotent, then $s_1 > s_2$, and, since the second polar decomposition yields the same $s_1, s_2, \mathbf{v}_1, \mathbf{v}_2$, the above development shows that the singular decomposition is unique.

Since $\mathbf{a} = \lambda \mathbf{a}_+$, by use of (3.37), (3.38) and (3.42), we get

$$|\det \mathbf{a}| = \det \mathbf{a}_+ = s_1 s_2. \tag{3.45}$$

This result corresponds to the fact that the determinant of a positive operator is the product of its eigenvalues. From (3.44) and (3.45), we have

$$s_1 \pm s_2 = \sqrt{2|\mathbf{a}|^2 \pm 2|\det \mathbf{a}|}. \tag{3.46}$$

For any $\mathbf{a}$ with singular decomposition (3.34), by use of (3.29) we get

$$\mathbf{a}^{(3)} := \{\mathbf{a}, \mathbf{a}, \mathbf{a}\} = s_1^3 \mathbf{v}_1 + s_2^3 \mathbf{v}_2, \tag{3.47}$$

implying that the cube of an element $\mathbf{a} \in \mathcal{S}^n$ can be calculated by cubing its singular numbers. Similarly, taking *any* odd power of $\mathbf{a}$ is equivalent to applying this odd power to its singular numbers.

3.1.6 The Peirce decomposition and the main identity

Let $\mathbf{v}$ be a minimal tripotent. Motivated by the discussion of the spectrum of the operator $D(\mathbf{v})$ on page 98, we define $P_1(\mathbf{v})$, $P_{1/2}(\mathbf{v})$ and $P_0(\mathbf{v})$ to be the projections onto the 1, 1/2 and 0 eigenspaces of $D(\mathbf{v})$, respectively. Thus,

$$P_1(\mathbf{v}) = P_{\mathbf{v}}, \;\; P_{1/2}(\mathbf{v}) = I - P_{\mathbf{v}} - P_{\overline{\mathbf{v}}}, \;\; P_0(\mathbf{v}) = P_{\overline{\mathbf{v}}}. \tag{3.48}$$

Then, from (3.27), we have

$$D(\mathbf{v}) = P_1(\mathbf{v}) + \frac{1}{2} P_{1/2}(\mathbf{v}). \tag{3.49}$$

Since

$$I = P_1(\mathbf{v}) + P_{1/2}(\mathbf{v}) + P_0(\mathbf{v}), \tag{3.50}$$

these projections induce a decomposition of $\mathcal{S}^n$ into the sum of the three eigenspaces:

$$\mathcal{S}^n = \mathcal{S}_1^n(\mathbf{v}) + \mathcal{S}_{1/2}^n(\mathbf{v}) + \mathcal{S}_0^n(\mathbf{v}). \tag{3.51}$$

This is called the *Peirce decomposition* of $\mathcal{S}^n$ with respect to a minimal tripotent $\mathbf{v}$.

Next, we want to derive the *main identity* of the triple product. Decompose $\mathbf{v}$ as in (3.30). We denote the mutually orthogonal norm 1 elements $2\mathbf{x}$

and $2\mathbf{y}$ in $\mathcal{S}^n_R$ by $\mathbf{u}_1$ and $\mathbf{u}_2$, respectively. Complete them to a TCAR basis B. From (3.22), the operator $\delta = i\tau I + \tau D(\mathbf{u}_1, \mathbf{u}_2)$ is an element of taut $(\mathcal{S}^n)$ for any real τ. Denote by $(w_1, w_2, \cdots, w_n)$ the coordinates of an arbitrary element $\mathbf{w} \in \mathcal{S}^n$ in the basis B. Then, writing $\mathbf{v} = (\frac{1}{2}, \frac{1}{2}i, 0, 0, \dots)$, we have

$$\delta(\mathbf{w}) = i\tau\mathbf{w} + \tau D(\mathbf{u}_1, \mathbf{u}_2)\mathbf{w} = i\tau(w_1 - iw_2, i(w_1 - iw_2), w_3, \cdots, w_n)$$

$$= i\tau(2P_1(\mathbf{v})\mathbf{w} + P_{1/2}(\mathbf{v})\mathbf{w}) = i2\tau D(\mathbf{v})\mathbf{w},$$

implying that

$$\exp(i\tau D(\mathbf{v})) \in \text{Taut}\,(\mathcal{S}^n) \tag{3.52}$$

is a triple product automorphism. Thus, for any $\mathbf{a}, \mathbf{b}, \mathbf{c} \in \mathcal{S}^n$, we have

$$\exp(i\tau D(\mathbf{v}))\{\mathbf{a}, \mathbf{b}, \mathbf{c}\} = \{\exp(i\tau D(\mathbf{v}))\mathbf{a}, \exp(i\tau D(\mathbf{v}))\mathbf{b}, \exp(i\tau D(\mathbf{v}))\mathbf{c}\}.$$

By differentiating both sides of this equation with respect to τ, substituting $\tau = 0$, using the linearity and conjugate linearity of the spin triple product, and dividing by i, we get

$$D(\mathbf{v})\{\mathbf{a}, \mathbf{b}, \mathbf{c}\} = \{D(\mathbf{v})\mathbf{a}, \mathbf{b}, \mathbf{c}\} - \{\mathbf{a}, D(\mathbf{v})\mathbf{b}, \mathbf{c}\} + \{\mathbf{a}, \mathbf{b}, D(\mathbf{v})\mathbf{c}\}. \tag{3.53}$$

We can use this last expression to describe the behavior of the triple product on the eigenspaces $\mathcal{S}^n_j(\mathbf{v})$. Suppose

$$\mathbf{a} \in \mathcal{S}^n_j(\mathbf{v}), \quad \mathbf{b} \in \mathcal{S}^n_k(\mathbf{v}), \quad \mathbf{c} \in \mathcal{S}^n_l(\mathbf{v}),$$

where $j, k, l \in \{1, \frac{1}{2}, 0\}$. Then, from (3.53), we get

$$D(\mathbf{v})\{\mathbf{a}, \mathbf{b}, \mathbf{c}\} = \{j\mathbf{a}, \mathbf{b}, \mathbf{c}\} - \{\mathbf{a}, k\mathbf{b}, \mathbf{c}\} + \{\mathbf{a}, \mathbf{b}, l\mathbf{c}\}$$

$$= (j - k + l)\{\mathbf{a}, \mathbf{b}, \mathbf{c}\}.$$

This implies that the vector $\{\mathbf{a}, \mathbf{b}, \mathbf{c}\}$ is an eigenvector of $D(\mathbf{v})$ and therefore is in the range of a Peirce projection. Thus,

$$\{\mathcal{S}^n_j(\mathbf{v}), \mathcal{S}^n_k(\mathbf{v}), \mathcal{S}^n_l(\mathbf{v})\} \subset \mathcal{S}^n_{j-k+l}(\mathbf{v}) \tag{3.54}$$

if $j - k + l \in \{1, \frac{1}{2}, 0\}$, and otherwise, $\{\mathcal{S}^n_j(\mathbf{v}), \mathcal{S}^n_k(\mathbf{v}), \mathcal{S}^n_l(\mathbf{v})\} = 0$. Equation (3.54) is called the *Peirce calculus* formula.

The main identity of the triple product is a generalization of (3.53), in which the minimal tripotent $\mathbf{v}$ is replaced with an arbitrary element $\mathbf{d} \in \mathcal{S}^n$. Use the singular decomposition of $\mathbf{d}$ as

$$\mathbf{d} = s_1\mathbf{v}_1 + s_2\mathbf{v}_2,$$

where $\mathbf{v}_1, \mathbf{v}_2$ are orthogonal minimal tripotents. The tripotent $\mathbf{v}_2$ belongs to the image of $P_0(\mathbf{v}_1)$, which implies that $\mathbf{v}_2 = \lambda\overline{\mathbf{v}}_1$ for some $\lambda \in C$ with $|\lambda| = 1$. This implies that for any $\mathbf{w} \in S^n$, we have

$$D(\mathbf{v}_1, \mathbf{v}_2)\mathbf{w} = \langle \mathbf{v}_1|\mathbf{v}_2\rangle\mathbf{w} + \langle \mathbf{w}|\mathbf{v}_2\rangle\mathbf{v}_1 - \langle \mathbf{w}|\overline{\mathbf{v}}_1\rangle\overline{\mathbf{v}}_2$$

$$= 0 + \overline{\lambda}\langle \mathbf{w}|\overline{\mathbf{v}}_1\rangle\mathbf{v}_1 - \overline{\lambda}\langle \mathbf{w}|\overline{\mathbf{v}}_1\rangle\mathbf{v}_1 = 0.$$

Thus,

$$D(\mathbf{d}) = D(s_1\mathbf{v}_1 + s_2\mathbf{v}_2, s_1\mathbf{v}_1 + s_2\mathbf{v}_2) = s_1^2 D(\mathbf{v}_1) + s_2^2 D(\mathbf{v}_2). \tag{3.55}$$

By using (3.49) and the Peirce decomposition, we can rewrite (3.55) as

$$D(\mathbf{d}) = s_1^2 P_1(\mathbf{v}_1) + \frac{s_1^2 + s_2^2}{2} P_{1/2}(\mathbf{v}_1) + s_2^2 P_0(\mathbf{v}_1). \tag{3.56}$$

This shows that the spectrum of the linear operator $D(\mathbf{d})$ is non-negative for each $\mathbf{d}$.

For any $\mathbf{a}, \mathbf{b}, \mathbf{c} \in S^n$, by use of (3.53) and (3.55), we get

$$D(\mathbf{d})\{\mathbf{a}, \mathbf{b}, \mathbf{c}\} = \{D(\mathbf{d})\mathbf{a}, \mathbf{b}, \mathbf{c}\} - \{\mathbf{a}, D(\mathbf{d})\mathbf{b}, \mathbf{c}\} + \{\mathbf{a}, \mathbf{b}, D(\mathbf{d})\mathbf{c}\}. \tag{3.57}$$

This is the *main identity* of the triple product.

3.2 Geometry of the spin factor

The bounded symmetric domains D_v and D_s, considered in the first two chapters, are Euclidean balls. The geometry of these balls is somewhat trivial. Any two points on the boundary can be mapped to each other by a rotation, and any two internal points can be mapped to each other by elements of the appropriate automorphism group. Thus, the only significant distinction is that between objects moving with the speed of light and objects moving with less than the speed of light. Such a model is too simple to describe the variety of different phenomena in our physical world. On the other hand, as we will see, the complex spin triple factor is a bounded symmetric domain with *non-trivial* geometry.

3.2.1 The norm of S^n.

For $\mathbf{a}$ with singular decomposition (3.34), we define a norm, called the *operator norm of* $\mathbf{a}$, by

$$\|\mathbf{a}\| = s_1. \tag{3.58}$$

From (3.42), we have

$$\|\mathbf{a}\| = |Re(\mathbf{a}_+)| + |Im(\mathbf{a}_+)|, \tag{3.59}$$

and from (3.46), we have

$$\|\mathbf{a}\| = \frac{1}{2}(\sqrt{2|\mathbf{a}|^2 + 2|\det \mathbf{a}|} + \sqrt{2|\mathbf{a}|^2 - 2|\det \mathbf{a}|}). \tag{3.60}$$

From (3.47), it follows that the operator norm satisfies the identity

$$\|\mathbf{a}^{(3)}\| = \|\mathbf{a}\|^3, \tag{3.61}$$

and hence is a natural norm for a set with a triple product. The above identity is the analog of the *star identity* $\|\mathbf{a}\mathbf{a}^*\| = \|\mathbf{a}\|^2$ in C^*-algebras.

The operator norm (3.58) coincides with the usual operator norm of a positive operator, defined to be the maximal eigenvalue (corresponding to the maximal singular value for the triple product). Note that from (3.56), it follows that

$$\|\mathbf{a}\|^2 = \|D(\mathbf{a})\|_{op}, \tag{3.62}$$

where $\|D(\mathbf{a})\|_{op}$ denotes the operator norm of $D(\mathbf{a})$. This identity can also be used to define the norm of $\mathbf{a} \in \mathcal{S}^n$.

Let $\mathbf{a}$ be any element of $\mathcal{S}^n$, and let μ be a complex number with $|\mu| = 1$. Let the polar decomposition of $\mathbf{a}$ with $\det \mathbf{a} \neq 0$ be given by (3.35). Then the polar decomposition of $\mu\mathbf{a}$ is $\mu\mathbf{a} = \mu\lambda\mathbf{a}_+$, implying that $(\mu\mathbf{a})_+ = \mathbf{a}_+$, and so

$$Re((\mu\,\mathbf{a})_+) = Re(\mathbf{a}_+), \quad Im((\mu\,\mathbf{a})_+) = Im(\mathbf{a}_+). \tag{3.63}$$

Thus, from (3.59) we get

$$\|\mu\,\mathbf{a}\| = \|\mathbf{a}\| \qquad \text{for any } \mu \in C,\ |\mu| = 1. \tag{3.64}$$

In particular,

$$\|\mathbf{a}_+\| = \|\mathbf{a}\|. \tag{3.65}$$

Moreover, for any complex number z, we have $(z\mathbf{a})_+ = |z|\mathbf{a}_+$, and from (3.59), we get

$$\|z\,\mathbf{a}\| = |z|\,\|\mathbf{a}\| \qquad \text{for any } z \in C. \tag{3.66}$$

In the next section, we will show that the operator norm satisfies the triangle inequality

$$\|\mathbf{a} + \mathbf{b}\| \le \|\mathbf{a}\| + \|\mathbf{b}\|. \tag{3.67}$$

To compare the Euclidean norm in C^n and the operator norm in $\mathcal{S}^n$, note that from the polar decomposition (3.35) and (3.59), we get

$$|\mathbf{a}|^2 = |\mathbf{a}_+|^2 = |Re(\mathbf{a}_+)|^2 + |Im(\mathbf{a}_+)|^2 \leq (|Re(\mathbf{a}_+)| + |Im(\mathbf{a}_+)|)^2 = \|\mathbf{a}\|^2,$$

and the equality $|\mathbf{a}| = \|\mathbf{a}\|$ holds if and only if $Im(\mathbf{a}_+) = 0$, implying that $\mathbf{a}$ is a multiple of a maximal tripotent. Since for any two real numbers x and y, we have $(x+y)^2 \leq 2(x^2+y^2)$, it follows that for any $\mathbf{a}$ in $\mathcal{S}^n$, we have

$$\|\mathbf{a}\|^2 = (|Re(\mathbf{a}_+)| + |Im(\mathbf{a}_+)|)^2 \leq 2(|Re(\mathbf{a}_+)|^2 + |Im(\mathbf{a}_+)|^2) = 2|\mathbf{a}|^2,$$

and the equality $\|\mathbf{a}\| = \sqrt{2}|\mathbf{a}|$ holds if and only if $|Re(\mathbf{a}_+)| = |Im(\mathbf{a}_+)|$, implying that $\mathbf{a}$ is a multiple of a minimal tripotent. Thus,

$$|\mathbf{a}| \leq \|\mathbf{a}\| \leq \sqrt{2}|\mathbf{a}|. \tag{3.68}$$

This implies that the operator norm is equivalent to the Euclidean norm on C^n. For $n = \infty$, the space C^n is a complex Hilbert space H, so we can define $\mathcal{S}^\infty$ to be equal as a set to H, with the triple product defined by (3.4) and norm defined by (3.59). From our observations, $\mathcal{S}^\infty$ will be norm closed.

3.2.2 The unit ball of the spin factor

We denote the unit ball of $\mathcal{S}^n$ by

$$D_{s,n} = \{\mathbf{a} \in \mathcal{S}^n : \ \|\mathbf{a}\| \leq 1\}. \tag{3.69}$$

The intersection of this ball with $\mathcal{S}^n_R$ is the Euclidean unit ball D^n_s of R^n. It is a symmetric domain with respect to the conformal group and was considered in Chapter 2. For example, D^3_s is the ball of s-velocities. The unit ball $D_{s,n}$ is our first example of a domain with non-trivial geometry. To gain an understanding of this geometry, we will consider two three-dimensional sections of $D_{s,n}$.

Let us consider first the three-dimensional section D_1 obtained by intersecting $D_{s,n}$ with the real subspace

$$M_1 = \{(x, y, iz, 0, ...) : \ x, y, z \in R\}. \tag{3.70}$$

Each element of $\mathbf{a} \in D_1$ is of the form $\mathbf{a} = (x, y, iz, 0, ...)$. From the definition of the determinant, we have

$$\det \mathbf{a} = x^2 + y^2 - z^2.$$

Hence, $\arg\det \mathbf{a}$ is either 1 or -1. If $x^2+y^2 > z^2$, then $\mathbf{a} = \mathbf{a}_+$, $Re(\mathbf{a}_+) = (x, y, 0, ...)$ and $Im(\mathbf{a}_+) = (0, 0, z, 0, ...)$. Thus, from (3.59), we get

$$||\mathbf{a}|| = \sqrt{x^2+y^2} + |z|. \tag{3.71}$$

If $x^2+y^2 < z^2$, then $\mathbf{a}_+ = -i\,\mathbf{a}$, $Re(\mathbf{a}_+) = (0, 0, z, 0, ...)$ and $Im(\mathbf{a}_+) = (-x, -y, 0, ...)$. Thus, from (3.59), we get that $||\mathbf{a}||$ is defined by (3.71). Thus,

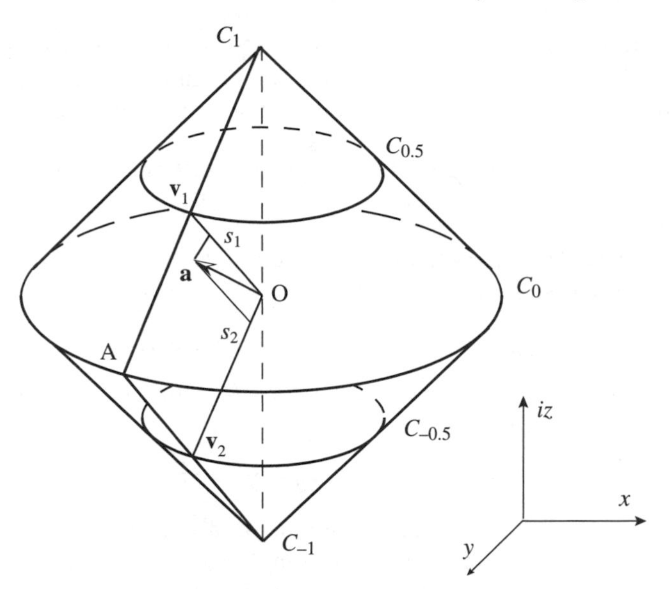

Fig. 3.1. The domain D_1 obtained by intersecting $D_{s,n}$ with the subspace $M_1 = \{(x, y, iz) : \ x, y, z \in R\}$. The domain is the intersection of two circular cones. The minimal tripotents belong to two circles $C_{0.5}$ and $C_{-0.5}$, whose respective equations are $iz = 0.5$ and $iz = -0.5$. The maximal tripotents are the two points $C_1 = (0, 0, i)$ and $C_{-1} = (0, 0, -i)$, as well as the points of the circle $C_0 : iz = 0$. The norm-exposed faces are either points or line segments.

$$D_1 = \{(x, y, iz, 0, ...) : \ \sqrt{x^2 + y^2} \leq 1 - |z|\},$$

which is a double cone (see Figure 3.1).

To locate the minimal tripotents $\mathbf{v}$ in D_1 we introduce polar coordinates r, θ in the x-y plane. Then $\mathbf{v} = (r\cos\theta, r\sin\theta, iz, 0, ...)$. The conditions $\det \mathbf{v} = 0$ and $|Re(\mathbf{v})| = |Im(\mathbf{v})| = 1/2$ lead to

$$\mathbf{v} = 1/2(\cos\theta, \sin\theta, \pm i, 0, ...), \tag{3.72}$$

implying that the minimal tripotents lie on two circles $C_{0.5}$ and $C_{-0.5}$ of radius 1/2. Maximal tripotents are multiples of a real vector of unit length. Thus, the maximal tripotents of D_1 are $C_1 = (0, 0, i, 0, ...)$, and $C_{-1} = (0, 0, -i, 0, ...)$ and the circle $C_0 = \{(\cos\theta, \sin\theta, 0, ...) : \theta \in R\}$ of radius 1.

We can now visualize the geometry of the singular decomposition. Let $\mathbf{a} = (r\cos\theta, r\sin\theta, iz, 0, ...)$ and let $r > z > 0$. Then $\mathbf{a}_+ = \mathbf{a}$. The minimal tripotents in the singular decomposition of $\mathbf{a}$ are obtained from (3.40), yielding

$$\mathbf{v}_1 = 1/2(\cos\theta, \sin\theta, i, 0, ...), \quad \mathbf{v}_2 = 1/2(\cos\theta, \sin\theta, -i, 0, ...).$$

These tripotents are the intersection of the plane through $\mathbf{a}$, C_1 and C_{-1} with the circles $C_{0.5}$ and $C_{-0.5}$ of minimal tripotents . The *singular numbers* of $\mathbf{a}$ are $s_1 = r + z$ and $s_2 = r - z$. Thus the singular decomposition is

$$\mathbf{a} = \frac{r+z}{2}(\cos\theta, \sin\theta, i, 0, ...) + \frac{r-z}{2}(\cos\theta, \sin\theta, -i, 0, ...).$$

See Figure 3.1.

Consider now the three-dimensional section D_2 obtained by intersecting $D_{s,n}$ with the real subspace

$$M_2 = \{(x+iy, z, 0, ...) : \ x, y, z \in R\}. \tag{3.73}$$

Each element of $\mathbf{a} \in D_2$ is of the form $\mathbf{a} = (x+iy, z, 0, ...)$. From the definition of the determinant, we have

$$\det \mathbf{a} = x^2 + 2ixy - y^2 + z^2.$$

Hence, arg det $\mathbf{a}$ can be any complex number of absolute value 1. This makes the calculation of the norm much more complicated. Let's consider the intersection of this ball with the basic two-dimensional planes.

If $z = 0$, then $\mathbf{a} = (x+iy, 0, ...)$. Since for such $\mathbf{a}$, we have $\mathbf{a}^{(3)} = (x^2+y^2)\mathbf{a}$, the norm $||\mathbf{a}|| = |x+iy| = \sqrt{x^2+y^2}$. This also follows from the fact that $\mathbf{a} = (x+iy)(1, 0, ...)$, is a multiple of a maximal tripotent $(1, 0, ...)$, and, since the operator norm of any tripotent is 1, we get $||\mathbf{a}|| = |x+iy|$. Thus, the intersection of D_2 with the x-y plane is a unit ball $x^2 + y^2 \leq 1$, with the boundary consisting of maximal tripotents.

If $y = 0$, then $\mathbf{a} = (x, z, 0, ...)$. Then $\det \mathbf{a} = x^2 + z^2$, so $\arg\det \mathbf{a} = 1$, and $\mathbf{a}_+ = \mathbf{a}$. Hence, $||\mathbf{a}|| = \sqrt{x^2+z^2}$. Thus, the intersection of D_2 with the x-z plane is a unit ball $x^2 + z^2 \leq 1$, with the boundary consisting of maximal tripotents.

If $x = 0$, then $\mathbf{a} = (iy, z, 0, ...)$. So $\det \mathbf{a} = -y^2+z^2$. Hence, $\arg\det \mathbf{a} = \pm 1$, and $\mathbf{a}_+$ is either $\mathbf{a}$ or $-i\mathbf{a}$. Thus, $||\mathbf{a}|| = |y| + |z|$. Thus, the intersection of D_2 with the y-z plane is a square rotated 45°. Here, we have four minimal tripotents $1/2(\pm i, \pm 1, 0, ...)$. The singular decomposition of an element in the y-z plane is a linear combination of these tripotents. But note that for any $\mathbf{a} \in D_2$ which is *not* in the y-z plane, the singular decomposition of $\mathbf{a}$ consists of tripotents *not* belonging to D_2. Figure 3.2 shows the domain D_2.

The geometry of a domain can also be understood from the structure of the *norm-exposed faces* or the *flat components* of the boundary of the domain. For this approach, we need to study the linear functionals, or the dual space, of $\mathcal{S}^n$.

3.3 The dual space of $\mathcal{S}^n$

Every normed linear space A over the complex numbers equipped with a norm has a *dual space*, denoted A^*, consisting of complex linear functionals,

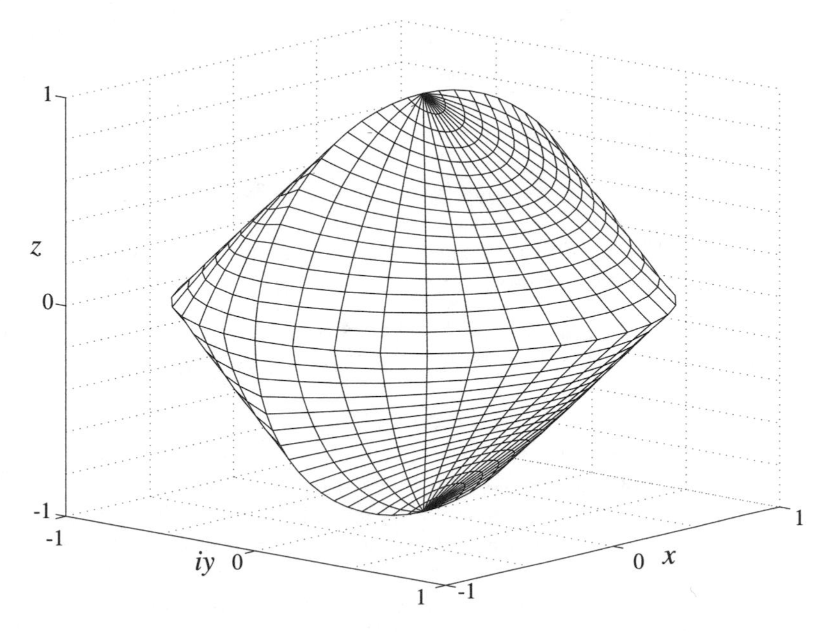

Fig. 3.2. The domain D_2 obtained by intersecting $D_{s,n}$ with the subspace $M_2 = \{(x+iy, z) : x, y, z \in R\}$.

i.e., linear maps from A to the complex numbers. We define a norm on A^* by

$$\|f\| = \sup\{|f(\mathbf{w})| : \mathbf{w} \in A, \|\mathbf{w}\| \leq 1\}. \tag{3.74}$$

In general, $A \subset A^{**}$, where A^{**} denotes the dual of A^*. If $A = A^{**}$, then A is called *reflexive*. In this case, A can be considered as the dual of A^*. Hence, we sometimes refer to A^* as the predual of A.

It is known that if A is finite dimensional or isomorphic to a Hilbert space, then A is reflexive. From (3.68), it follows that $\mathcal{S}^n$ is reflexive.

3.3.1 The norm on the dual of $\mathcal{S}^n$

The dual (or predual) of $\mathcal{S}^n$ is the set of complex linear functionals on $\mathcal{S}^n$. We denote it by $\mathcal{S}^n_*$. We use the inner product on C^n to define an imbedding of $\mathcal{S}^n$ into $\mathcal{S}^n_*$, as follows. For any element $\mathbf{a} \in \mathcal{S}^n$, we define a complex linear functional $\hat{\mathbf{a}} \in \mathcal{S}^n_*$ by

$$\hat{\mathbf{a}}(\mathbf{w}) = \langle \mathbf{w} | 2\mathbf{a} \rangle. \tag{3.75}$$

The coefficient 2 of $\mathbf{a}$ is needed to make the dual of a minimal tripotent have norm 1. This is a convenient normalization for all non-spin factors.

Conversely, for any $\mathbf{f} \in \mathcal{S}^n_*$, by the Riesz theorem, there is an element $\check{\mathbf{f}} \in \mathcal{S}^n$ such that for all $\mathbf{w} \in \mathcal{S}^n$,

$$\mathbf{f}(\mathbf{w}) = \langle \mathbf{w} | 2\check{\mathbf{f}} \rangle. \tag{3.76}$$

The norm on $\mathcal{S}^n_*$, called the *trace norm*, is defined by

$$\|\mathbf{f}\|_* = \sup\{|\mathbf{f}(\mathbf{w})| : \mathbf{w} \in \mathcal{S}^n, \ \|\mathbf{w}\| \le 1\}. \tag{3.77}$$

Let $\mathbf{v} \in \mathcal{S}^n$ be a minimal tripotent. We will show now that the functional $\hat{\mathbf{v}}$ has norm 1 and has value 1 on $\mathbf{v}$.

To do this, write $\mathbf{v} = \mathbf{x} + i\mathbf{y}$, where $\mathbf{x}, \mathbf{y} \in \mathcal{S}^n_R$ satisfy $\langle \mathbf{x} | \mathbf{y} \rangle = 0$ and $|\mathbf{x}| = |\mathbf{y}| = 1/2$. From the definition (3.75) of $\hat{\mathbf{v}}$, we have, for $\mathbf{w} \in \mathcal{S}^n$,

$$\hat{\mathbf{v}}(\mathbf{w}) = \langle \mathbf{w} | 2\mathbf{v} \rangle = \langle \mathbf{w} | (2\mathbf{x} + i2\mathbf{y}) \rangle. \tag{3.78}$$

Note that $2\mathbf{x}$ and $2\mathbf{y}$ are mutually orthogonal norm 1 vectors in C^n. Let $\mathbf{j} = 2\mathbf{x}$ and $\mathbf{k} = 2\mathbf{y}$. Then $P_{\mathbf{j}}\mathbf{w} = \langle \mathbf{w} | \mathbf{j} \rangle \mathbf{j}$ and $P_{\mathbf{k}}\mathbf{w} = \langle \mathbf{w} | \mathbf{k} \rangle \mathbf{k}$ are the orthogonal projections onto $\mathbf{j}$ and $\mathbf{k}$, respectively. Then we can rewrite (3.78) as

$$\hat{\mathbf{v}}(\mathbf{w}) = \langle \mathbf{w} | \mathbf{j} \rangle - i \langle \mathbf{w} | \mathbf{k} \rangle = |P_{\mathbf{j}}\mathbf{w}| - i |P_{\mathbf{k}}\mathbf{w}|. \tag{3.79}$$

Use the polar decomposition $\mathbf{w} = \lambda \mathbf{w}_+$ and decompose $\mathbf{w}_+$ as $\mathbf{w}_+ = Re(\mathbf{w}_+) + iIm(\mathbf{w}_+) = \mathbf{w}_1 + i\mathbf{w}_2$. Then, from (3.79), we obtain

$$|\hat{\mathbf{v}}(\mathbf{w})| = |\hat{\mathbf{v}}(\mathbf{w}_+)| = |\langle \mathbf{w}_1 | \mathbf{j} \rangle - i \langle \mathbf{w}_1 | \mathbf{k} \rangle + i \langle \mathbf{w}_2 | \mathbf{j} \rangle + \langle \mathbf{w}_2 | \mathbf{k} \rangle|,$$

and, hence,

$$|\hat{\mathbf{v}}(\mathbf{w})|^2 = (\langle \mathbf{w}_1 | \mathbf{j} \rangle + \langle \mathbf{w}_2 | \mathbf{k} \rangle)^2 + (\langle \mathbf{w}_2 | \mathbf{j} \rangle - \langle \mathbf{w}_1 | \mathbf{k} \rangle)^2.$$

Note that $\widetilde{P} = P_{\mathbf{j}} + P_{\mathbf{k}}$ is the orthogonal projection onto the plane Π generated by $\mathbf{j}$ and $\mathbf{k}$. Thus, for any $\mathbf{b} \in C^n$ we have, by the Pythagorean Theorem,

$$|\widetilde{P}\mathbf{b}|^2 = |P_{\mathbf{j}}\mathbf{b}|^2 + |P_{\mathbf{k}}\mathbf{b}|^2 \le |\mathbf{b}|^2.$$

Thus, we get

$$|\hat{\mathbf{v}}(\mathbf{w})|^2 = |\widetilde{P}\mathbf{w}_1|^2 + |\widetilde{P}\mathbf{w}_2|^2$$

$$+2(\langle \mathbf{w}_1 | \mathbf{j} \rangle \langle \mathbf{w}_2 | \mathbf{k} \rangle - \langle \mathbf{w}_2 | \mathbf{j} \rangle \langle \mathbf{w}_1 | \mathbf{k} \rangle).$$

In the basis $\mathbf{j}, \mathbf{k}$ of the plane Π, the coordinates of $\widetilde{P}\mathbf{w}_1$ and $\widetilde{P}\mathbf{w}_2$ are $(\langle \mathbf{w}_1 | \mathbf{j} \rangle, \langle \mathbf{w}_1 | \mathbf{k} \rangle)$ and $(\langle \mathbf{w}_2 | \mathbf{j} \rangle, \langle \mathbf{w}_2 | \mathbf{k} \rangle)$, respectively. Hence,

$$|\langle \mathbf{w}_1 | \mathbf{j} \rangle \langle \mathbf{w}_2 | \mathbf{k} \rangle - \langle \mathbf{w}_2 | \mathbf{j} \rangle \langle \mathbf{w}_1 | \mathbf{k} \rangle| = |(\widetilde{P}\mathbf{w}_1) \times (\widetilde{P}\mathbf{w}_2)| \le |\widetilde{P}\mathbf{w}_1| \, |\widetilde{P}\mathbf{w}_2|,$$

and so

$$|\hat{\mathbf{v}}(\mathbf{w})|^2 \leq (|\widetilde{P}\mathbf{w}_1| + |\widetilde{P}\mathbf{w}_2|)^2 \leq (|\mathbf{w}_1| + |\mathbf{w}_2|)^2 = \|\mathbf{w}\|^2, \tag{3.80}$$

implying that $\|\hat{\mathbf{v}}\|_* \leq 1$. On the other hand, from (3.24) we have

$$\hat{\mathbf{v}}(\mathbf{v}) = \langle \mathbf{v}|2\mathbf{v}\rangle = 2\langle \mathbf{v}|\mathbf{v}\rangle = 1, \tag{3.81}$$

and, hence, $\|\hat{\mathbf{v}}\|_* = 1$.

For an arbitrary element $\mathbf{f}$ in $\mathcal{S}^n_*$, let $\mathbf{a} = \check{\mathbf{f}}$. We use the singular decomposition $\mathbf{a} = s_1\mathbf{v}_1 + s_2\mathbf{v}_2$ to calculate the norm of $\mathbf{f}$. From (3.76), for any $\mathbf{w} \in D_{s,n}$, we have

$$\mathbf{f}(\mathbf{w}) = \langle \mathbf{w}|2\check{\mathbf{f}}\rangle = s_1\langle \mathbf{w}|2\mathbf{v}_1\rangle + s_2\langle \mathbf{w}|2\mathbf{v}_2\rangle = (s_1\hat{\mathbf{v}}_1 + s_2\hat{\mathbf{v}}_2)(\mathbf{w}),$$

and so

$$\mathbf{f} = s_1\hat{\mathbf{v}}_1 + s_2\hat{\mathbf{v}}_2. \tag{3.82}$$

Note that since $\|\hat{\mathbf{v}}_1\|_* = \|\hat{\mathbf{v}}_2\|_* = 1$, we have

$$|\mathbf{f}(\mathbf{w})| = |s_1\hat{\mathbf{v}}_1(\mathbf{w}) + s_2\hat{\mathbf{v}}_2(\mathbf{w})| \leq s_1\|\hat{\mathbf{v}}_1\|_* + s_2\|\hat{\mathbf{v}}_2\|_* \leq s_1 + s_2.$$

But for the tripotent $\mathbf{v}_1 + \mathbf{v}_2$ which has norm 1, we have

$$|\mathbf{f}(\mathbf{v}_1 + \mathbf{v}_2)| = |s_1\hat{\mathbf{v}}_1(\mathbf{v}_1 + \mathbf{v}_2) + s_2\hat{\mathbf{v}}_2(\mathbf{v}_1 + \mathbf{v}_2)| = s_1 + s_2. \tag{3.83}$$

Therefore,

$$\|\mathbf{f}\|_* = s_1 + s_2 = 2|Re(\check{\mathbf{f}}_+)|, \tag{3.84}$$

where s_1, s_2 are the singular numbers of $\check{\mathbf{f}}$. From (3.46) we get

$$\|\mathbf{f}\|_* = s_1 + s_2 = \sqrt{2|\check{\mathbf{f}}|^2 + 2|\det \check{\mathbf{f}}|}. \tag{3.85}$$

Now we can prove the triangle inequality (3.67) for the operator norm. Let $\mathbf{a}$ and $\mathbf{b}$ be arbitrary elements of $\mathcal{S}^n$. Use the singular decomposition to decompose $\mathbf{a} + \mathbf{b} = s_1\mathbf{v}_1 + s_2\mathbf{v}_2$ as a linear combination of two minimal, orthogonal tripotents $\mathbf{v}_1$, $\mathbf{v}_2$. Then, since $2\langle \mathbf{v}_1|\mathbf{v}_1\rangle = 1$, $\langle \mathbf{v}_1|\mathbf{v}_2\rangle = 0$ and $\|\hat{\mathbf{v}}_1\|_* = 1$, we have

$$\|\mathbf{a} + \mathbf{b}\| = s_1 = \langle s_1\mathbf{v}_1 + s_2\mathbf{v}_2|2\mathbf{v}_1\rangle = \hat{\mathbf{v}}_1(\mathbf{a} + \mathbf{b})$$

$$= \hat{\mathbf{v}}_1(\mathbf{a}) + \hat{\mathbf{v}}_1(\mathbf{b}) \leq \|\mathbf{a}\| + \|\mathbf{b}\|. \tag{3.86}$$

3.3.2 The facial structure of $D_{s,n}$

The geometry of a domain can be understood from the structure of the norm-exposed faces, or flat components, of the boundary of the domain. To define the notion of a norm-exposed face F of a domain D in a normed space X, we will use the concept of a *tangent hyperplane*. A hyperplane L in X is the parallel translation of the kernel of a linear map from X to R. It has real codimension 1. A hyperplane L is *tangent* to D if $L \cap D \subset \partial D$. The subset F of D is a *norm-exposed face* of D if

$$F = L \cap D \tag{3.87}$$

for some hyperplane L which is tangent to D. Any point of F is said to be *exposed* by L. For a Euclidean ball (like D_v or D_s), each boundary point is a norm-exposed face, and any norm-exposed face is a single boundary point. In the previous section, we introduced the domain $D_{s,n}$, the unit ball of $\mathcal{S}^n$. We are now ready to describe the norm-exposed faces of $D_{s,n}$.

Let $\mathbf{v}$ be a minimal tripotent. Since $\|\hat{\mathbf{v}}\|_* = 1$, for any $\mathbf{w} \in D_{s,n}$, we have $Re\, \hat{\mathbf{v}}(\mathbf{w}) \leq |\hat{\mathbf{v}}(\mathbf{w})| \leq 1$. Define a hyperplane

$$\Pi = \{\mathbf{w} \in \mathcal{S}^n : \ Re\, \hat{\mathbf{v}}(\mathbf{w}) = 1\}.$$

Suppose $\mathbf{w} \in \Pi \cap D_{s,n}$. Then $|\hat{\mathbf{v}}(\mathbf{w})| = 1$. But

$$1 = |\hat{\mathbf{v}}(\mathbf{w})| \leq \|\hat{\mathbf{v}}\|_* \cdot \|\mathbf{w}\| = \|\mathbf{w}\| \leq 1.$$

Thus, $\mathbf{w} \in \partial D_{s,n}$ and Π is a tangent hyperplane to $D_{s,n}$.

Suppose $Re\, \hat{\mathbf{v}}(\mathbf{w}) = 1$. Then, since $|\hat{\mathbf{v}}(\mathbf{w})| \leq \|\hat{\mathbf{v}}\|_* = 1$, we have $\hat{\mathbf{v}}(\mathbf{w}) = 1$. Decompose $\mathbf{w}$ using the Peirce decomposition with respect to $\mathbf{v}$ as

$$\mathbf{w} = P_1(\mathbf{v})\mathbf{w} + P_{1/2}(\mathbf{v})\mathbf{w} + P_0(\mathbf{v})\mathbf{w}.$$

From the definition (3.75) of $\hat{\mathbf{v}}$ and the definition (3.48) of $P_1(\mathbf{v})$, we have

$$1 = \hat{\mathbf{v}}(\mathbf{w}) = \langle \mathbf{w}|2\mathbf{v}\rangle = \langle \mathbf{w}|2P_1(\mathbf{v})\mathbf{v}\rangle = \langle P_1(\mathbf{v})\mathbf{w}|2\mathbf{v}\rangle = \hat{\mathbf{v}}(P_1(\mathbf{v})\mathbf{w}).$$

Thus, $P_1(\mathbf{v})\mathbf{w} = \mathbf{v}$. From (3.48), we get $P_0(\mathbf{v})\mathbf{w} = \lambda\overline{\mathbf{v}}$ for some constant λ.

Next, we will show that

$$P_1(\mathbf{v})\mathbf{w} = \mathbf{v} \text{ and } ||\mathbf{w}|| = 1 \text{ imply } P_{1/2}(\mathbf{v})\mathbf{w} = 0. \tag{3.88}$$

Let $\mathbf{a} = P_{1/2}(\mathbf{v})\mathbf{w}$. By use of the Peirce calculus (3.54) and the fact that $\overline{\mathbf{v}}$ is orthogonal to $\mathbf{v}$, we get

$$P_1(\mathbf{v})(\mathbf{w}^{(3)}) = \mathbf{v} + 2\{\mathbf{a}, \mathbf{a}, \mathbf{v}\} + \{\mathbf{a}, \lambda\overline{\mathbf{v}}, \mathbf{a}\}.$$

Since $\mathbf{a} = P_{1/2}(\mathbf{v})\mathbf{a}$, we have $\langle \mathbf{a}|\mathbf{v}\rangle = \langle \mathbf{a}|\overline{\mathbf{v}}\rangle = 0$. From the definition of the spin triple product, we have $\{\mathbf{a}, \mathbf{a}, \mathbf{v}\} = |\mathbf{a}|^2\mathbf{v}$ and $\{\mathbf{a}, \overline{\mathbf{v}}, \mathbf{a}\} = (\det \mathbf{a})\mathbf{v}$. Thus, by (3.44) and (3.45), we have

$$Re\,\hat{\mathbf{v}}(P_1(\mathbf{v})(\mathbf{w}^{(3)})) = Re\langle P_1(\mathbf{v})(\mathbf{w}^{(3)})|2\mathbf{v}\rangle$$

$$= Re(1 + 2|\mathbf{a}|^2 + \overline{\lambda}\det\mathbf{a}) \geq 1 + s_1^2 + s_2^2 - s_1 s_2 \geq 1,$$

where s_1, s_2 are the singular numbers of $\mathbf{a}$. This implies that $\|P_1(\mathbf{v})(\mathbf{w}^{(3)})\| \geq 1$. However, since $P_1(\mathbf{v})$ is a projection, we have

$$\|P_1(\mathbf{v})(\mathbf{w}^{(3)})\| \leq \|\mathbf{w}^{(3)}\| = 1.$$

Hence, we must have $\|P_1(\mathbf{v})(\mathbf{w}^{(3)})\| = 1$. This implies that $s_1 = s_2 = 0$, which, in turn, implies that $\mathbf{a} = 0$. This proves (3.88).

Since $\|\mathbf{w}\| = 1$, the constant λ satisfies $|\lambda| \leq 1$. Thus, the norm-exposed face $F_\mathbf{v}$, defined to be the intersection of Π with $D_{s,n}$, is

$$F_\mathbf{v} = \{\mathbf{v} + \lambda\overline{\mathbf{v}} : \ |\lambda| \leq 1\}, \tag{3.89}$$

which is a two-dimensional disc with center at the minimal tripotent $\mathbf{v}$ and of radius 1 in the operator norm. The boundary points of the face $F_\mathbf{v}$ correspond to $|\lambda| = 1$. Since such elements are the sum of two orthogonal minimal tripotents, they are maximal tripotents. On the other hand, for every maximal tripotent $\mathbf{u}$, there is a minimal tripotent $\mathbf{v}$ such that $\mathbf{u}$ belongs to the boundary of $F_\mathbf{v}$. Thus $\partial D_{s,n}$ consists of discs of real dimension 2 of radius 1, centered at a minimal tripotent, whose boundaries consist of maximal tripotents.

If $\mathbf{u}$ is a maximal tripotent, then we can write it as a sum $\mathbf{u} = \mathbf{v}_1 + \mathbf{v}_2$, where $\mathbf{v}_1$ and $\mathbf{v}_2$ are minimal tripotents. Define a hyperplane

$$\Pi_u = \{\mathbf{w} \in \mathcal{S}^n : \ Re\,(\hat{\mathbf{v}}_1 + \hat{\mathbf{v}}_2)(\mathbf{w}) = 2\}.$$

Then $(\hat{\mathbf{v}}_1 + \hat{\mathbf{v}}_2)(\mathbf{u}) = 2$, implying that $\mathbf{u} \in \Pi_u \cap D_{s,n}$. But if $(\hat{\mathbf{v}}_1 + \hat{\mathbf{v}}_2)(\mathbf{w}) = 2$, then $\hat{\mathbf{v}}_1(\mathbf{w}) = 1$ and $\hat{\mathbf{v}}_2(\mathbf{w}) = 1$, implying that $\mathbf{w} \in F_{\mathbf{v}_1} \cap F_{\mathbf{v}_2}$ and $\mathbf{w} = \mathbf{v}_1 + \mathbf{v}_2 = \mathbf{u}$. Thus, any maximal tripotent $\mathbf{u}$ is norm-exposed by Π_u, which is tangent to $D_{s,n}$ and is an *extreme point* of $D_{s,n}$.

It is not easy to visualize a surface $\partial D_{s,n}$ which is paved totally with two-dimensional discs. Let's take a look at the intersection of this surface with the three-dimensional subspace M_1 from Figure 3.1. In this figure, we see only one-dimensional sections of $F_\mathbf{v}$, which are intervals with center at $\mathbf{v}$. These intervals start at a vertex C_1 or C_{-1}, which are maximal tripotents, and end up on the circle C_0 consisting of maximal tripotents. The midpoint of the interval is a tripotent from one of the two circles $C_{0.5}$ and $C_{-0.5}$. In Figure 3.2, only four faces $F_\mathbf{v}$ have a one-dimensional intersection with the subspace M_2, and the rest either do not intersect the subspace or intersect it at a single point. The four one-dimensional faces are in the plane $x = 0$ and correspond to the four minimal tripotents $1/2(\pm i, \pm 1, 0, ...)$.

3.3.3 The unit ball in $\mathcal{S}_*^n$

The unit ball S_n in $\mathcal{S}_*^n$ is defined by

$$S_n = \{\mathbf{f} \in \mathcal{S}_*^n : \ \|\mathbf{f}\|_* \leq 1\}. \tag{3.90}$$

We call the ball S_n the *state space* of $\mathcal{S}^n$. Later we will show that this ball represents the state space of two-state quantum systems. The state space S_n has non-trivial geometry. To understand this geometry, we will examine two three-dimensional sections of this ball.

We consider first the three-dimensional section D_1^* consisting of those elements $\mathbf{f} \in S_n$ satisfying

$$2\check{\mathbf{f}} \in M_1 = \{(x, y, iz) : \quad x, y, z \in R\}. \tag{3.91}$$

From the definition of the determinant, we have $\det \check{\mathbf{f}} = 1/4(x^2 + y^2 - z^2)$. So arg det $\check{\mathbf{f}}$ is ± 1. If $x^2 + y^2 \geq z^2$, then $\check{\mathbf{f}}_+ = \check{\mathbf{f}}$ and $2Re(\check{\mathbf{f}}_+) = (x, y, 0, ...)$. Thus, from (3.84), we obtain

$$||\mathbf{f}||_* = 2|Re(\check{\mathbf{f}}_+)| = \sqrt{x^2 + y^2}. \tag{3.92}$$

If $x^2 + y^2 \leq z^2$, then $\check{\mathbf{f}}_+ = -i\check{\mathbf{f}}$ and $2Re(\check{\mathbf{f}}_+) = (0, 0, z, 0, ...)$. Thus, from (3.84), we get that $||\mathbf{f}||_* = |z|$. Thus,

$$D_1^* = \{(x, y, iz, 0, ...) : \ \max\{\sqrt{x^2 + y^2}, \ |z|\} \leq 1\},$$

which is a cylinder (see Figure 3.3).

To describe the functionals $\mathbf{f}$ in D_1^* which correspond to minimal tripotents $\mathbf{v} = \check{\mathbf{f}}$, we introduce polar coordinates r, θ in the x-y plane. For such functionals, by (3.72) we have

$$2\check{\mathbf{f}} = (\cos\theta, \sin\theta, \pm i, 0, ...), \tag{3.93}$$

yielding two circles C_1 and C_{-1} of radius 1. The functionals corresponding to multiples of maximal tripotents are multiples of a real vector of unit length. Thus, the norm 1 functionals corresponding to multiples of maximal tripotents are $A_1 = (0, 0, i, 0, ...)$, $A_2 = (0, 0, -i, ...)$, which are the centers of the two-dimensional discs of ∂S_n, and the points of the circle $C_0 = (\cos\theta, \sin\theta, 0, ...)$ of radius 1. See Figure 3.3.

Consider now the three-dimensional section D_2^* consisting of those elements $\mathbf{f} \in S_n$ satisfying

$$2\check{\mathbf{f}} \in M_2 = \{(x + iy, z, 0, ...) : \quad x, y, z \in R\}. \tag{3.94}$$

For these $\mathbf{f}$, we have $\det \check{\mathbf{f}} = 1/4(x^2 + 2ixy - y^2 + z^2)$. Hence, arg det $2\check{\mathbf{f}}$ can be any complex number of absolute value 1. This makes the calculation of

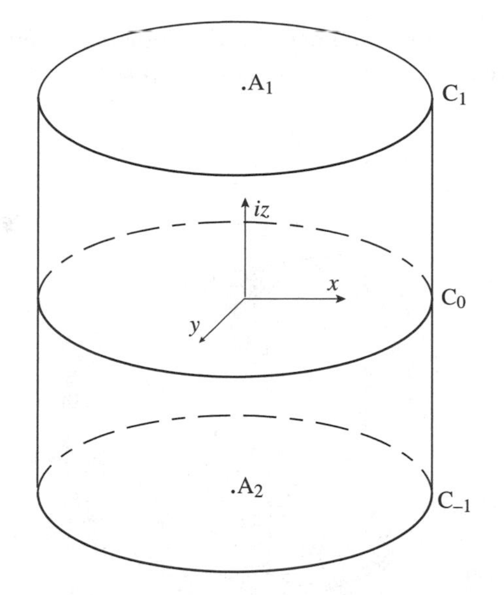

Fig. 3.3. The domain D_1^* obtained by intersecting the state space S_n with the subspace $M_1 = \{(x, y, iz, 0, ...) : x, y, z \in R\}$. This domain is a cylinder. The pure states, corresponding to minimal tripotents, are extreme points of the domain and belong to two unit circles $C_1 : iz = 1$ and $C_{-1} : iz = -1$. The functionals corresponding to maximal tripotents are $A_1 = (0, 0, i)$ and $A_2 = (0, 0, -i)$ and each point of the circle $C_0 : iz = 0$. They are centers of faces. The norm-exposed faces are either points, line segments or disks.

the norm much more complicated. Let's consider the intersection of this ball with the basic two-dimensional planes.

If $z = 0$, then $2\check{\mathbf{f}} = (x + iy)(1, 0, ...)$ is a multiple of a maximal tripotent. Note that if $\mathbf{a}$ is a multiple of a maximal tripotent, then $|Re\, \mathbf{a}_+| = |\mathbf{a}|$. So from (3.84), we get $||\mathbf{f}||_* = 2|Re\, \check{\mathbf{f}}_+| = |2\check{\mathbf{f}}| = |x+iy|$. Thus, the intersection of D_2^* with the x-y plane is a unit ball $x^2+y^2 \leq 1$, with the boundary consisting of functionals corresponding to multiples of maximal tripotents.

If $y = 0$, then $2\check{\mathbf{f}} = (x, z, 0, ...)$. Then $\det(2\check{\mathbf{f}}) = x^2+z^2$, so arg det $(2\check{\mathbf{f}}) = 1$, and $(2\check{\mathbf{f}})_+ = 2\check{\mathbf{f}}$. Hence, $|2\check{\mathbf{f}}| = \sqrt{x^2 + z^2}$. Thus, the intersection of D_2^* with the x-z plane is a unit ball $x^2 + z^2 \leq 1$, with the boundary consisting of maximal tripotents.

If $x = 0$, then $2\check{\mathbf{f}} = (iy, z, 0, ...)$. If $|z| \geq |y|$, then $\check{\mathbf{f}}_+ = \check{\mathbf{f}}$, and by (3.84), we get $||\mathbf{f}||_* = |z|$. If $|z| < |y|$, then $\check{\mathbf{f}}_+ = -i\check{\mathbf{f}}$ and $2Re\, \check{\mathbf{f}}_+ = (y, 0, ...)$. Then, by (3.84), we get $||\mathbf{f}||_* = |y|$ Thus, the intersection of D_2^* with the y-z plane is a square $\{(iy, z) : |y| \leq 1, |z| \leq 1\}$. Here, we have four functionals corresponding to the minimal tripotents $(\pm i, \pm 1, 0, ...)$. Figure 3.4 shows the ball D_2^*. It has the form of a pillow.

More information on the geometry of the domain S_n can be obtained from an analysis of its extreme points and norm-exposed faces. This will be done in the next section.

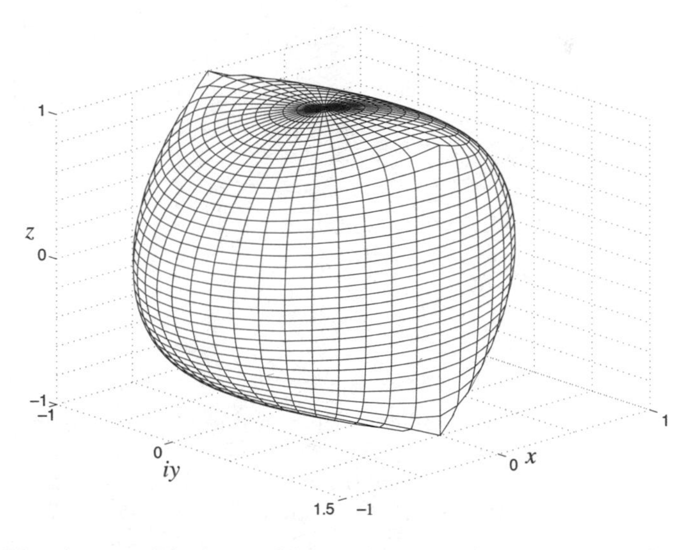

Fig. 3.4. The domain D_2^* obtained by intersecting the state space S_n with the subspace $M_2 = \{(x + iy, z, 0, ...) : x, y, z \in R\}$.

3.3.4 The geometry of the state space S_n

For any element $\mathbf{f}$ in S_*^n, there is a unique tripotent on which $\mathbf{f}$ attains its norm. This tripotent, denoted $s(\mathbf{f})$, is called the *support tripotent* of $\mathbf{f}$. From (3.83) and (3.84), it follows that

$$s(\mathbf{f}) = \mathbf{v}_1 + \mathbf{v}_2, \tag{3.95}$$

where $\mathbf{v}_1$, $\mathbf{v}_2$ are the the tripotents from the singular decomposition of $\check{\mathbf{f}}$. If $\check{\mathbf{f}}$ is not a multiple of a minimal tripotent, then $s_2 \neq 0$, and the support tripotent of $\mathbf{f}$ is a maximal tripotent. In this case, for any norm 1 element $\mathbf{f}$ of S_n, we have $s_1 + s_2 = 1$, and (3.82) then implies that

$$\mathbf{f} = s_1\hat{\mathbf{v}}_1 + s_2\hat{\mathbf{v}}_2, \quad s_1 + s_2 = 1. \tag{3.96}$$

This means that $\mathbf{f}$ is a convex combination of two norm 1 states. From the definition of an extreme point of a set, it follows that $\mathbf{f}$ is not an extreme point of S_n.

We will show now that $\hat{\mathbf{v}}$, where $\mathbf{v}$ is a minimal tripotent, is an extreme point of S_n. First, we will show that if $\mathbf{v}$ and $\mathbf{w}$ are minimal tripotents, then

$$\hat{\mathbf{v}}(\mathbf{w}) = 1 \Rightarrow \mathbf{v} = \mathbf{w}. \tag{3.97}$$

In this case, we have equality in (3.80). Thus $|\widetilde{P}\mathbf{w}_1| = |\mathbf{w}_1|$, $|\widetilde{P}\mathbf{w}_2| = |\mathbf{w}_2|$, which implies that $\widetilde{P}\mathbf{w}_1 = \mathbf{w}_1$, $\widetilde{P}\mathbf{w}_2 = \mathbf{w}_2$. Thus, writing $\mathbf{v} = \mathbf{x} + i\mathbf{y}$, there is a θ such that

$$\mathbf{w}_1 = \cos\theta\,\mathbf{x} - \sin\theta\,\mathbf{y}, \quad \mathbf{w}_2 = \sin\theta\,\mathbf{x} + \cos\theta\,\mathbf{y}.$$

Therefore,

$$\mathbf{w} = \mathbf{w}_1 + i\mathbf{w}_2 = \cos\theta(\mathbf{x} + i\mathbf{y}) + i\sin\theta(\mathbf{x} + i\mathbf{y}) = e^{i\theta}\mathbf{v}.$$

But from $\hat{\mathbf{v}}(\mathbf{w}) = 1$, we obtain $\theta = 0$, and so $\mathbf{v} = \mathbf{w}$.

Now let $\mathbf{f}$ be any norm 1 element in S_n with decomposition (3.96). Suppose $\mathbf{f}(\mathbf{v}) = 1$. We claim that

$$\mathbf{f}(\mathbf{v}) = \|\mathbf{f}\|_* = 1 \Longrightarrow \mathbf{f} = \hat{\mathbf{v}}. \tag{3.98}$$

Since

$$1 = |\mathbf{f}(\mathbf{v})| = |s_1\hat{\mathbf{v}}_1(\mathbf{v}) + s_2\hat{\mathbf{v}}_2(\mathbf{v})| \leq s_1 + s_2 = 1,$$

we have $\hat{\mathbf{v}}_1(\mathbf{v}) = 1$, and, by (3.97), $\mathbf{v}_1 = \mathbf{v}$. If $s_2 \neq 0$, then also $\hat{\mathbf{v}}_2(\mathbf{v}) = 1$ and $\mathbf{v}_2 = \mathbf{v}$, contradicting the fact that $\mathbf{v}_1$ is orthogonal to $\mathbf{v}_2$. Thus, $\mathbf{f} = \hat{\mathbf{v}}$ and (3.98) holds. Therefore, the minimal tripotent $\mathbf{v}$, considered as a linear functional on $\mathcal{S}^n_*$, exposes only $\hat{\mathbf{v}}$, implying that

$$\hat{\mathbf{v}} \text{ is a norm-exposed face} \tag{3.99}$$

in $\mathcal{S}^n_*$.

Suppose now that $\hat{\mathbf{v}}$ is a convex combination of two elements, say $\mathbf{f}_1$ and $\mathbf{f}_2$, of the state space S_n. Then $\hat{\mathbf{v}} = \alpha\mathbf{f}_1 + (1-\alpha)\mathbf{f}_2$, for some $0 < \alpha < 1$. Then

$$1 = \hat{\mathbf{v}}(\mathbf{v}) = \alpha\mathbf{f}_1(\mathbf{v}) + (1-\alpha)\mathbf{f}_2(\mathbf{v}) \leq \alpha + (1-\alpha) = 1.$$

Since $\mathbf{f}_1(\mathbf{v})$ and $\mathbf{f}_2(\mathbf{v})$ belong to the unit disc Δ of C and 1 is an extreme point of C, we get $\mathbf{f}_1(\mathbf{v}) = \mathbf{f}_2(\mathbf{v}) = 1$. Hence, $\mathbf{f}_1 = \hat{\mathbf{v}}$ and, similarly, $\mathbf{f}_2 = \hat{\mathbf{v}}$. This proves that for a minimal tripotent $\mathbf{v}$, the functional $\hat{\mathbf{v}}$ is an extreme point of the state space S_n. We call such a $\hat{\mathbf{v}}$ a *pure state.*

We say that a pair of maximal tripotents $\mathbf{u}$ and $\widetilde{\mathbf{u}}$ are *complementary* if there are $\mathbf{r}, \widetilde{\mathbf{r}} \in \mathcal{S}^n_R$ and $\lambda \in C$ such that $|\lambda| = 1$ and

$$\mathbf{u} = \lambda\mathbf{r}, \; \widetilde{\mathbf{u}} = i\lambda\widetilde{\mathbf{r}} \text{ and } \langle\mathbf{r}|\widetilde{\mathbf{r}}\rangle = 0. \tag{3.100}$$

It is easy to check that maximal tripotents $\mathbf{u}$ and $\widetilde{\mathbf{u}}$ are complementary if and only if $\{\mathbf{u}, \widetilde{\mathbf{u}}, \mathbf{u}\} = \widetilde{\mathbf{u}}$.

For any $\mathbf{a} \in \mathcal{S}^n$, use the polar decomposition (3.35) $\mathbf{a} = \lambda \mathbf{a}_+$, where $Re(\mathbf{a}_+)$ and $Im(\mathbf{a}_+)$ satisfy (3.38). If $Im(\mathbf{a}_+) \neq 0$ (meaning that $\mathbf{a}$ is not a multiple of a maximal tripotent), then $\mathbf{u} := \lambda Re(\mathbf{a}_+)/|Re(\mathbf{a}_+)|$ and $\widetilde{\mathbf{u}} := i\lambda Im(\mathbf{a}_+)/|Im(\mathbf{a}_+)|$ are maximal complementary tripotents. Thus, $\mathbf{a}$ can be decomposed as

$$\mathbf{a} = \alpha_1 \mathbf{u} + \alpha_2 \widetilde{\mathbf{u}}, \tag{3.101}$$

where $\alpha_1 = |Re(\mathbf{a}_+)|$ and $\alpha_2 = |Im(\mathbf{a}_+)|$. The decomposition (3.101) by maximal complementary tripotents $\mathbf{u}$ and $\widetilde{\mathbf{u}}$, with $\alpha_1 \geq \alpha_2 \geq 0$, is called *facial decomposition*. From the definition (3.59) of the norm, we get

$$\|\mathbf{a}\| = \alpha_1 + \alpha_2. \tag{3.102}$$

Thus, any $\mathbf{a} \in \mathcal{S}^n$ of norm one is a convex combination of two maximal tripotents which are extreme points of $D_{s,n}$. In Figure 3.1, for example, we see that any $\mathbf{a} \in \partial D_{s,n}$ which is not a maximal tripotent belongs to a line segment which connects two maximal tripotents. This implies that $\mathbf{a}$ is a convex combination of the end points of this line segment.

By (3.84), we have

$$\|\hat{\mathbf{a}}\|_* = 2|Re(\mathbf{a}_+)| = 2\alpha_1. \tag{3.103}$$

Note that the norm $\|\hat{\mathbf{u}}\|_*$ of the dual of a maximal tripotent $\mathbf{u}$ equals 2.

To describe the faces of S_n, recall that each element $\hat{\mathbf{v}} \in \mathcal{S}^n_*$ corresponding to a minimal tripotent $\mathbf{v}$ is an extreme point of S_n and a norm-exposed face. If $\mathbf{a}$ is not a minimal tripotent and $\|\hat{\mathbf{a}}\|_* = 1$, then from (3.103) and (3.101), it follows that

$$2\mathbf{a} = \mathbf{u} + \alpha \widetilde{\mathbf{u}},$$

where $\mathbf{u}$ and $\widetilde{\mathbf{u}}$ are complementary and $0 \leq \alpha \leq 1$. Since

$$\hat{\mathbf{a}}(\mathbf{u}) = \langle \mathbf{u} | \mathbf{u} + \alpha \widetilde{\mathbf{u}} \rangle = 1,$$

the maximal tripotent $\mathbf{u}$ is the support tripotent $s(\hat{\mathbf{a}})$ of $\hat{\mathbf{a}}$. Thus, $\hat{\mathbf{a}}$ belongs to a face defined by a maximal tripotent. To describe these faces, we now define, for each maximal tripotent $\mathbf{u}$, the set

$$\widehat{F}_{\mathbf{u}} = \{\mathbf{f} \in S_n : \; 2\check{\mathbf{f}} = \mathbf{u} + \alpha \widetilde{\mathbf{u}}\}, \tag{3.104}$$

where $\mathbf{u}$ and $\widetilde{\mathbf{u}}$ are complementary and $0 \leq \alpha \leq 1$. Then $\widehat{F}_{\mathbf{u}}$ is a face of S_n, exposed by $\mathbf{u}$, and consisting of all norm 1 functionals $\mathbf{f}$ in S_n with support $s(\mathbf{f}) = \mathbf{u}$. From (3.100), it follows that the set $\widehat{F}_{\mathbf{u}}$ is a Euclidean ball in $\mathcal{S}^n_*$, with center at $0.5\hat{\mathbf{u}}$, of real dimension $(n-1)$ and radius $0.5\|\widetilde{\mathbf{u}}\|_* = 1$. Thus, the boundary of the state space S_n is paved with faces in the form of $(n-1)$-dimensional balls. In Figure 3.3, we see the intersection of faces that are two-dimensional discs or one-dimensional line segments.

3.3.5 S_n as the state space of a two-state quantum system

A quantum system is called a *two-state system* if any measurement of the system has at most two possible outcomes. Examples of such systems are the polarization of photons and the spin of spin-half particles. If the measurement of some physical quantity has at most two distinct possible outcomes for some state ψ, then this state can be written as a convex combination of two states, corresponding to the two possible outcomes of the experiment. For a two-state system, each of these states must be a *pure (indecomposable) state*, called also an *atom*. Equation (3.96) shows that a similar property holds for elements of S_n.

The measuring process of quantum systems implies that each pure state ψ_0 has a filtering projection. This projection represents the process of transforming any state (incoming beam) into a multiple of ψ_0 of possibly smaller intensity. The Stern–Gerlach apparatus, after blocking the $|z-\rangle$ component of an incoming beam of electrons, is a filtering projection for the pure state $|z+\rangle$. If we have a beam of photons, we can use the R-projector based on a right-left polarization analyzer to create a filtering projection for photons with right circular polarization. In general, a measurement causes the system to move into an eigenstate of the observable that is being measured. Thus, the measuring process defines a projection, called a *filtering projection*, on the state space for each value that could be observed. Since applying the filtering a second time will not affect the output state of the system, the filtering maps are indeed projections.

Note that a pure state in S_n is given by $\hat{\mathbf{v}}$, with $\mathbf{v}$ a minimal tripotent. We can associate to $\hat{\mathbf{v}}$ a projection $P_1^*(\mathbf{v}) : S_n \to S_n$ defined by

$$(P_1^*(\mathbf{v})\mathbf{f})(\mathbf{w}) = \mathbf{f}(P_1(\mathbf{v})\mathbf{w}). \tag{3.105}$$

From the definition (3.48) of $P_1(\mathbf{v})$, we get

$$\mathbf{f}(P_1(\mathbf{v})\mathbf{w}) = \langle P_1(\mathbf{v})\mathbf{w}|2\check{\mathbf{f}}\rangle = \langle 2\langle \mathbf{w}|\mathbf{v}\rangle\mathbf{v}|2\check{\mathbf{f}}\rangle = \langle \mathbf{v}|2\check{\mathbf{f}}\rangle\langle \mathbf{w}|2\mathbf{v}\rangle = \mathbf{f}(\mathbf{v})\hat{\mathbf{v}}(\mathbf{w}),$$

implying that

$$P_1^*(\mathbf{v})\mathbf{f} = \mathbf{f}(\mathbf{v})\hat{\mathbf{v}}. \tag{3.106}$$

Since $\hat{\mathbf{v}}(\mathbf{v}) = 1$, the map $P_1^*(\mathbf{v})$ is a projection, and, since $||P_1^*(\mathbf{v})\mathbf{f}||_* \leq |\mathbf{f}(\mathbf{v})|\,||\hat{\mathbf{v}}||_* \leq ||\mathbf{f}||_*$, it is a contraction. This projection transforms any $\mathbf{f} \in S_n$ to a multiple of $\hat{\mathbf{v}}$ and behaves like a filtering projection.

Let P be a filtering projection. The norm $||P\psi||_*$ represents the probability that a beam in the state ψ, represented by a norm 1 element of S, will pass the filter. Another important property of a filtering projection is *neutrality*. A projection P is called *neutral* if

$$||P\mathbf{f}||_* = ||\mathbf{f}||_* \quad \text{implies} \quad P\mathbf{f} = \mathbf{f}.$$

This means that if the beam passes the filter definitely (with probability 1), then it is already in the range of the filter. For our set S_n, the filtering projection is defined by (3.106). If for some state $\mathbf{f} \in S_n$, with $\|\mathbf{f}\|_* = 1$, we have $\|P_1^*(\mathbf{v})(\mathbf{f})\|_* = |\mathbf{f}(\mathbf{v})| = 1$, then from (3.98) we have $\mathbf{f} = \lambda\hat{\mathbf{v}}$ with $|\lambda| = 1$. But for such $\mathbf{f}$ we have $P_1^*(\mathbf{v})(\mathbf{f}) = \mathbf{f}$, showing that $P_1^*(\mathbf{v})$ is a neutral projection.

For any two pure states ψ and ϕ, the *transition probability* $P_{\psi\to\phi}$ between ψ and ϕ is defined as the probability that a beam in the state ψ will pass the filter preparing the state ϕ. The transition probability on a state space must satisfy the *symmetry of transition probability property*, meaning that

$$P_{\psi\to\phi} = P_{\phi\to\psi}.$$

For any two minimal tripotents $\mathbf{u}$ and $\mathbf{v}$, we have $P_{\hat{\mathbf{u}}\to\hat{\mathbf{v}}} = ||P_1^*(\mathbf{v})\hat{\mathbf{u}}||_* = |\hat{\mathbf{u}}(\mathbf{v})|$. Since

$$|\hat{\mathbf{u}}(\mathbf{v})| = |\hat{\mathbf{v}}(\mathbf{u})|,$$

the transition probability is symmetric on S_n.

On a state space S, each filtering projection P has a unique complementary filtering projection, denoted by $P^\sharp$. If P prepares the state ψ, the projection $P^\sharp$ prepares the state complementary to ψ. If an observable has two possible values and the probability of getting the first value when the system is in state ψ is zero, then it will definitely have the second value and belong to the complementary state. The complementary projection $P^\sharp$ is contractive and neutral, like P. Moreover, the sum $P + P^\sharp$ is a contractive projection, which, in general, differs from the identity. The operator $S_P = 2(P+P^\sharp) - I$ on the state space S is a symmetry and fixes the state ψ and its complementary state. This property is called *facial symmetry.*

If $P = P_1^*(\mathbf{v})$, then the complementary filtering projection $P^\sharp$ is $P_1^*(\overline{\mathbf{v}})$, which we also denote by $P_0^*(\mathbf{v})$. This projection prepares the state associated to $\overline{\mathbf{v}}$, which is orthogonal to $\mathbf{v}$. Note that the complementary projection is contractive and neutral. The operator

$$S_{\mathbf{v}} = 2(P_1^*(\mathbf{v}) + P_0^*(\mathbf{v})) - I = P_1^*(\mathbf{v}) - P_{1/2}^*(\mathbf{v}) + P_0^*(\mathbf{v})$$

is a symmetry of S_n. From (3.49) and (3.52), it follows that

$$S_{\mathbf{v}} = \exp(i2\pi D(\mathbf{v}))$$

is a triple automorphism of $D_{s,n}$ and an isometry of S_n. Thus, S_n is facially symmetric.

It was shown in [35] that if the state space of a two-state quantum system is facially symmetric and satisfies the above-mentioned pure state properties, then it is isometric to the dual of a spin factor. The proof is based on the construction of a natural basis, called a *grid*, on a facially symmetric space. We turn now to the construction of grids.

3.3.6 $\mathcal{S}^4$ and Pauli matrices

Let $\mathbf{v} = \mathbf{x} + i\mathbf{y}$ be a minimal tripotent in $\mathcal{S}^4$. Then $|\mathbf{x}| = |\mathbf{y}| = 1/2$ and $\langle \mathbf{x}|\mathbf{y}\rangle = 0$. Let $\mathbf{u}_1 = 2\mathbf{x}$ and $\mathbf{u}_2 = 2\mathbf{y}$. Then $\mathbf{u}_1 = \mathbf{v} + \overline{\mathbf{v}}$ and $\mathbf{u}_2 = (\mathbf{v} - \overline{\mathbf{v}})/i$. In any TCAR basis $\{\mathbf{u}_1, \mathbf{u}_2, \mathbf{u}_3, \mathbf{u}_4\}$, we will have

$$\mathbf{v} = 0.5(1, i, 0, 0), \quad \overline{\mathbf{v}} = 0.5(1, -i, 0, 0). \tag{3.107}$$

Applying the Peirce decomposition with respect to $\mathbf{v}$, we have $\mathcal{S}^4 = \mathcal{S}^4_1(\mathbf{v}) + \mathcal{S}^4_{1/2}(\mathbf{v}) + \mathcal{S}^4_0(\mathbf{v})$.

Note that $\mathcal{S}^4_{1/2}(\mathbf{v})$ has dimension 2. Choose a minimal tripotent $\mathbf{w} \in \mathcal{S}^4_{1/2}(\mathbf{v})$. Without loss of generality, we may assume that

$$\mathbf{w} = 0.5(0, 0, 1, i), \quad \overline{\mathbf{w}} = 0.5(0, 0, 1, -i). \tag{3.108}$$

Then we can choose $\mathbf{u}_3 = \mathbf{w} + \overline{\mathbf{w}}$ and $\mathbf{u}_4 = (\mathbf{w} - \overline{\mathbf{w}})/i$. Let us calculate $\{\mathbf{w}, \mathbf{v}, \overline{\mathbf{w}}\}$. Since the dot product of $\mathbf{v}$ with both $\mathbf{w}$ and $\overline{\mathbf{w}}$ is zero, we have

$$\{\mathbf{w}, \mathbf{v}, \overline{\mathbf{w}}\} = \langle \mathbf{w}|\mathbf{v}\rangle \overline{\mathbf{w}} + \langle \overline{\mathbf{w}}|\mathbf{v}\rangle \mathbf{w} - \langle \mathbf{w}|\mathbf{w}\rangle \overline{\mathbf{v}} = -0.5\overline{\mathbf{v}}. \tag{3.109}$$

This leads us to the definition of an odd quadrangle. We say that in a space with a triple product, four elements $(\mathbf{v}, \mathbf{w}, \overline{\mathbf{v}}, \overline{\mathbf{w}})$ form an *odd quadrangle* if the following relations hold:

- $\mathbf{v}, \mathbf{w}, \overline{\mathbf{v}}, \overline{\mathbf{w}}$ are minimal tripotents,
- $\overline{\mathbf{v}}$ is algebraically orthogonal to $\mathbf{v}$ and $\overline{\mathbf{w}}$ is algebraically orthogonal to $\mathbf{w}$,
- the pairs $(\mathbf{v}, \mathbf{w}), (\mathbf{v}, \overline{\mathbf{w}}), (\mathbf{w}, \overline{\mathbf{v}})$ and $(\overline{\mathbf{w}}, \overline{\mathbf{v}})$ are *co-orthogonal* (the pair $(\mathbf{v}, \mathbf{w})$ is said to be co-orthogonal if $D(\mathbf{v})\mathbf{w} = 0.5\mathbf{w}$ and $D(\mathbf{w})\mathbf{v} = 0.5\mathbf{v}$),
- $\{\mathbf{w}, \mathbf{v}, \overline{\mathbf{w}}\} = -0.5\overline{\mathbf{v}}$ and $\{\mathbf{v}, \mathbf{w}, \overline{\mathbf{v}}\} = -0.5\overline{\mathbf{w}}$.

An example of an odd quadrangle is the following set of 2×2 matrices:

$$\mathbf{v} = \begin{pmatrix} 1 & 0 \\ 0 & 0 \end{pmatrix}, \overline{\mathbf{v}} = \begin{pmatrix} 0 & 0 \\ 0 & 1 \end{pmatrix}, \mathbf{w} = \begin{pmatrix} 0 & 1 \\ 0 & 0 \end{pmatrix}, \overline{\mathbf{w}} = \begin{pmatrix} 0 & 0 \\ -1 & 0 \end{pmatrix}, \tag{3.110}$$

where the triple product is

$$\{a, b, c\} = \frac{ab^*c + cb^*a}{2}. \tag{3.111}$$

Our TCAR basis becomes

$$\mathbf{u}_1 = \begin{pmatrix} 1 & 0 \\ 0 & 1 \end{pmatrix}, \mathbf{u}_2 = \begin{pmatrix} -i & 0 \\ 0 & i \end{pmatrix}, \mathbf{u}_3 = \begin{pmatrix} 0 & 1 \\ -1 & 0 \end{pmatrix}, \mathbf{u}_4 = \begin{pmatrix} 0 & -i \\ -i & 0 \end{pmatrix}.$$

Thus,

$$\mathbf{u}_1 = I, \mathbf{u}_2 = -i\sigma_3, \mathbf{u}_3 = -i\sigma_2, \mathbf{u}_4 = -i\sigma_1,$$

where σ_j denote the Pauli matrices.

Any element $\mathbf{a} = (a_1, a_2, a_3, a_4) \in S^4$ can be represented by a 2×2 matrix A as

$$\sum_{j=1}^{4} a_j \mathbf{u}_j = \begin{pmatrix} a_1 - ia_2 & a_3 - ia_4 \\ -a_3 - ia_4 & a_1 + ia_2 \end{pmatrix} = A.$$

Note that

$$\det A = a_1^2 + a_2^2 + a_3^2 + a_4^2 = \det \mathbf{a}, \tag{3.112}$$

providing another justification for the definition of the determinant in the complex spin factor.

3.3.7 The spin grid and TCAR bases in S^n

In the previous section, we studied the duality between minimal and maximal tripotents in the spin factor S^n. In Table 3.2 below, we summarize the dual properties of these objects.

TCAR bases consist of maximal tripotents. In physical applications, maximal tripotents correspond to physical quantities or components of such quantities. They are distinguished extreme points on the ball $D_{s,n}$ in the space representing observables. On the other hand, it is sometimes convenient to use a basis consisting of pure states. In quantum mechanics, for example, we often work with a basis for the state space built from eigenstates of a commuting family of observables. These states are, in general, pure states, and they correspond to minimal tripotents. This type of basis is called a *spin grid* of S^n.

We assume here that $n = 2m$ is an even number. We say that

$$\mathbf{v}_1, \mathbf{v}_2, ..., \mathbf{v}_n \text{ form a } \textit{spin grid} \text{ in } S^n$$

if for any $1 \leq j < k \leq m$, the elements $(\mathbf{v}_{2j-1}, \mathbf{v}_{2k-1}, \mathbf{v}_{2j}, \mathbf{v}_{2k})$ form an odd quadrangle. A spin grid in S^n is a basis consisting of minimal tripotents. The connection between a TCAR basis and a spin grid is as follows. If $\{\mathbf{u}_1, \mathbf{u}_2, ..., \mathbf{u}_n\}$ is a TCAR basis, then $\{\mathbf{v}_1, \mathbf{v}_2, ..., \mathbf{v}_n\}$ is a spin grid, where

$$\mathbf{v}_{2j-1} = 0.5(\mathbf{u}_{2j-1} + i\mathbf{u}_{2j}),\ \mathbf{v}_{2j} = 0.5(\mathbf{u}_{2j-1} - i\mathbf{u}_{2j}). \tag{3.113}$$

Conversely, if $\{\mathbf{v}_1, \mathbf{v}_2, ..., \mathbf{v}_n\}$ is a spin grid, then $\{\mathbf{u}_1, \mathbf{u}_2, ..., \mathbf{u}_n\}$ is a TCAR basis, where

$$\mathbf{u}_{2j-1} = \mathbf{v}_{2j-1} + \mathbf{v}_{2j},\ \mathbf{u}_{2j} = i(\mathbf{v}_{2j} - \mathbf{v}_{2j-1}). \tag{3.114}$$

Equation (3.113) explains why in quantum mechanics we encounter expressions like $a = \hat{x} + i\hat{p}_x$, $a^* = \hat{x} - i\hat{p}_x$, $J_+ = J_x + iJ_y$ and $J_- = J_x - iJ_y$.

For the case when n is odd, see Chapter 6.

Type	Ball $D_{s,n}$	Ball S_n	Basis	Decomposition	Quantum Mechanics
Maximal	Extreme point	Center (n-1)D face	TCAR	Facial	Observables
Minimal	Center 2D face	Extreme point	Spin grid	Spectral	Pure states

Table 3.2. The geometric and quantum mechanical properties of tripotents in $\mathcal{S}^n$.

3.4 The unit ball $D_{s,n}$ of $\mathcal{S}^n$ as a bounded symmetric domain

In this section, we will show that the unit ball $D_{s,n}$ of $\mathcal{S}^n$ is a bounded symmetric domain with respect to the group $Aut_a(D_{s,n})$ of analytic automorphisms of $D_{s,n}$. This domain is a good example with which to demonstrate various concepts that will be generalized in Chapter 5 to general bounded symmetric domains.

3.4.1 Complete analytic vector fields on $D_{s,n}$

Let D be a domain in a complex linear space X. A map $\xi : D \to X$ is called a *vector field*. We will say that a vector field ξ is *analytic* if, for any point $\mathbf{a} \in D$, there is a neighborhood of $\mathbf{a}$ in which ξ is the sum of a power series (see Chapter 5 for more details). A well-known theorem from the theory of differential equations states that if ξ is an analytic vector field on D and $\mathbf{a} \in D$, then the initial-value problem

$$\frac{d\mathbf{w}(\tau)}{d\tau} = \xi(\mathbf{w}(\tau)), \quad \mathbf{w}(0) = \mathbf{a} \tag{3.115}$$

has a unique solution $\mathbf{w_a}(\tau)$ for real τ in some neighborhood of $\tau = 0$. A field ξ is called a *complete analytic vector field* if the solution of the initial-value problem (3.115) exists for all real τ, all $\mathbf{a} \in D$, and all $\mathbf{w}(\tau) \in D$. In this case, for any fixed $\tau \in R$, the map $\phi_\tau : D \to D$, defined by

$$\phi_\tau(\mathbf{a}) = \mathbf{w_a}(\tau), \tag{3.116}$$

is analytic and is called the *flow generated by* ξ. If ξ is complete, then the flow ϕ_τ generated by ξ belongs to the group $Aut_a(D)$ of analytic automorphisms of D. We denote

$$\phi_\tau = \exp(\tau\xi), \tag{3.117}$$

where $\tau \in R$. By a result in [64], an analytic flow ξ on a bounded domain D is complete if and only if it is tangent to the boundary ∂D of the domain.

For each $\mathbf{a} \in S^n$, we define a vector field representing a generator of translation by

$$\xi_{\mathbf{a}}(\mathbf{w}) = \mathbf{a} - \{\mathbf{w}, \mathbf{a}, \mathbf{w}\}. \tag{3.118}$$

Note the similarity of $\xi_{\mathbf{a}}$ to the generators of velocity addition in Chapter 1 and the generators of s-velocity addition in Chapter 2. Note that $\xi_{\mathbf{a}}$ is a second-degree polynomial in $\mathbf{w}$ and, thus, an analytic vector field. See Figure 3.5 for an example of $\xi_{\mathbf{a}}(\mathbf{w})$, where $\mathbf{w}$ ranges over the intersection of $D_{s,n}$ with the two-dimensional real span of the two minimal tripotents $\mathbf{v}_1$ and $\mathbf{v}_2$ from the singular decomposition of $\mathbf{a}$. We will show now that $\xi_{\mathbf{a}}$ is tangent on

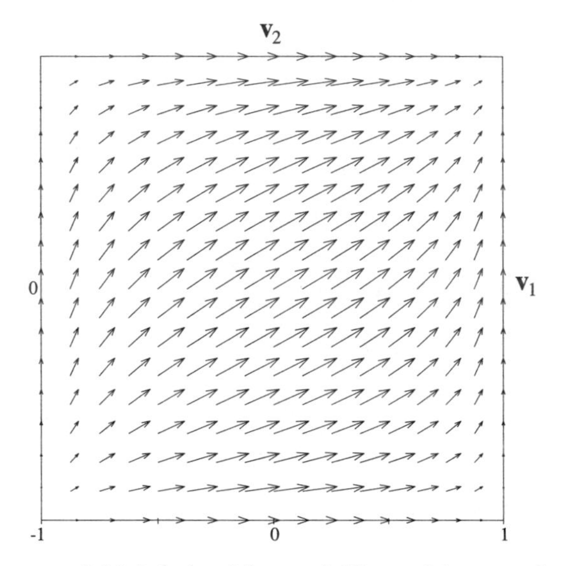

Fig. 3.5. The vector field $\xi_{\mathbf{a}}(\mathbf{w})$, with $\mathbf{a} = 0.17\mathbf{v}_1 + 0.1\mathbf{v}_2$, on the intersection of $D_{s,n}$ with $\text{span}_R\{\mathbf{v}_1, \mathbf{v}_2\}$. Note that the flow is tangent on $\partial D_{s,n}$.

the boundary of $D_{s,n}$. Thus, it generates a complete analytic flow on $D_{s,n}$.

Let $\mathbf{w} \in \partial D_{s,n}$. Then, from section 3.3.2, it follows that $\mathbf{w}$ belongs to a norm-exposed face $F_{\mathbf{v}}$, corresponding to a minimal tripotent $\mathbf{v}$. This face is exposed by the hyperplane $\Pi = \{\mathbf{w} : Re\, \hat{\mathbf{v}}(\mathbf{w}) = 1\}$. We will show that $\xi_{\mathbf{a}}(\mathbf{w})$ is parallel to the hyperplane, meaning that

$$Re\, \hat{\mathbf{v}}(\xi_{\mathbf{a}}(\mathbf{w})) = 0. \tag{3.119}$$

From (3.89), it follows that $\mathbf{w} = \mathbf{v} + \lambda\overline{\mathbf{v}}$, with $|\lambda| \leq 1$. Thus,

$$\{\mathbf{w}, \mathbf{a}, \mathbf{w}\} = \{\mathbf{v} + \lambda\overline{\mathbf{v}}, \mathbf{a}, \mathbf{v} + \lambda\overline{\mathbf{v}}\}$$

$$= \{\mathbf{v}, \mathbf{a}, \mathbf{v}\} + \lambda^2\{\overline{\mathbf{v}}, \mathbf{a}, \overline{\mathbf{v}}\} + 2\lambda\{\mathbf{v}, \mathbf{a}, \overline{\mathbf{v}}\}.$$

Since $\mathbf{v}$ and $\overline{\mathbf{v}}$ are orthogonal minimal tripotents, we have

$$\{\mathbf{w},\mathbf{a},\mathbf{w}\} = 2\langle \mathbf{v}|\mathbf{a}\rangle\mathbf{v} + 2\lambda^2\langle \overline{\mathbf{v}}|\mathbf{a}\rangle\overline{\mathbf{v}} + 2\lambda\langle \mathbf{v}|\mathbf{a}\rangle\overline{\mathbf{v}}$$

$$+2\lambda\langle \overline{\mathbf{v}}|\mathbf{a}\rangle\mathbf{v} - \lambda\overline{\mathbf{a}}.$$

Recall that $\hat{\mathbf{v}}(\mathbf{v}) = 1$, $\hat{\mathbf{v}}(\overline{\mathbf{v}}) = 0$ and $\hat{\mathbf{v}}(\mathbf{a}) = 2\langle \mathbf{a}|\mathbf{v}\rangle$. Thus,

$$\hat{\mathbf{v}}(\xi_{\mathbf{a}}(\mathbf{w})) = 2\langle \mathbf{a}|\mathbf{v}\rangle - 2\langle \mathbf{v}|\mathbf{a}\rangle - 2\lambda\langle \overline{\mathbf{v}}|\mathbf{a}\rangle + 2\lambda\langle \overline{\mathbf{a}}|\mathbf{v}\rangle$$

$$= 4iIm(\langle \mathbf{a}|\mathbf{v}\rangle),$$

implying (3.119). Thus, the analytic vector field $\xi_{\mathbf{a}}$ defined by (3.118) is tangent on $\partial D_{s,n}$ and, thus, is a complete analytic vector field.

3.4.2 Decomposition of the translations on $D_{s,n}$

Let $\mathbf{a} \in \mathcal{S}^n$. Let $\mathbf{a} = s_1\mathbf{v}_1 + s_2\mathbf{v}_2$ be the singular decomposition of $\mathbf{a}$. Using the linearity of the triple product, we can decompose the generator of translation $\xi_{\mathbf{a}}$, defined by (3.118), as

$$\xi_{\mathbf{a}} = \mathbf{a} - \{\mathbf{w},\mathbf{a},\mathbf{w}\} = \xi_{s_1\mathbf{v}_1} + \xi_{s_2\mathbf{v}_2}. \tag{3.120}$$

The vector fields $\xi_{s_1\mathbf{v}_1}$ and $\xi_{s_2\mathbf{v}_2}$ represent generators of translations in the directions of two orthogonal minimal tripotents. We will show that these vector fields *commute*, meaning that

$$[\xi_{s_1\mathbf{v}_1}, \xi_{s_2\mathbf{v}_2}] = 0\,. \tag{3.121}$$

Figure 3.6 shows the decomposition of the vector field $\xi_{\mathbf{a}}(\mathbf{w})$ from Figure 3.5 into the sum of the vector fields $\xi_{s_1\mathbf{v}_1}$ and $\xi_{s_2\mathbf{v}_2}$.

Recall from section 1.5.2 that the Lie bracket of two vector fields δ and ξ is defined as

$$[\delta,\xi](\mathbf{w}) = \frac{d\delta}{d\mathbf{w}}(\mathbf{w})\xi(\mathbf{w}) - \frac{d\xi}{d\mathbf{w}}(\mathbf{w})\delta(\mathbf{w}), \tag{3.122}$$

where $\mathbf{w} \in \mathcal{S}^n$ and $\frac{d\delta}{d\mathbf{w}}(\mathbf{w})\xi(\mathbf{w})$ denotes the derivative of δ at the point $\mathbf{w}$ in the direction of the vector $\xi(\mathbf{w})$. Let us calculate first $\frac{d\xi_{s_1\mathbf{v}_1}}{d\mathbf{w}}(\mathbf{w})\xi_{s_2\mathbf{v}_2}(\mathbf{w})$. Using the orthogonality of $\mathbf{v}_1$ and $\mathbf{v}_2$, we have

$$\frac{d\xi_{s_1\mathbf{v}_1}}{d\mathbf{w}}(\mathbf{w})\xi_{s_2\mathbf{v}_2}(\mathbf{w}) = -2\{\mathbf{w}, s_1\mathbf{v}_1, (s_2\mathbf{v}_2 - \{\mathbf{w}, s_2\mathbf{v}_2, \mathbf{w}\})\}$$

$$= 2s_1s_2\{\mathbf{w}, \mathbf{v}_1, \{\mathbf{w}, \mathbf{v}_2, \mathbf{w}\}\}.$$

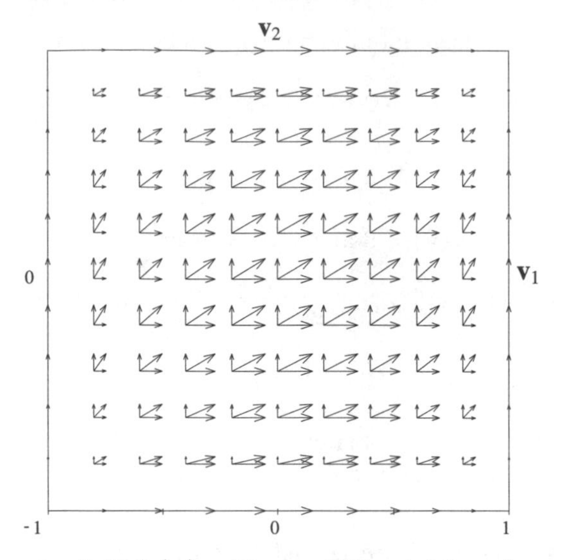

Fig. 3.6. The vector field $\xi_{\mathbf{a}}(\mathbf{w})$, with $\mathbf{a} = 0.17\mathbf{v}_1 + 0.1\mathbf{v}_2$, decomposed into the sum of the vector fields $\xi_{0.17\mathbf{v}_1}$ and $\xi_{0.1\mathbf{v}_2}$. Compare to Figure 3.5.

From (3.28), it follows that there is a constant λ such that $|\lambda| = 1$ and $\mathbf{v}_2 = \lambda\overline{\mathbf{v}}_1$. Thus,

$$\frac{d\xi_{s_1\mathbf{v}_1}}{d\mathbf{w}}(\mathbf{w})\xi_{s_2\mathbf{v}_2}(\mathbf{w}) = 2s_1s_2\overline{\lambda}\{\mathbf{w}, \mathbf{v}_1, \{\mathbf{w}, \overline{\mathbf{v}}_1, \mathbf{w}\}\}.$$

Similarly, we get

$$\frac{d\xi_{s_2\mathbf{v}_2}}{d\mathbf{w}}(\mathbf{w})\xi_{s_1\mathbf{v}_1}(\mathbf{w}) = 2s_1s_2\overline{\lambda}\{\mathbf{w}, \overline{\mathbf{v}}_1, \{\mathbf{w}, \mathbf{v}_1, \mathbf{w}\}\}.$$

Thus, in order to prove (3.121), it is enough to show that

$$\{\mathbf{w}, \mathbf{v}_1, \{\mathbf{w}, \overline{\mathbf{v}}_1, \mathbf{w}\}\} = \{\mathbf{w}, \overline{\mathbf{v}}_1, \{\mathbf{w}, \mathbf{v}_1, \mathbf{w}\}\}.$$

By the definition of the triple product, both sides of the equation are equal to

$$4\langle\mathbf{w}|\mathbf{v}_1\rangle\langle\mathbf{w}|\overline{\mathbf{v}}_1\rangle\mathbf{w} - \langle\mathbf{w}|\overline{\mathbf{w}}\rangle(\langle\mathbf{w}|\overline{\mathbf{v}}_1\rangle\overline{\mathbf{v}}_1 + \langle\mathbf{w}|\mathbf{v}_1\rangle\mathbf{v}_1 + 1/2),$$

proving (3.121).

Since the exponent of the sum of commuting vector fields is the product of the exponents of each field, we have

$$\exp(\xi_{\mathbf{a}}) = \exp(\xi_{s_2\mathbf{v}_2})\exp(\xi_{s_1\mathbf{v}_1}), \tag{3.123}$$

implying that any translation $\exp(\xi_{\mathbf{a}})$ can be decomposed as a product of translations $\exp(\xi_{s\mathbf{v}})$ defined by multiples of minimal tripotents.

3.4.3 The homogeneity of $D_{s,n}$

Since $D_{s,n}$ is the unit ball in the operator norm, the reflection map $\mathbf{w} \to -\mathbf{w}$ is an analytic symmetry on $D_{s,n}$ which fixes only the origin. Clearly, $D_{s,n}$ is bounded. Thus, in order to show that $D_{s,n}$ is a bounded symmetric domain, it is enough to show that it is homogeneous. This means that for a given $\mathbf{b} \in D_{s,n}$, we must find an analytic automorphism $\varphi_{\mathbf{b}}$ of $D_{s,n}$ such that $\varphi_{\mathbf{b}}(\mathbf{0}) = \mathbf{b}$. We will construct an element $\mathbf{a}$ of $D_{s,n}$ such that $\exp(\xi_{\mathbf{a}})(\mathbf{0}) = \mathbf{b}$.

The first step is to calculate $\exp(\xi_{s\mathbf{v}})(\mathbf{0})$, where s is a positive constant and $\mathbf{v}$ is a minimal tripotent. To do this, we have to solve the initial-value problem (3.115), which in our case is

$$\frac{d\mathbf{w}(\tau)}{d\tau} = s\mathbf{v} - s\{\mathbf{w}(\tau), \mathbf{v}, \mathbf{w}(\tau)\}, \quad \mathbf{w}(0) = \mathbf{0}. \tag{3.124}$$

But if we take

$$\mathbf{w}(\tau) = \tanh(s\tau)\mathbf{v}$$

(as in the solutions of the similar initial-value problem in Chapter 1), then

$$\frac{d\mathbf{w}(\tau)}{d\tau} = \frac{s}{\cosh^2(s\tau)}\mathbf{v} = s(1 - \tanh^2(s\tau))\mathbf{v} = \xi_{s\mathbf{v}}(\mathbf{w}(\tau)),$$

implying that $\mathbf{w}(\tau)$ is a solution of the initial-value problem (3.124). Thus, by (3.117), we have

$$\exp(\xi_{s\mathbf{v}})(\mathbf{0}) = \tanh(s)\mathbf{v}. \tag{3.125}$$

Now let $\mathbf{b} = s_1\mathbf{v}_1 + s_2\mathbf{v}_2$ be an arbitrary element of $D_{s,n}$. Then the function

$$\mathbf{w}(\tau) = \tanh(\tau \tanh^{-1}(s_1))\mathbf{v}_1 + \tanh(\tau \tanh^{-1}(s_2))\mathbf{v}_2$$

satisfies the initial-value problem

$$\frac{d\mathbf{w}(\tau)}{d\tau} = \xi_{\mathbf{a}}(\mathbf{w}(\tau)) = \mathbf{a} - \{\mathbf{w}(\tau), \mathbf{a}, \mathbf{w}(\tau)\}, \quad \mathbf{w}(0) = \mathbf{0}, \tag{3.126}$$

where

$$\mathbf{a} = \tanh^{-1}(s_1)\mathbf{v}_1 + \tanh^{-1}(s_2)\mathbf{v}_2.$$

Thus,

$$\exp(\xi_{\mathbf{a}})(\mathbf{0}) = \mathbf{b}.$$

This establishes the homogeneity of $D_{s,n}$. In Figure 3.7, we see $\mathbf{w}(\tau)$ for the vector field $\xi_{\mathbf{a}}(\mathbf{w})$ from Figure 3.5.

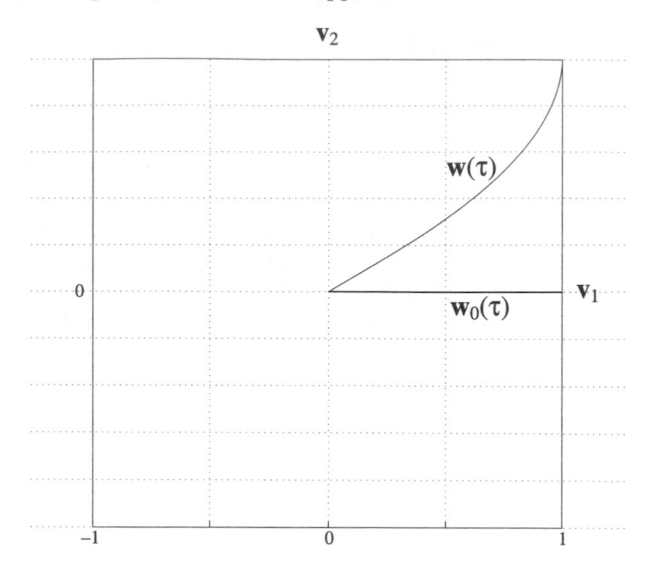

Fig. 3.7. The trajectory $\mathbf{w}(\tau)$ for the vector field $\xi_{\mathbf{a}}$, with $\mathbf{a} = 0.17\mathbf{v}_1 + 0.1\mathbf{v}_2$, from Figure 3.5. The trajectory starts at the origin and ends up at the maximal tripotent $\mathbf{v}_1 + \mathbf{v}_2$. Such behavior is stable under small perturbations of $\mathbf{a}$. The trajectory $\mathbf{w}_0(\tau)$, corresponding to the vector field $\xi_{\mathbf{v}_1}$, ends up at the minimal tripotent $\mathbf{v}_1$. This trajectory is unstable under small perturbations of $\mathbf{v}_1$.

3.4.4 The group $Aut_a(D_{s,n})$

The group $Aut_a(D_{s,n})$ of all analytic automorphisms of $D_{s,n}$ consists of translations $\varphi_{\mathbf{b}}$, which are the exponents of complete analytic vector fields $\xi_{\mathbf{a}}$, and rotations, which are linear isometries of $D_{s,n}$. We will denote the group of all linear isometries of $D_{s,n}$ by K, since this is a compact subgroup of $D_{s,n}$. We will show now that K is exactly the group $\mathrm{Taut}(\mathcal{S}^n)$ of triple product automorphisms of $\mathcal{S}^n$, defined earlier and characterized by (3.20).

Let $T \in \mathrm{Taut}(\mathcal{S}^n)$ be a triple product automorphism. Then, for any tripotent $\mathbf{u}$, we have

$$T(\mathbf{u}) = T(\{\mathbf{u}, \mathbf{u}, \mathbf{u}\}) = \{T(\mathbf{u}), T(\mathbf{u}), T(\mathbf{u})\}),$$

implying that $T(\mathbf{u})$ is also a tripotent. Since algebraic orthogonality of tripotents is defined by the triple product, the map T preserves algebraically orthogonal tripotents. Thus, for an arbitrary element $\mathbf{a} \in \mathcal{S}^n$ with singular decomposition $\mathbf{a} = s_1\mathbf{v}_1 + s_2\mathbf{v}_2$, the singular decomposition of $T(\mathbf{a})$ is

$$T(\mathbf{a}) = s_1 T(\mathbf{v}_1) + s_2 T(\mathbf{v}_2).$$

From the definition (3.58) of the norm on $\mathcal{S}^n$, we have $||T(\mathbf{a})|| = s_1 = ||\mathbf{a}||$, implying that T is an isometry and, hence, belongs to K.

Conversely, suppose T is a linear isometry of $\mathcal{S}^n$ preserving the unit ball $D_{s,n}$. In this case, the dual T^* is an isometry on $\mathcal{S}^n_*$ preserving the state space S_n. The linear map T must take extreme points of $D_{s,n}$, which are maximal tripotents, to extreme points of $\mathcal{S}^n$. So maximal tripotents are mapped to maximal tripotents. The same argument for T^* implies that the minimal tripotents which correspond to extreme points of S_n are mapped to minimal tripotents. Since two minimal tripotents are algebraically orthogonal if and only if their sum is a maximal tripotent, the map T preserves algebraic orthogonality of minimal tripotents. Thus, for any element $\mathbf{a} \in \mathcal{S}^n$ with singular decomposition $\mathbf{a} = s_1\mathbf{v}_1 + s_2\mathbf{v}_2$, the singular decomposition of $T(\mathbf{a})$ is

$$T(\mathbf{a}) = s_1 T(\mathbf{v}_1) + s_2 T(\mathbf{v}_2).$$

Thus,

$$T(\mathbf{a}^{(3)}) = T(s_1^3\mathbf{v}_1 + s_2^3\mathbf{v}_2) = s_1^3 T(\mathbf{v}_1) + s_2^3 T(\mathbf{v}_2) = T(\mathbf{a})^{(3)}.$$

By polarization, this implies that T is a triple product automorphism. Thus, we have shown that

$$K = \mathrm{Taut}(\mathcal{S}^n).$$

3.5 The Lorentz group representations on $\mathcal{S}^n$

In section 1.5.4 of Chapter 1, we saw that the transformation of the electromagnetic field strength E, B from one inertial system to another preserves the complex quantity F^2, where $F = E + icB$. If we consider F as an element of $\mathcal{S}^3$, then $F^2 = \det F$. Thus, if we take $\mathcal{S}^3$ as the space representing the set of all possible electromagnetic field strengths, then the Lorentz group acts on $\mathcal{S}^3$ by linear transformations which preserve the determinant. This leads us to study the Lie group of determinant-preserving linear maps on $\mathcal{S}^3$ (and, in general, on $\mathcal{S}^n$) and the Lie algebra of this group.

3.5.1 The determinant-preserving group Dinv $(\mathcal{S}^n)$ and its Lie algebra

Let Dinv $(\mathcal{S}^n)$ be the group of all invertible linear maps $\mathcal{S}^n \to \mathcal{S}^n$ which preserve the determinant. We introduce a complex bilinear symmetric form Bl on $\mathcal{S}^n$ by

$$Bl(\mathbf{a}, \mathbf{b}) = a_1 b_1 + a_2 b_2 + \cdots + a_n b_n, \tag{3.127}$$

where $\mathbf{a} = (a_1, a_2, \ldots, a_n)$ and $\mathbf{b} = (b_1, b_2, \ldots, b_n)$. Then

$$\mathrm{Dinv}\,(\mathcal{S}^n) = \{g \in GL(\mathcal{S}^n) : Bl(g\mathbf{a}, g\mathbf{a}) = Bl(\mathbf{a}, \mathbf{a}) \text{ for all } \mathbf{a} \in \mathcal{S}^n\}. \tag{3.128}$$

We denote the Lie algebra of Dinv $(\mathcal{S}^n)$ by dinv $(\mathcal{S}^n)$. If $g(t)$ is a smooth curve in Dinv $(\mathcal{S}^n)$, with $g(0) = I$, the identity map on $\mathcal{S}^n$, then $X := g'(0) \in$ dinv $(\mathcal{S}^n)$. Since $g(t) \in$ Dinv $(\mathcal{S}^n)$, from (3.128) we have

$$Bl(g(t)\mathbf{a}, g(t)\mathbf{a}) = Bl(\mathbf{a}, \mathbf{a}) \text{ for all } \mathbf{a} \in \mathcal{S}^n.$$

Differentiating this by t and substituting $t = 0$, we obtain

$$Bl(X\mathbf{a}, \mathbf{a}) + Bl(\mathbf{a}, X\mathbf{a}) = 0,$$

and so $Bl(X\mathbf{a}, \mathbf{a}) = 0$ for all $\mathbf{a} \in \mathcal{S}^n$. By polarization, for every $\mathbf{a}, \mathbf{b} \in \mathcal{S}^n$ we obtain

$$Bl(X\mathbf{a}, \mathbf{b}) + Bl(\mathbf{a}, X\mathbf{b}) = 0. \tag{3.129}$$

Note that from the definition (3.127) of the bilinear form Bl, for any element $\mathbf{u}_j$ of a TCAR basis B and any element $\mathbf{c} \in \mathcal{S}^n$, we have

$$Bl(\mathbf{c}, \mathbf{u}_j) = Bl(\mathbf{u}_j, \mathbf{c}) = \langle \mathbf{c} | \mathbf{u}_j \rangle,$$

implying that

$$\langle X\mathbf{u}_j | \mathbf{u}_k \rangle + \langle X\mathbf{u}_k | \mathbf{u}_j \rangle = 0$$

for any $1 \leq j, k \leq n$. Thus, the matrix $[x_{jk}]$ of X with respect to the basis B is antisymmetric.

Hence dinv $(\mathcal{S}^n) \subset A_n(C)$, where $A_n(C)$ denotes the space of all $n \times n$ complex antisymmetric matrices, a Cartan factor of type 2. Using the triple product on $\mathcal{S}^n$ and (3.22), we can express this Lie algebra as

$$\text{dinv}\,(\mathcal{S}^n) = \{\sum_{k<l} d_{kl} D(\mathbf{u}_k, \mathbf{u}_l) : \; d_{kl} \in C\}. \tag{3.130}$$

The Lie group Dinv $(\mathcal{S}^n)$ consists of matrix exponents of elements of X from dinv $(\mathcal{S}^n)$. If $X \in$ dinv $(\mathcal{S}^n)$ is real, then its exponent is an orthogonal matrix. Thus, $O(n) \subset$ Dinv $(\mathcal{S}^n)$. Since the trace of X is zero, the determinant of $\exp(X)$ equals 1. Thus, Dinv $(\mathcal{S}^n)$ is a subgroup of $SL(n, C)$.

The problem of characterizing the linear maps which preserve the determinant has a long history dating back to Frobenius in 1897. Frobenius showed that the linear maps of the $n \times n$ complex matrices which preserve the determinant are of one of the forms $A \mapsto PAQ$ or $A \mapsto PA^tQ$, where A^t is the transpose of A and P, Q are matrices with $\det PQ = 1$. The group Dinv $(\mathcal{S}^n)$ coincides with the pseudo-orthogonal group $SO(n, C)$,which consists of linear maps of C^n having determinant 1 and preserving the form $z_1^2 + \cdots + z_n^2$.

3.5.2 Representation of the Lorentz Lie algebra in dinv $(\mathcal{S}^3)$

We consider the complex spin triple factor $\mathcal{S}^3$ as a representation space for electromagnetic field strength. As mentioned at the beginning of this section, the Lorentz group preserves the complex value $F^2 = (E + icB)^2 = \det F$. Thus, the Lorentz group acts by elements of Dinv $(\mathcal{S}^3)$ and the generators of this group are elements of dinv $(\mathcal{S}^3)$. We can thus define a representation π_s^3 from the Lorentz Lie algebra into dinv $(\mathcal{S}^3)$.

To determine this representation, it is enough to define π_s^3 on the generators of rotation J_k and the generators of boosts K_k, for $k = 1, 2, 3$. Let $B = \{\mathbf{u}_1, \mathbf{u}_2, \mathbf{u}_3\}$ be a TCAR basis of $\mathcal{S}^3$. Similar to the representation π_c from section 2.4.5 of Chapter 2, we define

$$\pi_s^3(J_1) = D(\mathbf{u}_2, \mathbf{u}_3), \;\; \pi_s^3(J_2) = D(\mathbf{u}_3, \mathbf{u}_1), \;\; \pi_s^3(J_3) = D(\mathbf{u}_1, \mathbf{u}_2). \tag{3.131}$$

Using the definition (3.4) of the spin triple product and the definition (3.7) of the D operator, the matrix of $D(\mathbf{u}_2, \mathbf{u}_3)$, for example, in the basis B is

$$\pi_s^3(J_1) = D(\mathbf{u}_2, \mathbf{u}_3) = \begin{pmatrix} 0 & 0 & 0 \\ 0 & 0 & 1 \\ 0 & -1 & 0 \end{pmatrix}, \tag{3.132}$$

which is an antisymmetric matrix, and, by (3.130), an element of dinv $(\mathcal{S}^3)$. This matrix is the same as the matrix of momentum J_1 of rotation about the $\mathbf{u}_1$-axis.

To define the representation π_s^3 on the generators of boosts K_k as elements of dinv $(\mathcal{S}^3)$, it is natural to try to use the remaining three dimensions of the six real dimensions of dinv $(\mathcal{S}^3)$. By (3.130), the operators $iD(\mathbf{u}_2, \mathbf{u}_3)$,$iD(\mathbf{u}_3, \mathbf{u}_1)$ and $iD(\mathbf{u}_1, \mathbf{u}_2)$ are also elements of dinv $(\mathcal{S}^3)$. As shown in section 1.5.3, page 39, the generators of the Lorentz group satisfy the following commutation relations:

$$[J_1, J_2] = -J_3, \;\; [J_2, J_3] = -J_1, \;\; [J_3, J_1] = -J_2, \tag{3.133}$$

$$[J_1, K_1] = 0, \;\; [J_1, K_2] = -K_3, \;\; [J_1, K_3] = K_2, \tag{3.134}$$

$$[J_2, K_1] = K_3, \;\; [J_2, K_2] = 0, \;\; [J_2, K_3] = -K_1, \tag{3.135}$$

$$[J_3, K_1] = -K_2, \;\; [J_3, K_2] = K_1, \;\; [J_3, K_3] = 0, \tag{3.136}$$

$$[K_1, K_2] = J_3, \;\; [K_2, K_3] = J_1, \;\; [K_3, K_1] = J_2. \tag{3.137}$$

We define

$$\pi_s^3(K_1) = i\pi_s^3(J_1),\ \ \pi_s^3(K_2) = i\pi_s^3(J_2),\ \ \pi_s^3(K_3) = i\pi_s^3(J_3). \tag{3.138}$$

For example,

$$\pi_s^3(K_1) = i\pi_s^3(J_1) = \begin{pmatrix} 0 & 0 & 0 \\ 0 & 0 & i \\ 0 & -i & 0 \end{pmatrix}. \tag{3.139}$$

It is straightforward to verify that the images of J_k, K_k under π_s^3 also satisfy the relations (3.133)–(3.137). Thus, (3.131) and (3.138) define a *linear* representation π_s^3 of the Lorentz Lie algebra into $\operatorname{dinv}(\mathcal{S}^3)$. In fact, the range of π_s^3 is the entire Lie algebra $\operatorname{dinv}(\mathcal{S}^3)$.

3.5.3 Representation of the Lorentz group in Dinv ($\mathcal{S}^3$)

To define a representation of the Lorentz group into $\operatorname{Dinv}(\mathcal{S}^3)$, we compute the exponents of the matrices, defined above, representing the generators in $\operatorname{dinv}(\mathcal{S}^3)$. We use the usual formula for the exponent of a matrix A:

$$\exp(A) = \sum_{j=0}^{\infty} \frac{A^j}{j!}. \tag{3.140}$$

If A represents a generator of rotation $\pi_s^3(J_j)$, then from (3.132), $A^3 = -A$ and the exponent is a matrix of rotation. For example, using the notation (1.176) for the rotation operator, we obtain

$$\mathcal{R}_\varphi^{\mathbf{i}} = \exp(\varphi\pi_s^3(J_1)) = \begin{pmatrix} 1 & 0 & 0 \\ 0 & \cos\varphi & \sin\varphi \\ 0 & -\sin\varphi & \cos\varphi \end{pmatrix}. \tag{3.141}$$

If A represents a generator of a boost $\pi_s^3(K_j)$, then from (3.139), $A^3 = A$ and the exponent is a matrix of hyperbolic rotation. For example, for the boost in the x-direction, we have

$$\mathcal{B}_\varphi^{\mathbf{i}} = \exp(\varphi\pi_s^3(K_1)) = \begin{pmatrix} 1 & 0 & 0 \\ 0 & \cosh\varphi & i\sinh\varphi \\ 0 & -i\sinh\varphi & \cosh\varphi \end{pmatrix}. \tag{3.142}$$

We now compute the eigenvalues and eigenvectors of the operators $\mathcal{R}_\varphi^{\mathbf{i}}, \mathcal{B}_\varphi^{\mathbf{i}}$. It is obvious that $\mathbf{i} = (1,0,0)$ is an eigenvalue corresponding to the eigenvector 1 for both operators. By a direct calculation, we find that the other two eigenvectors are $\mathbf{v} = 0.5(0,1,i)$ and $\overline{\mathbf{v}} = 0.5(0,1,-i)$. The eigenvector $\mathbf{v}$ corresponds to the eigenvalue $e^{i\varphi}$ for the operator $\mathcal{R}_\varphi^{\mathbf{i}}$ and to the eigenvalue $(\cosh\varphi + \sinh\varphi)$ for the operator $\mathcal{B}_\varphi^{\mathbf{i}}$. Similarly, the eigenvector $\overline{\mathbf{v}}$ corresponds to the eigenvalue $e^{-i\varphi}$ for the operator $\mathcal{R}_\varphi^{\mathbf{i}}$ and to the eigenvalue $(\cosh\varphi - \sinh\varphi)$ for the operator $\mathcal{B}_\varphi^{\mathbf{i}}$.

Note that $\mathbf{v}$ and $\overline{\mathbf{v}}$ are orthogonal minimal tripotents and that $\mathbf{i}$ belongs to the Peirce 1/2 part of each of them. Thus, the Peirce decomposition

$$\mathcal{S}^3 = \mathcal{S}^3_1(\mathbf{v}) + \mathcal{S}^3_{1/2}(\mathbf{v}) + \mathcal{S}^3_0(\mathbf{v}) \tag{3.143}$$

is a decomposition of $\mathcal{S}^3$ into the eigenspaces of both $\mathcal{R}^{\mathbf{i}}_\varphi$ and $\mathcal{B}^{\mathbf{i}}_\varphi$. Note also that $\mathbf{v}$ and $\overline{\mathbf{v}}$ are photon helicity eigenstates (see [71] v.I, p.359) corresponding to circular polarization.

3.5.4 The electromagnetic field and $\mathcal{S}^3$.

The transformation of the electromagnetic field intensity E', B' from an inertial system K' to an inertial system K, moving with velocity $\mathbf{v} = (v, 0, 0)$ with respect to K', was discussed in Chapter 1 and is given by (1.121) as

$$E_1 = E'_1,\ E_2 = \frac{E'_2 - vB'_3}{\sqrt{1-\frac{v^2}{c^2}}},\ E_3 = \frac{E'_3 + vB'_2}{\sqrt{1-\frac{v^2}{c^2}}}, \tag{3.144}$$

and

$$B_1 = B'_1,\ B_2 = \frac{B'_2 + \frac{v}{c^2}E'_3}{\sqrt{1-\frac{v^2}{c^2}}},\ B_3 = \frac{B'_3 - \frac{v}{c^2}E'_2}{\sqrt{1-\frac{v^2}{c^2}}}, \tag{3.145}$$

where E and B have coordinates E_1, E_2, E_3 and B_1, B_2, B_2, respectively.

It was shown in section 1.5.4 that if $F = E + icB$, then

$$F^2 = F_1^2 + F_2^2 + F_3^2 = |E|^2 - c^2|B|^2 + 2ic\langle E|B\rangle \tag{3.146}$$

is invariant under the transition from one inertial frame to another. The above transformations of electric and magnetic fields take the form

$$F_1 = F'_1,\ F_2 = F'_2\cosh\varphi + iF'_3\sinh\varphi,\ F_3 = -iF'_2\sinh\varphi + F'_3\cosh\varphi.$$

Hence, the matrix of this transformation is $\exp(\pi^3_s(K_1))$, which is (3.142) where $\tanh\varphi = v/c$. Thus, it is natural to represent the electromagnetic field intensity as a vector F in C^3, in which case the Lorentz group acts by linear maps which are described by matrices of type (3.141) and (3.142) and preserve the value of F^2.

Consider now elements of the spin factor $\mathcal{S}^3$ as representing the vector $F = E + icB$ of the electromagnetic field strength. Let us understand first the physical meaning of multiples of minimal tripotents. Using singular decomposition, each element can be decomposed as a linear combination of two orthogonal multiples of minimal tripotents. The dual of such a tripotent is an extreme point of the state space. Since the Lorentz group representation π^3_s acts on $\mathcal{S}^3$ by operators which preserve the determinant, it follows that a multiple of a minimal tripotent remains a multiple of a minimal tripotent in any inertial system.

In the current interpretation, a multiple $k\mathbf{v}$ of a minimal tripotent $\mathbf{v}$ can be decomposed as $k\mathbf{v} = \mathbf{y} + i\mathbf{z}$, representing an electric field $E = \mathbf{y}$ and a magnetic field $cB = \mathbf{z}$, in which E and B are perpendicular and $|E| = c|B|$. Motion in such a field was described in Case 2 of section 2.6.3 of Chapter 2. From (3.144) and (3.145), it follows that only in this case will a charged particle moving with the speed of light in the direction $E \times B$ (the x-direction in our example) experience no field. Thus, it is natural to associate with a coherent *ray of photons* (which has no charge of its own and thus zero electromagnetic field in its own frame) a field of this type, where $E \times B$ is the direction of propagation of the ray. Such a ray is an eigenstate of both the rotation $\mathcal{R}_{\varphi}^{\mathbf{i}}$ about the x-axis and the boost $\mathcal{B}_{\varphi}^{\mathbf{i}}$ in the x-direction. It is also a helicity eigenstate and has definite circular polarization.

If two multiples of minimal tripotents are orthogonal to each other, then each one is the complex conjugate of the other. This means that the photons which they are representing move in opposite directions and have the same wavelength and complementary polarization.

A multiple $k\mathbf{u}$ of a maximal tripotent $\mathbf{u} = e^{i\varphi}\mathbf{r}$ represents an electric field $E = k\cos\varphi\,\mathbf{r}$ and magnetic field $B = c^{-1}k\sin\varphi\,\mathbf{r}$. This means that the electric and magnetic fields are parallel. Motion in such a field was studied in section 1.5.6 of Chapter 1.

A general field $F = E + icB$ can be decomposed by singular decomposition (3.34) as a sum of multiples of two orthogonal minimal tripotents. By the above interpretation, the field can be represented as a combination of two rays of photons moving in opposite directions with the complementary polarization.

3.5.5 Spin 1 representations of the Lorentz group on Dinv ($\mathcal{S}^4$)

The three-dimensional linear representation π_s^3 was defined on the rotation group and extended to the full Lorentz group. In this subsection, we construct a four-dimensional linear representation π_s^4 of the Lorentz group which also extends the linear representation π_s^3 of the rotation group. We will show that π_s^4 is equivalent to the standard representation of the Lorentz group on space-time.

We shall again let $J_1, J_2, J_3, K_1, K_2, K_3$ be the standard infinitesimal generators of the Lorentz Lie algebra. We also let $\mathbf{u}_0, \mathbf{u}_1, \mathbf{u}_2, \mathbf{u}_3$ denote a TCAR basis of $\mathcal{S}^4$ and define a representation π_s^4 from the Lorentz Lie algebra to the set of operators on $\mathcal{S}^4$. For ease of notation, we shall write D_{jk} instead of $D(\mathbf{u}_j, \mathbf{u}_k)$, for $j, k \in \{0, 1, 2, 3\}$, $j \neq k$. Note that here, $D(\mathbf{u}_2, \mathbf{u}_3)$ acts on the space spanned by $\mathbf{u}_0, \mathbf{u}_1, \mathbf{u}_2, \mathbf{u}_3$, whereas in (3.131), it acts on the span of $\mathbf{u}_1, \mathbf{u}_2, \mathbf{u}_3$. We define π_s^4 by

$$\pi_s^4(J_1) = D_{23},\ \pi_s^4(J_2) = D_{31},\ \pi_s^4(J_3) = D_{12}, \tag{3.147}$$

$$\pi_s^4(K_1) = iD_{01},\ \pi_s^4(K_2) = iD_{02},\ \pi_s^4(K_3) = iD_{03}. \tag{3.148}$$

To show that π_s^4 is a representation of the Lorentz group, it is enough to check that it satisfies the commutation relations (3.133)–(3.137). To do this, we need the following generalization of the main identity (3.53). By use of polarization, we can rewrite this identity as

$$D(\mathbf{x},\mathbf{y})\{\mathbf{a},\mathbf{b},\mathbf{c}\} = \{D(\mathbf{x},\mathbf{y})\mathbf{a},\mathbf{b},\mathbf{c}\} - \{\mathbf{a},D(\mathbf{y},\mathbf{x})\mathbf{b},\mathbf{c}\} + \{\mathbf{a},\mathbf{b},D(\mathbf{x},\mathbf{y})\mathbf{c}\}, \tag{3.149}$$

which is equivalent to

$$D(\mathbf{x},\mathbf{y})D(\mathbf{a},\mathbf{b})\mathbf{c} = D(D(\mathbf{x},\mathbf{y})\mathbf{a},\mathbf{b})\mathbf{c} - D(\mathbf{a},D(\mathbf{y},\mathbf{x})\mathbf{b})\mathbf{c} + D(\mathbf{a},\mathbf{b})D(\mathbf{x},\mathbf{y})\mathbf{c}.$$

This implies that

$$[D(\mathbf{x},\mathbf{y}),D(\mathbf{a},\mathbf{b})] = D(D(\mathbf{x},\mathbf{y})\mathbf{a},\mathbf{b}) - D(\mathbf{a},D(\mathbf{y},\mathbf{x})\mathbf{b}). \tag{3.150}$$

Using this identity and the TCAR relations (3.14)–(3.16), it is easy to verify the commutation relations. For example,

$$[\pi_s^4(J_1),\pi_s^4(K_2)] = i[D_{23},D_{02}]$$

$$= iD(D(\mathbf{u}_2,\mathbf{u}_3)\mathbf{u}_0,\mathbf{u}_2) - iD(\mathbf{u}_0,D(\mathbf{u}_3,\mathbf{u}_2)\mathbf{u}_2) = -iD_{03} = -\pi_s^4(K_3).$$

An additional example is

$$[\pi_s^4(K_1),\pi_s^4(K_2)] = -[D_{01},D_{02}]$$

$$= -D(D(\mathbf{u}_0,\mathbf{u}_1)\mathbf{u}_0,\mathbf{u}_2) + D(\mathbf{u}_0,D(\mathbf{u}_1,\mathbf{u}_0)\mathbf{u}_2) = D_{12} = \pi_s^4(J_3).$$

It is straightforward to check the remaining relations. Thus, π_s^4 is a representation of the Lie algebra of the Lorentz group into dinv $(\mathcal{S}^4)$.

The representation of the Lorentz group by elements of Dinv $(\mathcal{S}^4)$ is obtained by taking the exponent of the basic elements of π_s^4. For example, $\exp(\varphi\pi_s^4(K_1))$ with respect to the basis $\{\mathbf{u}_0,\mathbf{u}_1,\mathbf{u}_2,\mathbf{u}_3\}$ is, by use of (3.140),

$$\exp(\varphi\pi_s^4(K_1)) = \begin{pmatrix} \cosh\varphi & i\sinh\varphi & 0 & 0 \\ -i\sinh\varphi & \cosh\varphi & 0 & 0 \\ 0 & 0 & 1 & 0 \\ 0 & 0 & 0 & 1 \end{pmatrix}. \tag{3.151}$$

Similarly, we have

$$\exp(\varphi\pi_s^4(J_1)) = \begin{pmatrix} 1 & 0 & 0 & 0 \\ 0 & 1 & 0 & 0 \\ 0 & 0 & \cos\varphi & \sin\varphi \\ 0 & 0 & -\sin\varphi & \cos\varphi \end{pmatrix}.$$

3.5.6 Invariant subspaces of π_s^4

Unlike the representation π_s^3, which is onto dinv$(\mathcal{S}^3)$, the representation π_s^4 is not onto dinv$(\mathcal{S}^4)$. This follows from the fact that dinv$(\mathcal{S}^4)$, which is identical to $A_4(C)$, is of complex dimension 6, while the Lorentz group is of real dimension 6. Therefore, we want to find a subspace M_1 of $\mathcal{S}^4$ which is invariant under the representation π_s^4. Then the representation of the Lie algebra of the Lorentz group is by elements of dinv$(\mathcal{S}^4)$, which are vector fields generating flows on $\mathcal{S}^4$ that preserve the determinant and map the subspace M_1 into itself.

It is easy to check that the subspace

$$M_1 = \{(x^0, x^1, x^2, x^3) = x^\nu \mathbf{u}_\nu \in \mathcal{S}^4 : \ x^0 \in R, \ x^1, x^2, x^3 \in iR\} \tag{3.152}$$

is invariant under π_s^4. To see this, it is enough to check that the maps $\exp(\varphi\pi_s^4(J_k))$ and $\exp(\varphi\pi_s^4(K_k))$, for $k = 1, 2, 3$, keep M_1 invariant. We can attach the following meaning to the subspace M_1. Let M be the Minkowski space representing the space-time coordinates (t, x, y, z) of an event in an inertial system. Define a map $\Psi : M \to M_1$ by

$$\Psi(t, x, y, z) = ct\mathbf{u}_0 - ix\mathbf{u}_1 - iy\mathbf{u}_2 - iz\mathbf{u}_3. \tag{3.153}$$

We use the minus sign for the space coordinates in order that the resulting Lorentz transformations will have their usual form. Note that

$$\det(\Psi(t, x, y, z)) = (ct)^2 - x^2 - y^2 - z^2 = s^2,$$

where s is the space-time interval.

Any map $T \in$ Dinv$(\mathcal{S}^4)$ which maps M_1 into itself generates an interval-preserving map

$$\Lambda = \Psi^{-1} T \Psi \tag{3.154}$$

from M to M. Thus, any map T from π_s^4 generates by (3.154) a space-time Lorentz transformation. For example, if $T = \exp\varphi\pi_s^4(K_1)$, then

$$\Lambda(t, x, y, z) = \begin{pmatrix} \cosh\varphi & c^{-1}\sinh\varphi & 0 & 0 \\ c\sinh\varphi & \cosh\varphi & 0 & 0 \\ 0 & 0 & 1 & 0 \\ 0 & 0 & 0 & 1 \end{pmatrix} \begin{pmatrix} t \\ x \\ y \\ z \end{pmatrix},$$

which is the usual *Lorentz space-time transformation* for the boost in the x-direction, where $\tanh\varphi = \mathbf{v}/c$, and $\mathbf{v}$ is the relative velocity between the systems. Conversely, any space-time Lorentz transformation Λ generates a transformation $T = \Psi\Lambda\Psi^{-1}$ on M_1 which can be extended linearly to a map on $\mathcal{S}^4$ which belongs to π_s^4. Thus, the usual Lorentz space-time transformation is equivalent to a representation π_s^4, and π_s^4 can be considered as an extension of the usual representation of the Lorentz group from space-time to $\mathcal{S}^4$.

In addition to the subspace M_1, the representation π_s^4 preserves the subspace

$$M_2 = \{(p_0, p_1, p_2, p_3) = p_\nu \mathbf{u}_\nu \in \mathcal{S}^4 : \ p_0 \in iR, \ \ p_1, p_2, p_3 \in R\}, \tag{3.155}$$

which is complementary to M_1. We can attach the following meaning to the subspace M_2. Let $\widetilde{M}$ be the Minkowski space representing the four-vector momentum $(p_0, p_1, p_2, p_3) = m_0(c\gamma, \gamma\mathbf{v})$, where m_0 is the rest-mass and $(\gamma, \gamma\mathbf{v}/c)$ is the four-velocity of the object. Define a map $\widetilde{\Psi} : \widetilde{M} \to M_2$ by

$$\widetilde{\Psi}(p_0, p_1, p_2, p_3) = i\, p_0\mathbf{u}_0 + p_1\mathbf{u}_1 + p_2\mathbf{u}_2 + p_3\mathbf{u}_3. \tag{3.156}$$

Note that

$$\det(\widetilde{\Psi}(p_0, p_1, p_2, p_3)) = -p_0^2 + p_1^2 + p_2^2 + p_3^2 = -(E/c)^2 + \mathbf{p}^2 = -(m_0 c)^2$$

is invariant under the Lorentz transformations.

Any linear map T on $\mathcal{S}^4$ which maps M_2 into itself generates a map

$$T|_{\widetilde{M}} = \widetilde{\Psi}^{-1} T \widetilde{\Psi} \tag{3.157}$$

from $\widetilde{M}$ to $\widetilde{M}$. If $T \in \mathrm{Dinv}\,(\mathcal{S}^4)$, then the map $T|_{\widetilde{M}}$ is a Lorentz transformation on the four-vector momentum space. Thus, any map T from π_s^4 generates a Lorentz transformation on the four-vector momentum space. For example, if $T = \exp(\varphi\pi_s^4(K_1))$, then $\widetilde{\Lambda} := T|_{\widetilde{M}} = \exp(\varphi\pi_s^4(K_1))|_{\widetilde{M}}$ is

$$\widetilde{\Lambda}\begin{pmatrix} p_0 \\ p_1 \\ p_2 \\ p_3 \end{pmatrix} = \begin{pmatrix} \cosh\varphi & \sinh\varphi & 0 & 0 \\ \sinh\varphi & \cosh\varphi & 0 & 0 \\ 0 & 0 & 1 & 0 \\ 0 & 0 & 0 & 1 \end{pmatrix}\begin{pmatrix} p_0 \\ p_1 \\ p_2 \\ p_3 \end{pmatrix}. \tag{3.158}$$

Conversely, any four-vector momentum Lorentz transformation $\widetilde{\Lambda}$ generates a transformation $T = \widetilde{\Psi}\widetilde{\Lambda}\widetilde{\Psi}^{-1}$ on M_2 which can be extended linearly to a map T on $\mathcal{S}^4$ which belongs to π_s^4. Thus, the usual Lorentz four-vector momentum transformation is equivalent to a representation π_s^4, and π_s^4 can be considered as an extension of the usual representation of the Lorentz group from four-vector momentum space to $\mathcal{S}^4$.

3.5.7 Transformation of the electromagnetic field tensor under π_s^4

We can now use (3.157) to define the restriction of the generator of a boost $\pi_s^4(K_1)$ in the x-direction on the four-momentum space $\widetilde{M}$. We have

$$\pi_s^4(K_1)|_{\widetilde{M}}(p_0, p_1, p_2, p_3) = \widetilde{\Psi}^{-1}\pi_s^4(K_1)\widetilde{\Psi}(p_0, p_1, p_2, p_3) =$$

$$\widetilde{\Psi}^{-1} i D_{01}(ip_0\mathbf{u}_0 + p_1\mathbf{u}_1 + p_2\mathbf{u}_2 + p_3\mathbf{u}_3) = \widetilde{\Psi}^{-1}(ip_1\mathbf{u}_0 + p_0\mathbf{u}_1) = (p_1, p_0, 0, 0).$$

In matrix notation, we get

$$\pi_s^4(K_1)|_{\widetilde{M}}\begin{pmatrix}p_0\\p_1\\p_2\\p_3\end{pmatrix}=\begin{pmatrix}0&1&0&0\\-1&0&0&0\\0&0&0&0\\0&0&0&0\end{pmatrix}\begin{pmatrix}p_0\\p_1\\p_2\\p_3\end{pmatrix}.$$

Thus, $\pi_s^4(K_1)|_{\widetilde{M}}$ acts like the electromagnetic tensor for an electric field in the x-direction. Such a field generates a boost in the x-direction.

A similar calculation for $\pi_s^4(J_1)|_{\widetilde{M}}$ gives

$$\pi_s^4(J_1)|_{\widetilde{M}}\begin{pmatrix}p_0\\p_1\\p_2\\p_3\end{pmatrix}=\begin{pmatrix}0&0&0&0\\0&0&0&0\\0&0&0&1\\0&0&-1&0\end{pmatrix}\begin{pmatrix}p_0\\p_1\\p_2\\p_3\end{pmatrix},$$

which is the action of the magnetic field. Thus, the electromagnetic strength tensor F for an electromagnetic field $\mathbf{E},\mathbf{B}$ on $\widetilde{M}$ is represented by

$$F=E_k\pi_s^4(K_k)|_{\widetilde{M}}+cB_k\pi_s^4(J_k)|_{\widetilde{M}}=\begin{pmatrix}0&E_1&E_2&E_3\\-E_1&0&cB_3&-cB_2\\-E_2&-cB_3&0&cB_1\\-E_3&cB_2&-cB_1&0\end{pmatrix}. \qquad (3.159)$$

This is very natural, since the electric field generates boosts and the magnetic field generates rotations.

We are now able to obtain the transformation of the electromagnetic strength tensor F under the Lorentz transformations $\widetilde{\Lambda}$ on the four-vector momentum. The transformation of this tensor can be calculated by use of the formula

$$F'=\widetilde{\Lambda}F\widetilde{\Lambda}^{-1}, \qquad (3.160)$$

which stems from the following commutative diagram:

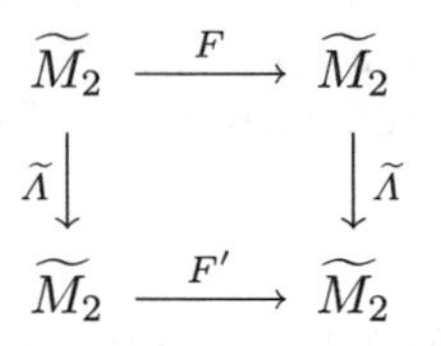

Thus, for example, let $\widetilde{\Lambda}=\exp\varphi\pi_s^4(K_1)|_{\widetilde{M}}$ be the boost in the x-direction with speed v, and set $\tanh\varphi=v/c$. Then, by use of (3.158), the electromagnetic field tensor F' after the boost is

$$\begin{pmatrix}0&E_1'&E_2'&E_3'\\-E_1'&0&cB_3'&-cB_2'\\-E_2'&-cB_3'&0&cB_1'\\-E_3'&cB_2'&-cB_1'&0\end{pmatrix}=$$

$$\begin{pmatrix} \cosh\varphi & \sinh\varphi & 0 & 0 \\ \sinh\varphi & \cosh\varphi & 0 & 0 \\ 0 & 0 & 1 & 0 \\ 0 & 0 & 0 & 1 \end{pmatrix} \begin{pmatrix} 0 & E_1 & E_2 & E_3 \\ -E_1 & 0 & cB_3 & -cB_2 \\ -E_2 & -cB_3 & 0 & cB_1 \\ -E_3 & cB_2 & -cB_1 & 0 \end{pmatrix} \begin{pmatrix} \cosh\varphi & -\sinh\varphi & 0 & 0 \\ -\sinh\varphi & \cosh\varphi & 0 & 0 \\ 0 & 0 & 1 & 0 \\ 0 & 0 & 0 & 1 \end{pmatrix}.$$

This coincides with the transformation of the electromagnetic field strength given by (3.144) and (3.145).

3.5.8 The relativistic phase space

In classical mechanics, we use the phase space, consisting of position and momentum, to describe the state of a system. The results from the previous subsection suggest that $\mathcal{S}^4$ can serve as a relativistic analog of the phase space (by representing space-time and four-momentum on it). Note that any relativistically invariant multiple of four-momentum can be used instead of four-momentum. For instance, we may use four-velocity instead of four-momentum.

In order to allow transformations under which the subspaces M_1 and M_2 are *not* invariant, we have to multiply the four-momentum by a universal constant that will make the units of M_1 equal to the units of M_2. Therefore, we propose two embeddings of the relativistic phase space $\mathcal{S}^4$:

1) The space-momentum embedding:

$$\Omega(t, x, y, z, E/c, p_1, p_2, p_3) = \tag{3.161}$$

$$(ct + \varsigma iE/c)\mathbf{u}_0 + (\varsigma p_1 - ix)\mathbf{u}_1 + (\varsigma p_2 - iy)\mathbf{u}_2 + (\varsigma p_3 - iz)\mathbf{u}_3,$$

where the universal constant ς transforms momentum into length.

2) The space-velocity embedding:

$$\widetilde{\Omega}(t, x, y, z, \gamma, \gamma \mathbf{v}_1/c, \gamma \mathbf{v}_2/c, \gamma \mathbf{v}_3/c) = \tag{3.162}$$

$$(ct + i\varpi\gamma)\mathbf{u}_0 + (\varpi\gamma \mathbf{v}_1/c - ix)\mathbf{u}_1 + (\varpi\gamma \mathbf{v}_2/c - iy)\mathbf{u}_2 + (\varpi\gamma \mathbf{v}_3/c - iz)\mathbf{u}_3,$$

where the universal constant ϖ transforms velocity into length.

We now explain the physical meaning of the tripotent $\mathbf{u} = 0.5(\mathbf{u}_0 + i\mathbf{u}_1)$ in the range of Ω. Since for any λ of modulus 1, $\lambda\mathbf{u}$ is also a tripotent, the coordinates of $\lambda\mathbf{u}$ satisfy

$$ct + \varsigma iE/c = -i(\varsigma p_1 - ix), \quad y = z = p_2 = p_3 = 0.$$

This implies that $ct = -x$ and $E/c = -p_1$. Since $E = m_0\gamma c^2$ and $\mathbf{p} = m_0\gamma\mathbf{v}$, the second equation becomes $c = -v_1$. This is also consistent with the first equation. Thus, the minimal tripotent $\mathbf{u} = 0.5(\mathbf{u}_0 + i\mathbf{u}_1)$ and, similarly, the minimal tripotent $\overline{\mathbf{u}} = 0.5(\mathbf{u}_0 - i\mathbf{u}_1)$, orthogonal to $\mathbf{u}$, represent particles moving in the direction of the x-axis with velocity $\mp c$, respectively. Such particles have rest-mass $m_0 = 0$ and spin 0, and their momentum is expressed in terms of their wavelength. Thus,

$$0.5(\mathbf{u}_0 \pm i\mathbf{u}_1) \Longrightarrow m_0 = 0,\ x = \mp ct,\ p_1 = \mp E/c,\ y = z = p_2 = p_3 = 0. \tag{3.163}$$

Consider now the minimal tripotent $\mathbf{u} = 0.5(\mathbf{u}_2 + i\mathbf{u}_3)$ in the range of Ω. The coordinates of this tripotent satisfy

$$\varsigma p_2 - iy = -i(\varsigma p_3 - iz),\quad t = E = p_1 = x = 0.$$

This implies that $\varsigma p_2 = -z$ and $\varsigma p_3 = y$. Since $\mathbf{p} = m_0\gamma\mathbf{v} = m_0\frac{d\mathbf{v}}{d\tau}$, the state corresponding to $\hat{\mathbf{u}}$ describes a rotation about the x-axis with constant angular velocity. This state behaves like a particle in a magnetic field in the x-direction. But in this case, the meaning of $t = E = 0$ is not clear. Perhaps this state must be considered as a limiting state.

The representation π_s^4 of the Lorentz group has several advantages over π_s^3. In general, a representation of the Lorentz Lie algebra consists of generators of rotations, interpreted as magnetic fields, and generators of boosts, interpreted as electric fields. As we saw in Chapter 1 (1.106), page 39, electric and magnetic fields commute if and only if they are parallel. This implies that the representations of K_k and J_k must commute for each k. Commuting, in general, is described mathematically by orthogonality. Thus, it is preferable to represent the Lie algebra of the Lorentz group by a rank 2 symmetric domain, which is a spin factor. Only for dimension 4 is the Lie algebra dinv $(\mathcal{S}^n)$ a spin factor. Moreover, under the representation π_s^4, for any direction, the generator of a rotation about this direction and the generator of a boost in this direction, defined by (3.147) and (3.148), are orthogonal tripotents. In the next section, we will encounter yet another advantage of π_s^4: modelling spin 1/2 particles on $\mathcal{S}^4$.

3.6 Spin-$\frac{1}{2}$ representation in dinv $(\mathcal{S}^4)$

3.6.1 The representation π^+ on $\mathcal{S}^4$

The elements D_{jk} from dinv $(\mathcal{S}^4)$, together with the triple product (3.111), form a spin grid, defined earlier on page 122. For example, the elements $(D_{01}, D_{02}, D_{23}, D_{31})$ form one of the odd quadrangles of this grid. For elements of an odd quadrangle, one may define a "sharp" operation which sends each minimal tripotent to its orthogonal one. If we choose a TCAR basis in the spin factor dinv $(\mathcal{S}^4)$, then this operation is complex conjugation, since in a TCAR basis, the complex conjugate of a minimal tripotent is an orthogonal minimal tripotent. The sharp operation is defined as follows:

$$D_{01}^\sharp = D_{23},\ D_{02}^\sharp = D_{31},\ D_{03}^\sharp = D_{12},\ D_{23}^\sharp = D_{01},\ D_{31}^\sharp = D_{02},\ D_{12}^\sharp = D_{03}$$

and is then extended to all of dinv $(\mathcal{S}^4)$ conjugate linearly.

Under the representation π_s^4, constructed in the previous section, $\pi_s^4(J_k)$ and $\pi_s^4(K_k)$ are minimal tripotents of dinv $(\mathcal{S}^4)$. Recall that equation (3.114) transforms a spin grid, which consists of minimal tripotents, into a TCAR basis, which consists of maximal tripotents. In a similar way, we now use the conjugation $\sharp$ to construct, from π_s^4, two representations π^+ and π^-, of the Lorentz group using maximal tripotents. The representation π^+ is defined to be the self-adjoint, or real part of π_s^4 on J_1, J_2, J_3 with respect to $\sharp$, and the skew-adjoint, or imaginary part, of π_s^4 with respect to $\sharp$ on K_1, K_2, K_3. To be precise, we define

$$\pi^+(J_k) = \frac{1}{2}(\pi_s^4(J_k) + \pi_s^4(J_k)^\sharp),$$

$$\pi^+(K_k) = \frac{1}{2}(\pi_s^4(K_k) - \pi_s^4(K_k)^\sharp).$$

Note that $\pi^+(K_k) = i\pi^+(J_k)$ for $k = 1, 2, 3$.

For example, the matrices of $\pi^+(J_1), \pi^+(J_2)$ and $\pi^+(J_3)$ in the basis $\mathbf{u}_0, \mathbf{u}_1, \mathbf{u}_2, \mathbf{u}_3$ are the following multiples of 4×4 orthogonal matrices:

$$\pi^+(J_1) = \frac{1}{2}(D_{01} + D_{23}) = \frac{1}{2}\begin{pmatrix} 0 & 1 & 0 & 0 \\ -1 & 0 & 0 & 0 \\ 0 & 0 & 0 & 1 \\ 0 & 0 & -1 & 0 \end{pmatrix},$$

$$\pi^+(J_2) = \frac{1}{2}(D_{02} + D_{31}) = \frac{1}{2}\begin{pmatrix} 0 & 0 & 1 & 0 \\ 0 & 0 & 0 & -1 \\ -1 & 0 & 0 & 0 \\ 0 & 1 & 0 & 0 \end{pmatrix},$$

$$\pi^+(J_3) = \frac{1}{2}(D_{03} + D_{12}) = \frac{1}{2}\begin{pmatrix} 0 & 0 & 0 & 1 \\ 0 & 0 & 1 & 0 \\ 0 & -1 & 0 & 0 \\ -1 & 0 & 0 & 0 \end{pmatrix}.$$

In order to check that $\pi^+(J_k)$ satisfy the commutation relations (3.133), we calculate the multiplication table for these elements. Direct calculation establishes Table 3.3 for $\pi^+(J_k)$. The constant $\frac{1}{2}$ is necessary if π^+ is to satisfy the commutation relations (3.133). Since $\pi^+(K_k) = i\pi^+(J_k)$, the commutation relations (3.134)–(3.137) also hold, and so π^+ is a representation of the Lorentz group. Note that for the representation π_s^4, the generators of the Lie

	$2\pi^+(J_1)$	$2\pi^+(J_2)$	$2\pi^+(J_3)$
$2\pi^+(J_1)$	$-I$	$-2\pi^+(J_3)$	$2\pi^+(J_2)$
$2\pi^+(J_2)$	$2\pi^+(J_3)$	$-I$	$-2\pi^+(J_1)$
$2\pi^+(J_3)$	$-2\pi^+(J_2)$	$2\pi^+(J_1)$	$-I$

Table 3.3. Multiplication table for $\pi^+(J_k)$

algebra of the Lorentz group are minimal tripotents, while for the representation π^+, the generators are maximal tripotents in $\operatorname{dinv}(\mathcal{S}^4)$. From Table 3.3 and the fact that $\pi^+(J_k)^* = -\pi^+(J_k)$, it follows that

$$\{2\pi^+(J_1), 2\pi^+(J_1), 2\pi^+(J_2)\} =$$

$$\frac{2\pi^+(J_1)\cdot 2\pi^+(J_1)^*\cdot 2\pi^+(J_2) + 2\pi^+(J_2)\cdot 2\pi^+(J_1)^*\cdot 2\pi^+(J_1)}{2} = 2\pi^+(J_2)$$

and

$$\{2\pi^+(J_1), 2\pi^+(J_2), 2\pi^+(J_1)\} = 2\pi^+(J_1)\cdot 2\pi^+(J_2)^*\cdot 2\pi^+(J_1) = -2\pi^+(J_2).$$

Similar identities hold for the other indexes. Thus, the elements

$$\{2\pi^+(J_1), 2\pi^+(J_2), 2\pi^+(J_3)\} \quad \text{satisfy TCAR,} \tag{3.164}$$

as elements of the spin factor $\operatorname{dinv}(\mathcal{S}^4) = A_4(C)$.

Notice also that $\pi^+(J_1), \pi^+(J_2), \pi^+(J_3)$ each have two distinct eigenvalues, namely $\pm\frac{1}{2}$. Note that if j denotes the spin value of a representation, then the number of distinct eigenvalues is $2j+1$. In the above representation, then, we have $j = 1/2$, implying that this is a spin $\frac{1}{2}$ representation. This is confirmed also by direct calculation of the exponent of the generators of rotations. Since $\pi^+(J_k) = -\frac{1}{4}I$, the matrix exponent, defined by (3.140), is

$$\exp(\varphi(\pi^+(J_k))) = \cos\frac{\varphi}{2}I + \sin\frac{\varphi}{2}(\pi^+(J_k)).$$

For example, the matrix of the operator $R_3(\varphi) = \exp(\varphi\pi^+(J_3))$ in the TCAR basis $\{\mathbf{u}_0, \mathbf{u}_1, \mathbf{u}_2, \mathbf{u}_3\}$ is

$$R_3(\varphi) = \exp(\varphi\pi^+(J_3)) = \begin{pmatrix} \cos\frac{\varphi}{2} & 0 & 0 & \sin\frac{\varphi}{2} \\ 0 & \cos\frac{\varphi}{2} & \sin\frac{\varphi}{2} & 0 \\ 0 & -\sin\frac{\varphi}{2} & \cos\frac{\varphi}{2} & 0 \\ -\sin\frac{\varphi}{2} & 0 & 0 & \cos\frac{\varphi}{2} \end{pmatrix}.$$

This shows that the angle of rotation in the representation is half of the actual angle of rotation.

3.6.2 Invariant subspaces of π^+

We will show now that the representation π^+ is a direct sum of two copies of the standard spin $\frac{1}{2}$ representation in terms of the Pauli spin matrices σ_k, $k = 1, 2, 3$. To do this, it suffices to identify the invariant subspaces under the action of the rotations (exponents of $\pi^+(J_k)$) and boosts (exponents of $\pi^+(K_k)$) under this representation. Since $\pi^+(K_k) = i\pi^+(J_k)$, it suffices to consider only the restriction of π^+ to rotations $\exp(\pi^+(J_k))$. Moreover, since $(\pi^+(J_k))^2 = -I$, the $\exp(\pi^+(J_k))$ belongs to the linear span of I and $\pi^+(J_k)$ for $k = 1, 2, 3$. Thus, it is enough to find the subspaces of $\mathcal{S}^4$ which are invariant under all $\pi^+(J_k)$.

Let us replace the TCAR basis $\mathbf{u}_0, \mathbf{u}_1, \mathbf{u}_2, \mathbf{u}_3$ in $\mathcal{S}^4$, which consists of maximal tripotents, with a basis $\mathbf{v}_1, \mathbf{v}_2, \mathbf{v}_{-1}, \mathbf{v}_{-2}$ of minimal tripotents which form an odd quadrangle, as in section 3.3.6. This basis is defined by

$$\mathbf{v}_{\pm 1} = 0.5(\mathbf{u}_0 \pm i\mathbf{u}_3), \quad \mathbf{v}_{\pm 2} = 0.5(\mathbf{u}_2 \pm i\mathbf{u}_1).$$

Using the definition of $\pi^+(J_k)$ and the TCAR relations (3.14)–(3.16), we get

$$\pi^+(J_1)\mathbf{v}_{\pm 1} = \frac{1}{4}(D_{01} + D_{23})(\mathbf{u}_0 \pm i\mathbf{u}_3) = \frac{1}{4}(-\mathbf{u}_1 \pm i\mathbf{u}_2) = \pm\frac{1}{2}i\mathbf{v}_{\pm 2},$$

$$\pi^+(J_1)\mathbf{v}_{\pm 2} = \frac{1}{4}(D_{01} + D_{23})(\mathbf{u}_2 \pm i\mathbf{u}_1) = \frac{1}{4}(-\mathbf{u}_3 \pm i\mathbf{u}_0) = \pm\frac{1}{2}i\mathbf{v}_{\pm 1},$$

$$\pi^+(J_2)\mathbf{v}_{\pm 1} = \frac{1}{4}(D_{02} + D_{31})(\mathbf{u}_0 \pm i\mathbf{u}_3) = \frac{1}{4}(-\mathbf{u}_2 \mp i\mathbf{u}_1) = -\frac{1}{2}\mathbf{v}_{\pm 2},$$

$$\pi^+(J_2)\mathbf{v}_{\pm 2} = \frac{1}{4}(D_{02} + D_{31})(\mathbf{u}_2 \pm i\mathbf{u}_1) = \frac{1}{4}(\mathbf{u}_0 \pm i\mathbf{u}_3) = \frac{1}{2}\mathbf{v}_{\pm 1},$$

$$\pi^+(J_3)\mathbf{v}_{\pm 1} = \frac{1}{4}(D_{03} + D_{12})(\mathbf{u}_0 \pm i\mathbf{u}_3) = \frac{1}{4}(-\mathbf{u}_3 \pm i\mathbf{u}_0) = \pm\frac{1}{2}i\mathbf{v}_{\pm 1},$$

$$\pi^+(J_3)\mathbf{v}_{\pm 2} = \frac{1}{4}(D_{03} + D_{12})(\mathbf{u}_2 \pm i\mathbf{u}_1) = \frac{1}{4}(\mathbf{u}_1 \mp i\mathbf{u}_2) = \mp\frac{1}{2}i\mathbf{v}_{\pm 2}.$$

Thus, in the basis $\mathbf{v}_1, \mathbf{v}_2, \mathbf{v}_{-1}, \mathbf{v}_{-2}$ of minimal tripotents, the matrices of $\pi^+(J_k)$ are

$$\pi^+(J_1) = \frac{1}{2}\begin{pmatrix} 0 & i & 0 & 0 \\ i & 0 & 0 & 0 \\ 0 & 0 & 0 & -i \\ 0 & 0 & -i & 0 \end{pmatrix},$$

$$\pi^+(J_2) = \frac{1}{2}\begin{pmatrix} 0 & 1 & 0 & 0 \\ -1 & 0 & 0 & 0 \\ 0 & 0 & 0 & 1 \\ 0 & 0 & -1 & 0 \end{pmatrix},$$

$$\pi^+(J_3) = \frac{1}{2}\begin{pmatrix} i & 0 & 0 & 0 \\ 0 & -i & 0 & 0 \\ 0 & 0 & -i & 0 \\ 0 & 0 & 0 & i \end{pmatrix}.$$

This shows that the following two subspaces

$$\Upsilon_1 = \mathrm{sp}_C\{\mathbf{v}_1, \mathbf{v}_2\} \quad \text{and} \quad \Upsilon_2 = \mathrm{sp}_C\{\mathbf{v}_{-1}, \mathbf{v}_{-2}\}$$

are invariant under the representation π^+. Note that from section 3.5.8, it follows that Υ_1 and Υ_2 represent differently polarized states in parallel electric and magnetic fields in the z-direction.

As we have seen, the representation π^+ leaves the two two-dimensional complex subspaces Υ_1 and Υ_2 invariant, and thus we obtain two two-dimensional representations of the Lorentz group. These representations are related to the Pauli matrices as follows:

$$\pi^+(J_1)|_{\Upsilon_1} = \frac{1}{2}\begin{pmatrix} 0 & i \\ i & 0 \end{pmatrix} = i\sigma_1, \quad \pi^+(J_1)|_{\Upsilon_2} = \frac{1}{2}\begin{pmatrix} 0 & -i \\ -i & 0 \end{pmatrix} = -i\sigma_1. \tag{3.165}$$

Similarly,

$$\pi^+(J_2)|_{\Upsilon_1} = \pi^+(J_2)|_{\Upsilon_2} = \frac{1}{2}\begin{pmatrix} 0 & 1 \\ -1 & 0 \end{pmatrix} = i\sigma_2, \tag{3.166}$$

and

$$\pi^+(J_3)|_{\Upsilon_1} = \frac{1}{2}\begin{pmatrix} i & 0 \\ 0 & -i \end{pmatrix} = i\sigma_3, \quad \pi^+(J_3)|_{\Upsilon_2} = \frac{1}{2}\begin{pmatrix} -i & 0 \\ 0 & i \end{pmatrix} = -i\sigma_3. \tag{3.167}$$

Hence, π^+ defines the usual spin $\frac{1}{2}$ representation on the subspace Υ_1 via the Pauli matrices. This means that $\xi_1\mathbf{v}_1 + \xi_2\mathbf{v}_2$ forms a *spinor*. On the subspace Υ_2, the representation π^+ acts by complex conjugation on the usual spin $\frac{1}{2}$ representation. Hence $\eta_1\mathbf{v}_{-1} + \eta_2\mathbf{v}_{-2}$ forms a *dotted spinor*. This is similar to the action of the Lorentz group on Dirac bispinors, and so the basis $\{\mathbf{v}_1, \mathbf{v}_2, \mathbf{v}_{-1}, \mathbf{v}_{-2}\}$ of $\mathcal{S}^4$ can serve as a basis for bispinors. Note that the TCAR basis $\{\mathbf{u}_0, \mathbf{u}_1, \mathbf{u}_2, \mathbf{u}_3\}$, on the other hand, is a basis for four-vectors. For a possible interpretation of the basis in Υ_1, Υ_2, see section 3.5.8.

3.6.3 The representation π^-

In addition to the representation π^+ on $\mathcal{S}^4$ by maximal tripotents in $\mathrm{dinv}(\mathcal{S}^4)$, we define a related representation π^-, which is the skew-adjoint part of π^4_s

with respect to $\sharp$ on J_1, J_2, J_3 and the self-adjoint part of π_s^4 with respect to $\sharp$ on K_1, K_2, K_3. Thus,

$$\pi^-(J_k) = \frac{1}{2}(\pi_s^4(J_k) - \pi_s^4(J_k)^\sharp), \quad \pi^-(K_k) = \frac{1}{2}(\pi_s^4(K_k) + \pi_s^4(K_k)^\sharp).$$

As for π^+, we have $\pi^-(K_k) = i\pi^-(J_k)$ for any $k = 1, 2, 3$. The connection between the representations π^+ and π^- is the same as obtained on the electromagnetic tensor by lowering the indices (see [51]).

The matrices of $\pi^-(J_1), \pi^-(J_2)$ and $\pi^-(J_3)$ in the basis $\mathbf{u}_0, \mathbf{u}_1, \mathbf{u}_2, \mathbf{u}_3$ are

$$\pi^-(J_1) = \frac{1}{2}(D_{23} - D_{01}) = \frac{1}{2}\begin{pmatrix} 0 & -1 & 0 & 0 \\ 1 & 0 & 0 & 0 \\ 0 & 0 & 0 & 1 \\ 0 & 0 & -1 & 0 \end{pmatrix},$$

$$\pi^-(J_2) = \frac{1}{2}(D_{31} - D_{02}) = \frac{1}{2}\begin{pmatrix} 0 & 0 & -1 & 0 \\ 0 & 0 & 0 & -1 \\ 1 & 0 & 0 & 0 \\ 0 & 1 & 0 & 0 \end{pmatrix},$$

$$\pi^-(J_3) = \frac{1}{2}(D_{12} - D_{03}) = \frac{1}{2}\begin{pmatrix} 0 & 0 & 0 & -1 \\ 0 & 0 & 1 & 0 \\ 0 & -1 & 0 & 0 \\ 1 & 0 & 0 & 0 \end{pmatrix}.$$

Comparing these matrices with the corresponding matrices for the representation π^+ from section 3.6.1, we see that π^- can be obtained from π^+ by space inversion or by time inversion. Conversely, π^+ can be obtained from π^- by space or time inversion.

The operators $\pi^-(J_k)$ have a multiplication table similar to Table 3.3. Thus, the $\pi^-(J_k)$ satisfy the commutation relations (3.133). The commutation relations (3.134)–(3.137) hold also for π^-. Hence, π^- is a representation of the Lorentz group. Also here, the elements

$$\{2\pi^-(J_1), 2\pi^-(J_3), 2\pi^-(J_3)\} \text{ satisfy TCAR,} \tag{3.168}$$

as elements of the spin factor dinv $(\mathcal{S}^4) = A_4(C)$. Moreover,

$$\{2\pi^+(J_1), 2\pi^+(J_2), 2\pi^+(J_3), 2\pi^-(K_1), 2\pi^-(K_2), 2\pi^-(K_3)\} \tag{3.169}$$

is a TCAR basis of dinv $(\mathcal{S}^4) = A_4(C)$.

3.6.4 The representation π^- in the basis of minimal tripotents

Let us represent the matrices of $\pi^-(J_k)$ in the basis of minimal tripotents $\mathbf{v}_{\pm 1}, \mathbf{v}_{\pm 2}$ defined above. To do this, we have to calculate the action of these matrices on this basis. Modifying the calculations in section 3.6.2, we obtain

$$\pi^-(J_1)\mathbf{v}_{\pm 1} = \pm\frac{1}{2}i\mathbf{v}_{\mp 2}, \ \pi^-(J_1)\mathbf{v}_{\pm 2} = \mp\frac{1}{2}i\mathbf{v}_{\mp 1},$$

$$\pi^-(J_2)\mathbf{v}_{\pm 1} = \frac{1}{2}\mathbf{v}_{\mp 2}, \ \pi^-(J_2)\mathbf{v}_{\pm 2} = -\frac{1}{2}\mathbf{v}_{\mp 1},$$

$$\pi^-(J_3)\mathbf{v}_{\pm 1} = \mp\frac{1}{2}i\mathbf{v}_{\pm 1}, \ \pi^-(J_3)\mathbf{v}_{\pm 2} = \mp\frac{1}{2}i\mathbf{v}_{\pm 2}.$$

Hence, the two subspaces

$$\widetilde{\Upsilon}_1 = \mathrm{sp}_C\{\mathbf{v}_{-2}, \mathbf{v}_1\} \quad \text{and} \quad \widetilde{\Upsilon}_2 = \mathrm{sp}_C\{\mathbf{v}_{-1}, \mathbf{v}_2\}$$

are invariant under the representation π^-. Note that $\widetilde{\Upsilon}_1, \widetilde{\Upsilon}_2$ are obtained from the same spin grid that was used for defining the invariant subspaces Υ_1, Υ_2 of the representation π^+. In both cases, the invariant subspaces are obtained by partitioning the set of four elements of the spin grid into two pairs of non-orthogonal tripotents. Both possible partitions are realized in the representations π^+ and π^-.

In the basis $\mathbf{v}_1, \mathbf{v}_2, \mathbf{v}_{-1}, \mathbf{v}_{-2}$ of minimal tripotents, the matrices of $\pi^-(J_k)$ become

$$\pi^-(J_1) = \frac{1}{2}\begin{pmatrix} 0 & 0 & 0 & i \\ 0 & 0 & -i & 0 \\ 0 & -i & 0 & 0 \\ i & 0 & 0 & 0 \end{pmatrix},$$

$$\pi^-(J_2) = \frac{1}{2}\begin{pmatrix} 0 & 0 & 0 & -1 \\ 0 & 0 & 1 & 0 \\ 0 & -1 & 0 & 0 \\ 1 & 0 & 0 & 0 \end{pmatrix},$$

$$\pi^-(J_3) = \frac{1}{2}\begin{pmatrix} -i & 0 & 0 & 0 \\ 0 & -i & 0 & 0 \\ 0 & 0 & i & 0 \\ 0 & 0 & 0 & i \end{pmatrix}.$$

The representation π^- is also a spin $\frac{1}{2}$ representation. The restriction of π^- to the invariant subspaces $\widetilde{\Upsilon}_1$ and $\widetilde{\Upsilon}_2$ leads to the same Pauli spin matrices as in (3.165)–(3.167). Thus, the representation π^- is a direct sum of two complex conjugate copies of the spin $\frac{1}{2}$ two-dimensional representation given by the Pauli spin matrices. Hence, π^- is a representation of the Lorentz group on the Dirac bispinors.

By direct verification, we can show that the commutant of $\{\pi^+(J_k) : k = 1,2,3\}$ is $\mathrm{sp}_C[\{\pi^-(J_k) : k = 1,2,3\} \cup \{I\}]$, which, when restricted to real scalars, is a four-dimensional associative algebra isomorphic to the quaternions. This can be seen by examining the multiplication Table 3.3. The commutant of $\{\pi^-(J_k) : k = 1,2,3\}$ is $\mathrm{sp}_C[\{\pi^+(J_k) : k = 1,2,3\} \cup \{I\}]$, which is also, after restriction to real scalars, isomorphic to the quaternions. Thus, the two representations π^+ and π^- commute.

The above construction of the representations π^+ and π^- from the representation π_s^4 can also be done via the *Hodge operator*, also called the star operator. Borrowing the definition from the theory of differential forms, we define

$$*D_{jk} = \epsilon_{jklm} g^{jj} g^{kk} D_{lm},$$

where $\{j,k,l,m\} = \{0,1,2,3\}$; g^{pq} is the Lorentz metric: $g^{00} = 1,\ g^{kk} = -1$ for $k = 1,2,3$, and $g^{pq} = 0$ if $p \neq q$; and ϵ_{jklm} is the signature of the permutation $(j,k,l,m) \mapsto (0,1,2,3)$. Specifically,

$$*D_{01} = D_{23},\ *D_{02} = D_{31},\ *D_{03} = D_{12},$$

$$*D_{23} = -D_{01},\ *D_{31} = -D_{02},\ *D_{12} = -D_{03}.$$

The representation π^+ is then the skew-adjoint part of π_s^4 with respect to the Hodge operator, and the representation π^- is the self-adjoint part of π_s^4 with respect to the Hodge operator.

3.6.5 Action of the representations π^+ and π^- on dinv $(\mathcal{S}^4)$

We now lift the representations π^+ and π^- from $\mathcal{S}^4$ to an action on dinv $(\mathcal{S}^4)$, which can be identified with the spin factor $\mathcal{S}^6$. For a basis, we choose the TCAR basis given by (3.169).

Fix an action Λ on $\mathcal{S}^4$. For any linear operator T on $\mathcal{S}^4$, we define, as in (3.160), a transformation $\Phi(\Lambda)$ by

$$\Phi(\Lambda)T = \Lambda T \Lambda^{-1}.$$

From the definition of Dinv $(\mathcal{S}^4)$, it follows that if $\Lambda, T \in \mathrm{Dinv}\,(\mathcal{S}^4)$, then also $\Phi(\Lambda)T \in \mathrm{Dinv}\,(\mathcal{S}^4)$.

We define the action of the rotations of π^+ on dinv $(\mathcal{S}^4)$ by $R_k(\varphi) = \Phi(\exp(\varphi\pi^+(J_k)))$. Similarly, we define the action of the boosts of π^+ on dinv $(\mathcal{S}^4)$ by $B_k(\varphi) = \Phi(\exp(\varphi\pi^+(K_k)))$. With respect to the basis (3.169) of dinv $(\mathcal{S}^4)$, we get

$$R_1(\varphi) = \begin{pmatrix} 1 & 0 & 0 & 0 \\ 0 & \cos\varphi & \sin\varphi & 0 \\ 0 & -\sin\varphi & \cos\varphi & 0 \\ 0 & 0 & 0 & I_3 \end{pmatrix},$$

$$B_1(\varphi) = \begin{pmatrix} 1 & 0 & 0 & 0 \\ 0 & \cosh\varphi & i\sinh\varphi & 0 \\ 0 & -i\sinh\varphi & \cosh\varphi & 0 \\ 0 & 0 & 0 & I_3 \end{pmatrix},$$

$$R_2(\varphi) = \begin{pmatrix} \cos\varphi & 0 & -\sin\varphi & 0 \\ 0 & 1 & 0 & 0 \\ \sin\varphi & 0 & \cos\varphi & 0 \\ 0 & 0 & 0 & I_3 \end{pmatrix},$$

$$B_2(\varphi) = \begin{pmatrix} \cosh\varphi & 0 & -i\sinh\varphi & 0 \\ 0 & 1 & 0 & 0 \\ i\sinh\varphi & 0 & \cosh\varphi & 0 \\ 0 & 0 & 0 & I_3 \end{pmatrix},$$

$$R_3(\varphi) = \begin{pmatrix} \cos\varphi & \sin\varphi & 0 & 0 \\ -\sin\varphi & \cos\varphi & 0 & 0 \\ 0 & 0 & 1 & 0 \\ 0 & 0 & 0 & I_3 \end{pmatrix},$$

$$B_3(\varphi) = \begin{pmatrix} \cosh\varphi & i\sinh\varphi & 0 & 0 \\ -i\sinh\varphi & \cosh\varphi & 0 & 0 \\ 0 & 0 & 1 & 0 \\ 0 & 0 & 0 & I_3 \end{pmatrix}.$$

This coincides with the spin 1 representation π_s^3 from section 3.5.3 on the spin factor $\mathcal{S}^3$, which is the complex span of $\{\pi^+(J_1), \pi^+(J_2), \pi^+(J_3)\} \in \operatorname{dinv}(\mathcal{S}^4)$.

3.7 Summary of the representations of the Lorentz group on $\mathcal{S}^3$ and $\mathcal{S}^4$

Let us summarize the various representations of the Lorentz group on the spin factors $\mathcal{S}^3$ and $\mathcal{S}^4$. It is important to understand the function of the minimal and maximal tripotents under each representation. We will also give a possible interpretation for the pure states represented by norm 1 functionals corresponding to minimal tripotents. Since for any k, the generator of rotation J_k commutes with the generator of boosts K_k, there are two ways to represent them on $\operatorname{dinv}(\mathcal{S}^n)$. The first one, based on results of section 3.4.2, is to represent them by a pair of orthogonal minimal tripotents. The other way is to define $\pi(K_k) = i\pi(J_k)$.

3.7.1 π_s^3

The representation π_s^3 is a representation of the action of the Lorentz group on $\mathcal{S}^3$. The generators of this representation are elements of $\text{dinv}(\mathcal{S}^3) = A_3(C)$, which as a space is equal to C^3. The triple product is that of a domain of rank 1, of type I, corresponding to 1×3 complex matrices. The generators of rotation are represented by the basis elements of $A_3(C)$. For example, $\pi_s^3(J_1) = D_{23}$ and $\pi_s^3(K_1) = i\pi_s^3(J_1)$. We can associate with the elements of $\mathcal{S}^3$ an electromagnetic field strength $F = E + icB$, and the Lorentz group properly transforms the components of the field under rotations and boosts. A choice of a TCAR basis in $\mathcal{S}^3$ corresponds to the choice of the space axes. The real part of the decomposition with respect to this basis is related to the strength of the electric field, while the complex part of the decomposition is related to the strength of the magnetic field. The extreme points of the state space (unit ball of the dual ball) correspond to helicity eigenstates of a photon.

3.7.2 π_s^4

The representation π_s^4 is a representation of the Lorentz group on $\mathcal{S}^4$. The generators of this representation are elements of $\text{dinv}(\mathcal{S}^4) = A_4(C) = \mathcal{S}^6$. The factor $\mathcal{S}^4$ is the only one among the spin factors $\mathcal{S}^n$ for which the space $\text{dinv}(\mathcal{S}^n)$ is also a spin factor. This allows us to define different representations for the generators of the Lorentz group, based on two types of bases in $\mathcal{S}^6$ — the TCAR basis and the spin grid.

The representation π_s^4 represents the generators of the Lorentz group by minimal tripotents from a spin grid in $\text{dinv}(\mathcal{S}^4) = \mathcal{S}^6$. For example, $\pi_s^4(J_1) = D_{23}$ and $\pi_s^4(K_1) = iD_{01}$. This representation has two invariant subspaces M_1 and M_2, each of which can be associated with four-vectors. The restriction of the representation to the invariant subspaces acts in the same way as the Lorentz group acts on four-vectors. For example, if we interpret M_1 as the space-time continuum, then the restriction of π_s^4 to M_1 is the usual Lorentz space-time transformations.

Similarly, we may interpret M_2 as four-momentum. In this case, the restriction of π_s^4 to M_2 is the usual Lorentz transformations of the four-momentum, and we can interpret the real span of the generators of the group as the electromagnetic field tensor. We propose to interpret $\mathcal{S}^4$ as the relativistic phase space, by representing space-time in M_1 and four-momentum in M_2. At this point, the meaning of the pure states in this representation is not so clear.

Another possibility is to add to the four-momentum description on one invariant subspace an analog of a four-angular-momentum on the other invariant subspace. In this case, the conserved quantity would be J^2, which is a function of the spin of the particle and the zero-component connected with the angular energy of the particle. This interpretation corresponds to the interpretation of electromagnetic field strength $F = E + icB$, where E is the generator of the changes of momentum, and B is the generator for angular momentum changes.

3.7.3 π^+ and π^-

The representations π^+ and π^-, like π_s^4, are representations of the Lorentz group on $\mathcal{S}^4$. The generators of the representation π_s^4 are minimal tripotents in $\operatorname{dinv}(\mathcal{S}^4) = \mathcal{S}^6$. The generators of the representations π^+ and π^- are also represented by tripotents of $\operatorname{dinv}(\mathcal{S}^4) = \mathcal{S}^6$, but they each use only three out of the six elements of a TCAR basis, which consists of maximal tripotents. For example $\pi^+(J_1) = \frac{1}{2}(D_{23} + D_{01})$ and $\pi^-(J_1) = \frac{1}{2}(D_{23} - D_{01})$. For any $k = 1, 2, 3$, we have $\pi^\pm(K_k) = i\pi^\pm(J_k)$. Switching between the representations π^+ and π^- is equivalent to space or time reversal.

Both representations π^+ and π^- have two two-dimensional invariant subspaces. To describe these subspaces, one selects a spin grid basis in $\mathcal{S}^4$, consisting of minimal tripotents. The invariant subspaces are the span of two non-orthogonal tripotents of the grid. There are two possible ways to partition the grid into two such pairs. For each of the two representations, one of the partitions is invariant. We have seen that the action of π^+ and π^- on the invariant subspaces is the action of the Lorentz group on spinors or dotted spinors. This implies that the space $\mathcal{S}^4$ with the spin grid basis can be considered as the Dirac bispinors representing the relativistic state of an electron, while the representations π^+ and π^- properly represent the action of the Lorentz group, including space inversion, on the Dirac bispinors.

The representations π^+ and π^- on $\mathcal{S}^4$ induce an action on $\operatorname{dinv}(\mathcal{S}^4) = \mathcal{S}^6$. The latter can be decomposed into two subspaces, each of them equal to $\mathcal{S}^3$, which are related by space reversal. Both representations induce spin 1 representations on each of these subspaces. More precisely, the group acts on each subspace like the representation π_s^3, which is the way the Lorentz group acts on the electromagnetic field strength $F = E + icB$. Thus, for instance, we can use this space to represent the state of a photon. A major advantage to our approach is that one mathematical object can be used to represent both spin 1 and spin 1/2 particles and can thus serve as a model for supersymmetry.

Another interesting observation is that adding an additional symmetry with respect to the Hodge operator (or the sharp operation) to the representation π_s^4, which is an extension to $\mathcal{S}^4$ of the regular Lorentz group representation on four-vectors, leads to the representations π^+ and π^- on the bispinors. It will be interesting to find the physical meaning of this observation.

3.8 Notes

The complex spin factor as a bounded symmetric domain was introduced be E. Cartan [12]. Another approach to the spinors, which is closer to the one used in this chapter, can be found in Dirac [17]. The connection of the spin domain and the triple product can be found in [69]. The possibility of embedding the spin factor as operators on a Hilbert space is described in [7] and [8].

The approach presented in this chapter is further development of the results of [37], which is a continuation of [15].

In this chapter, we use ideas from Transmission Line Theory (TLT) to show that the classical domains are bounded symmetric domains. TLT studies the transformation of signals in a transmission line with two inputs and two outputs. The theory treats the line as a linear black box. Analysis of the line provides explicit formulas for the analytic automorphisms of the domain which map the origin of the domain to any other given point.

Transmission Line Theory is similar to *scattering theory* in Quantum Field Theory. In scattering theory, the inputs consist of the states of particles before interaction and are considered to be elements of a complex Hilbert space. The inputs are transformed by an interaction, considered as a black box, into output states. Also in this case, it is natural to consider two types of Hilbert spaces H and K. For example, the Hilbert space H may be used to represent bosons, while K represents fermions. Another possibility is to have H represent states of particles and K represent antiparticles. As in Transmission Line Theory, the *S-matrix* is assumed to be linear and even unitary. Until now, we are not aware of any use of the fact that the existence of such a transformation establishes the homogeneity of the domain associated with the above model.

In this chapter, we will also study the Peirce decomposition for JC^*-triples. This provides a model of filtering projections in the quantum measuring process. We will define functions of operators in $L(H,K)$ and describe their derivatives, which are needed for perturbation theory. We will describe the geometry of the unit ball of both a JC^*-triple and its dual. We will also define a natural basis for the classical domains and, for the infinite dimensional case, define the type of closure needed to ensure that the closure of the span of the basis is the entire domain.

4.1 The classical domains and operators between Hilbert spaces

4.1.1 Definition of the classical domains

Let H and K be finite or infinite-dimensional complex Hilbert spaces, endowed with the usual Euclidean norm, denoted by $|\cdot|$. Let a be a linear map from H to K. As usual, we define the *operator norm* of a by

$$\|a\| = \sup\{|a(\xi)| \ : \ \ \xi \in H, \ |\xi| \leq 1\}. \tag{4.1}$$

We say that a is *bounded* if $\|a\| < \infty$. We denote by $L(H,K)$ the space of all bounded linear operators from H to K, with the above operator norm. In the particular case $H = K$, we write $L(H)$ instead of $L(H,H)$.

A *Type I domain* is the unit ball

$$D_1 = \{a \in L(H,K) : \|a\| \leq 1\} \tag{4.2}$$

of $L(H,K)$. In the finite-dimensional case, where, say, $\dim H = n$ and $\dim K = m$, we can identify $a \in L(H,K)$ with an $m \times n$ complex matrix by choosing an orthonormal basis in each Hilbert space. Hence, the dimension of $L(H,K)$ is $m \cdot n$.

The domains of type II and III are subspaces of $L(H)$. To define these domains, we need the idea of conjugation. A conjugate-linear map $Q : H \to H$ is called a *conjugation* if $\|Q\| \leq 1$ and $Q^2 = I$, the identity map on H. Let Q be a conjugation on H. For $a \in L(H)$, we define $a^t \in L(H)$ by

$$a^t = Qa^*Q, \tag{4.3}$$

where a^* is the operator adjoint to a. Recall that the adjoint a^* of an operator $a \in L(H,K)$ is defined as an operator $a^* \in L(K,H)$ such that

$$\langle \xi | a^* \eta \rangle = \langle a\, \xi | \eta \rangle \tag{4.4}$$

for all $\xi \in H,\ \eta \in K$.

For the finite-dimensional case, if $\dim H = n$ and $\dim K = m$, then, by choosing an orthonormal basis $\{\mathbf{e}_1, \mathbf{e}_2, ..., \mathbf{e}_n\}$ of H and $\{\mathbf{f}_1, \mathbf{f}_2, ..., \mathbf{f}_m\}$ of K, we can identify any $a \in L(H,K)$ with an $m \times n$ complex matrix. The entries a_{kj} of this matrix are given by

$$a_{kj} = \langle \mathbf{f}_k | a\mathbf{e}_j \rangle.$$

Then, from (4.4), the entries $(a^*)_{jk}$ of the adjoint operator satisfy $(a^*)_{jk} = \overline{a_{kj}}$, where the bar denotes complex conjugation. If $H = K$ is a finite-dimensional Hilbert space and Q is complex conjugation, the matrix of a^t is the transpose of the matrix of a. To see this, consider the case when H is of dimension 2. Then

$$a^t \begin{pmatrix} z_1 \\ z_2 \end{pmatrix} = Qa^*Q \begin{pmatrix} z_1 \\ z_2 \end{pmatrix} = Q \begin{pmatrix} \overline{a_{11}} & \overline{a_{21}} \\ \overline{a_{12}} & \overline{a_{22}} \end{pmatrix} \begin{pmatrix} \overline{z_1} \\ \overline{z_2} \end{pmatrix} = \begin{pmatrix} a_{11} & a_{21} \\ a_{12} & a_{22} \end{pmatrix} \begin{pmatrix} z_1 \\ z_2 \end{pmatrix}.$$

Thus, the matrix representing the operator a^t, defined by (4.3), is the transpose of the matrix representing a.

A *Type II domain* has the form

$$D_2 = \{a \in L(H) : a^t = -a,\ \|a\| \leq 1\}, \tag{4.5}$$

and a *Type III domain* has the form

$$D_3 = \{a \in L(H) : a^t = a,\ \|a\| \leq 1\}. \tag{4.6}$$

The domains D_2 and D_3 are the intersections of the domain D_1 of $L(H)$ with the subspaces

$$E_2 = \{a \in L(H) :\ \ a^t = -a\ \} \tag{4.7}$$

and

$$E_3 = \{a \in L(H) : \ a^t = a \}, \tag{4.8}$$

respectively, where the transpose a^t is defined by (4.3) and depends on the conjugation Q. If Q is complex conjugation with respect to some basis and $\dim H = n$, then the domain D_2 of type II is the unit ball in the operator norm of the space $A_n(C)$ of $n \times n$ complex antisymmetric matrices, while the domain D_3 of type III is the unit ball in the operator norm of the space $S_n(C)$ of $n \times n$ complex symmetric matrices.

Note that the velocity ball D_v from chapter 1 is the real part of a type I domain, where H represents the time and K the space displacement of a motion with constant velocity. We showed that D_v is a BSD with respect to the group $Aut_p(D_v)$ of projective automorphisms of D_v. For another example, take the unit ball $D_1^{n,1}$ in the space $M_{n,1}(C)$ of all $n \times 1$ matrices over C, which can be identified with $L(C, C^n)$, the linear maps from C to C^n. It is a bounded symmetric domain with respect to the group $Aut_a(D_1^{n,1})$ of analytic automorphisms of $D_1^{n,1}$. In fact, as we will show in the next section, *every* Type I domain D is a bounded symmetric domain with respect to $Aut_a(D)$. Note that D_v is the real part of the domain $D_1^{3,1}$, and $Aut_p(D_v)$ is the restriction of $Aut_a(D_1^{3,1})$ to this real part.

4.1.2 The JC^*-triples

The space $L(H, K)$ is closed under the map $a \to aa^*a$. To see this, note first that aa^*a maps H to K, as seen from the following diagram:

$$H \xrightarrow{\ a\ } K \xrightarrow{\ a^*\ } H \xrightarrow{\ a\ } K.$$

Moreover, since the product of linear operators is linear and the product of bounded operators is bounded, the operator aa^*a belongs to $L(H, K)$.

Note that since $(ab^*c)^t = c^t(b^t)^*a^t$, both subspaces E_2 and E_3 are invariant under the map $a \to aa^*a$. Thus, all three classical domains D_1, D_2 and D_3 are unit balls of a subspace E of $L(H, K)$ which is closed under the map $a \to aa^*a$.

A subspace E of $L(H, K)$ which is closed under the map $a \to aa^*a$ is called a *JC^*-triple*. The name derives from the fact that any JC^*-triple admits a triple product. Our next goal is to derive the triple product on a JC^*-triple E. For any $a, b \in E$ and complex number λ, the operator

$$(a + \lambda b)(a^* + \overline{\lambda} b^*)(a + \lambda b) \in E.$$

Expanding this expression, we get

$$aa^*a + \lambda(aa^*b + ba^*a) + \overline{\lambda} ab^*a + \lambda^2 ba^*b + |\lambda|^2(ab^*b + bb^*a) + \lambda|\lambda|^2 bb^*b \in E.$$

Since λ is arbitrary, we must have

$$aa^*b + ba^*a \in E \quad \text{and} \quad ab^*a \in E.$$

From this, it follows that for any $a, b, c \in E$ and any complex number λ, we have

$$(a + \lambda b)(a^* + \overline{\lambda} b^*)c + c(a^* + \overline{\lambda} b^*)(a + \lambda b) \in E,$$

implying that the coefficient $ab^*c + cb^*a$ of $\overline{\lambda}$ also belongs to E. Thus, we have shown that the JC^*-triple $E \subseteq L(H, K)$ is closed under the map $\{\cdot, \cdot, \cdot\} : E \times E \times E \to L(H, K)$ defined by

$$\{a, b, c\} = \frac{ab^*c + cb^*a}{2}. \tag{4.9}$$

This map is called the *triple product* on the JC^*-triple E.

Since $L(H, K)$ and E_2, E_3 are JC^*-triples, they are closed under the triple product (4.9). Therefore, for any pair of elements $a, b \in E$, we can define the following two real linear maps from E to E. The first one, denoted by $D(a, b)$, is also a complex linear map and is defined by

$$D(a, b)z = \{a, b, z\} = \frac{ab^*z + zb^*a}{2}. \tag{4.10}$$

We write $D(a)$ instead of $D(a, a)$. The second map, $Q(a, b)$, is a conjugate linear map defined by

$$Q(a, b)z = \{a, z, b\} = \frac{az^*b + bz^*a}{2}. \tag{4.11}$$

We write $Q(a)$ instead of $Q(a, a)$. Note, in particular, that

$$Q(a)z = Q(a, a)z = \{a, z, a\} = az^*a. \tag{4.12}$$

4.1.3 The absolute value operator of $a \in L(H, K)$

The theory of self-adjoint operators is well known to physicists, since they represent observables of quantum systems. The theory of non–self-adjoint operators, however, has fewer applications and is less familiar to scientists. To be able to show that the classical domains are bounded symmetric domains, we need to review some of the basic theory of non–self-adjoint operators. If $H \neq K$, an operator $a \in L(H, K)$ cannot be self-adjoint. Nevertheless, using the triple product on $L(H, K)$, we can still obtain results for non–self-adjoint operators similar to those for self-adjoint operators.

Recall that a linear operator $b \in L(H)$ is *positive* if

$$\langle b\xi | \xi \rangle \geq 0 \quad \text{for all} \quad \xi \in H. \tag{4.13}$$

If b is positive, then b^* is also positive, since $\langle b^*\xi | \xi \rangle = \overline{\langle \xi | b^*\xi \rangle} = \overline{\langle b\xi | \xi \rangle} \geq 0$. Note that a positive operator has non-negative spectrum. Let $a \in L(H, K)$. Since for any $\xi \in H$ we have

$$\langle a^* a\xi|\xi\rangle = \langle \xi|a^* a\xi\rangle = \langle a\xi|a\xi\rangle = |a\xi|^2 \geq 0,$$

the operator $a^*a \in L(H)$ is a positive operator.

It is well known that any positive operator $b \in L(H)$ has a spectral decomposition, defined as follows. For each $0 \leq \lambda \leq \|b\|$, define an orthogonal projection $p_\lambda(b)$ from H onto a subspace $H_\lambda(b)$, with the following properties:

$$\lambda_1 \leq \lambda_2 \Rightarrow H_{\lambda_1}(b) \subseteq H_{\lambda_2}(b) \quad \text{and} \quad p_{\lambda_1}(b)p_{\lambda_2}(b) = p_{\lambda_1}(b), \tag{4.14}$$

$$\xi \in H_\lambda(b) \Rightarrow \langle b\xi|\xi\rangle \leq \lambda, \tag{4.15}$$

and

$$H_\lambda(b) \text{ is the maximal subspace safisfying (4.15).} \tag{4.16}$$

The family of projections $p_\lambda(b)$ is called the *spectral family* of b.

For any partition $0 = \lambda_0 < \lambda_1 < \cdots < \lambda_n = \|b\|$, we define an integral sum

$$S_n = \sum_{j=1}^{n} \lambda_j \triangle_j p_\lambda(b),$$

where $\triangle_j p_\lambda(b) = p_{\lambda_j}(b) - p_{\lambda_{j-1}}(b)$ is an orthogonal projection. Moreover, for any $j \neq k$, we have

$$\triangle_j p_\lambda(b) \triangle_k p_\lambda(b) = \delta_j^k \triangle_j p_\lambda(b), \tag{4.17}$$

where δ_j^k denotes the Kronecker delta. A family $\{\triangle_j p_\lambda(b)\}$ satisfying (4.17) is called an algebraically orthogonal family of orthogonal projections. It is obvious that S_n is a positive operator belonging to $L(H)$. The limit of S_n as $\triangle\lambda \to 0$ is equal to b, and we write

$$b = \int_{\lambda=0}^{\|b\|} \lambda \, d p_\lambda(b) \, . \tag{4.18}$$

Equation (4.18) is called the *spectral decomposition* of the positive operator b.

If $\dim H < \infty$, the spectral decomposition becomes a discrete decomposition

$$b = \sum_j \lambda_j p_j, \tag{4.19}$$

where λ_j are the eigenvalues of b, and p_j is the orthogonal projection from H onto the subspace consisting of all λ_j-eigenvectors of b. Formula (4.19) also holds when $\dim H = \infty$, if the spectrum of b is discrete.

Fix a positive operator b with spectral decomposition (4.18). For any real-valued function f which is defined on the interval $[0, \|b\|]$, we define the operator $f(b) \in L(H)$ by

$$f(b) = \int_{\lambda=0}^{\|b\|} f(\lambda)\, dp_\lambda(b). \tag{4.20}$$

This definition of a *function of an operator* is consistent with the operations on operators. For example, if $p(x)$ is a polynomial in x, then $f(b)$ is equal to the operator $p(b)$ obtained by substitution of b into the polynomial p. It is known that if f is positive on $[0, \|b\|]$ and monotone increasing, then

$$\|f(b)\| = f(\|b\|). \tag{4.21}$$

We end this subsection with the definition of the absolute value operator. For any $a \in L(H, K)$, apply the function $f(x) = \sqrt{x}$ to the positive operator $b = a^*a$ and define

$$|a| = f(b) = (a^*a)^{1/2}. \tag{4.22}$$

The operator $|a|$ is also positive and is called the *absolute value operator* of a. Note that if $u \in L(K)$ is a unitary operator, then

$$|ua| = ((ua)^*ua)^{1/2} = (a^*u^*ua)^{1/2} = (a^*a)^{1/2} = |a|. \tag{4.23}$$

In other words, left multiplication of an operator by a unitary operator does not change its absolute value.

4.1.4 Polar decomposition in $L(H, K)$

Let $a \in L(H, K)$. In order to define the polar decomposition of a, we have to define first its right and left supports. To do this, we first define the *kernel* of a, denoted $\ker a$, by

$$H_1 = \ker a = \{\xi \in H : \ a\xi = 0\}.$$

Next, decompose H into a direct sum $H = H_1 \oplus H_2$. The summand H_2 is called the *right support space* of a. We denote the orthogonal projection from H onto H_2 by $r(a)$. Any element $\xi \in H$ can be decomposed as $\xi = \xi_1 + \xi_2$, where $\xi_1 \in H_1$ and $\xi_2 \in H_2$. Then

$$a\xi = a(\xi_1 + \xi_2) = a\xi_1 + a\xi_2 = a\xi_2 = a\, r(a)\xi, \tag{4.24}$$

which shows that $a = a\, r(a)$. The projection $r(a)$ is called the *right support projection* of a. Note that $r(a)$ is the smallest projection satisfying $a = a\, r(a)$, where the order between projections is defined by inclusion of their images.

The left support of a is defined similarly. First define the image of a, denoted $\operatorname{Im} a$, by

$$\operatorname{Im} a = \{a\xi : \xi \in H\}.$$

The orthogonal projection from K onto Im a, denoted $l(a)$, is called the *left support projection* of a. Since $a\xi = l(a)a\xi$ for any $\xi \in H$, we have

$$a = l(a)a.$$

Note that $l(a)$ is the smallest projection satisfying this identity.

For arbitrary $\xi \in H$, using the self-adjointness of $|a|$, we have

$$\big|\, |a|\xi\big|^2 = \langle |a|\xi |\, |a|\xi\rangle = \langle \xi |\, |a|^2\xi\rangle = \langle \xi | a^*a\xi\rangle = \langle a\xi | a\xi\rangle = |a\xi|^2. \tag{4.25}$$

It is easy to show that the image $\widetilde{H} = \{\eta = |a|\xi : \xi \in H\}$ of $|a|$ is dense in H_2. Thus, there is a unique map $v \in L(H,K)$ which vanishes on H_1, and on H_2 continuously extends the map

$$v(\eta) = a\xi \ \text{ for } \ \eta \in \widetilde{H}, \quad \eta = |a|\xi.$$

From (4.25), it follows that v is an isometry from H_2, the support subspace of a, to the image Im a of a and vanishes on the kernel H_1 of a. The map v is called the *support tripotent* of a. The decomposition

$$a = v|a| \tag{4.26}$$

is called the *polar decomposition* of a.

Since, for any $\eta \in \widetilde{H}$, we have $|v\eta| = |\eta|$, it follows that

$$\langle \eta|\eta\rangle = \langle v\eta|v\eta\rangle = \langle \eta|v^*v\eta\rangle.$$

This implies that v^*v is the identity on $\widetilde{H}$ and, by continuity, also on H_2. From this, it follows that

$$\{v,v,v\} = v\,v^*v = v. \tag{4.27}$$

An operator v satisfying this identity is called a *partial isometry* and a *tripotent* of the triple product (4.9). Thus, the polar decomposition decomposes an operator a into the product of a non-negative operator $|a|$ (its absolute value operator) and a partial isometry v (its support tripotent).

Since v^*v is the identity on H_2 and vanishes on H_1, the orthogonal complement of H_2, the map v^*v is an orthogonal projection from H onto H_2. Thus,

$$a = av^*v, \quad v^*v = r(a),$$

where $r(a)$ is the right support projection of a. Note that $r(a)$ is both the right and left support of $|a|$. Thus,

$$v^*v|a| = |a| \ \Rightarrow a = v|a| = vv^*v|a| = (vv^*)v|a| = vv^*a,$$

implying that the left support $l(a)$ of a is equal to vv^*.

From the polar decomposition, it follows that

$$a^{(3)} = \{a, a, a\} = v|a|\,|a|v^*\,v|a| = v\,|a|^3. \tag{4.28}$$

Since multiplication by an isometry v on the support $r(v)$ does not change the norm, and since the function $f(x) = x^3$ is positive and monotone on $[0, \|a\|]$, from (4.21) it follows that

$$\|a^{(3)}\| = \|a\|^3. \tag{4.29}$$

This is the so-called *star identity.* It connects the norm and the triple product.

4.1.5 Singular decomposition

Let $a \in L(H, K)$. The spectral values of $|a|$ are called the *singular values* of a. The geometric meaning of the polar decomposition for the finite-dimensional case is as follows. The positive operator $|a|$ has a set of eigenvectors $\mathbf{e}_1, \mathbf{e}_2, ..., \mathbf{e}_n$, corresponding to the eigenvalues $s_1, s_2, ..., s_n$, which are the singular values of a. We may and will assume that

$$s_1 \geq s_2 \geq \cdots \geq s_n.$$

By a well-known theorem, the eigenvectors of a self-adjoint operator (for example, the positive operator $|a|$) corresponding to different eigenvalues are orthogonal. Thus, we may assume that $\mathbf{e}_1, \mathbf{e}_2, ..., \mathbf{e}_n$ form an orthonormal basis of H. The operator $|a|$ transforms any vector $c^j\mathbf{e}_j$ into $c^j s_j \mathbf{e}_j$, which amounts to changing the scale in each direction $\mathbf{e}_j$ by a factor s_j. Such an operator transforms the unit ball of H into an ellipsoid with semi-axes $\mathbf{e}_j$ of length s_j. Then the support tripotent from the polar decomposition rotates the vectors $\mathbf{e}_1, \mathbf{e}_2, ..., \mathbf{e}_n$ to the vectors $\mathbf{v}_1, \mathbf{v}_2, ..., \mathbf{v}_n$ which are the semi-axes of the ellipsoidal image of the unit ball under the map a (see Figure 4.1).

To introduce a singular decomposition of an operator $a \in L(H, K)$, we will apply the spectral decomposition (4.18) to the positive operator $|a|$ on the Hilbert space H. This yields

$$|a| = \int_{s=0}^{\|a\|} s\, d p_s(|a|). \tag{4.30}$$

Apply now the polar decomposition (4.26) to obtain a decomposition

$$a = v\,|a| = \int_{s=0}^{\|a\|} s\, d(v\, p_s(|a|)) = \int_{s=0}^{\|a\|} s\, d\, v_s(a), \tag{4.31}$$

where, for any $0 \leq s \leq \|a\|$, the partial isometry $v_s(a) = v\, p_s(|a|) \in L(H, K)$. We call the decomposition (4.31) the *singular decomposition* of a.

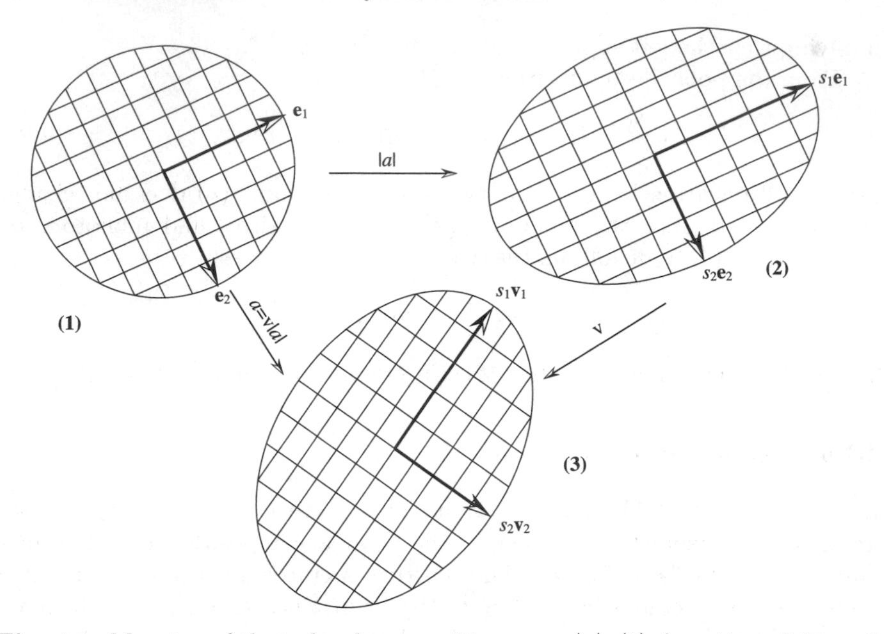

Fig. 4.1. Meaning of the polar decomposition $a = v\,|a|$. (1) A section of the unit ball with a grid on it. We choose the directions of the grid parallel to the eigenvectors $\mathbf{e}_1, \mathbf{e}_2$ of the operator $|a|$. (2) the transform $|a|$ of this grid. The semi-axes of the ellipse are $\mathbf{e}_1, \mathbf{e}_2$, multiplied by the singular values s_1, s_2 of a. (3) the grid from part (1) transformed by a, which is equal (by the polar decomposition) to rotation of the grid from part (2) by the tripotent v.

If the spectrum of $|a|$ is discrete, then, by (4.19), the singular decomposition becomes

$$a = \sum_j s_j v_j, \tag{4.32}$$

where the s_j's are the *singular numbers* of a and the v_j's are an algebraically orthogonal family of tripotents, a concept we define now. A family $\{v_j\}$ of tripotents is said to be *algebraically orthogonal* if their right and left supports are mutually orthogonal, meaning that for any $j \neq k$, we have

$$r(v_j)r(v_k) = 0 \text{ and } l(v_j)l(v_k) = 0. \tag{4.33}$$

Note that if the tripotents v_j and v_k are algebraically orthogonal, then

$$\{v_j, v_k, d\} = 0 \tag{4.34}$$

for any $d \in L(H, K)$.

Note that $v_s^*(a)v_s(a) = p_s(|a|)$ is the right support $r(v_s(a))$ of $v_s(a)$ from the decomposition (4.31). Note also that $v_{\|a\|}(a)$ is equal to the support tripotent v of a. From (4.14), it follows that the family $v_s(a)$ is an increasing family of partial isometries in the following sense. Let $0 \leq s \leq t \leq \|a\|$. Then the projection $p_t(|a|)$ can be written as a sum of two algebraically orthogonal projections $p_t(|a|) = p_s(|a|) + (p_t(|a|) - p_s(|a|))$. This implies that the partial isometry $v_t(a)$ can be written as a sum $v_t(a) = v_s(a) + v(p_t(|a|) - p_s(|a|))$ of two algebraically orthogonal tripotents. In this case, we say that $v_t(a)$ is *larger* (or *extends*) $v_s(a)$ and that $v_s(a)$ is *dominated* by $v_t(a)$, and we write $v_s(a) \preceq v_t(a)$. The increasing family $v_s(a)$ of tripotents is called the *singular family* of a.

4.1.6 Functions of operators $a \in L(H, K)$

We can now define functions of an element $a \in L(H, K)$ with singular decomposition (4.31). Let f be a continuous function on the interval $[0, \|a\|\,]$. Define

$$f(a) = \int_{s=0}^{\|a\|} f(s)\, d\, v_s(a). \tag{4.35}$$

If $f(x) = x^3$ and $|a|$ has a discrete spectrum, implying that a has the singular decomposition (4.32), then, using (4.34), we get

$$aa^*a = \{a, a, a\} = \left\{ \left(\sum s_j v_j\right), \left(\sum s_k v_k\right), \left(\sum s_l v_l\right) \right\}$$

$$= \sum s_j^3 \{v_j, v_j, v_j\} = \sum s_j^3 v_j = f(a) := a^3. \tag{4.36}$$

Note that if $|a|$ does not have a discrete spectrum, we can modify our argument and show that (4.36) holds in this case as well.

Similarly, if $f(x)$ is a polynomial of odd degree $2n + 1$, then

$$f(x) = \sum_{k=0}^{n} c_k x^{2k+1} = x \sum_{k=0}^{n} c_k x^{2k} = x g(x^2), \tag{4.37}$$

where g is an n-th degree polynomial. Then

$$f(a) = \sum_{k=1}^{n} c_k a^{2n+1} = a g(a^* a) = g(aa^*) a.$$

Any continuous function f on the interval $[0, \|a\|\,]$ can be extended to the interval $[-\|a\|, \|a\|\,]$ by setting $f(-x) = -f(x)$ and obtained as a limit of odd polynomials f_n. If we denote by g_n the polynomial corresponding to f_n in (4.37) and let $g = \lim g_n$, then $f(x) = x g(x^2)$ and

$$f(a) = a g(a^* a) = g(aa^*) a. \tag{4.38}$$

4.2 Classical domains are BSDs

In this section, we show that domains of types I, II, and III and, even more generally, any unit ball D of a JC^*-triple are, in fact, bounded symmetric domains with respect to the analytic automorphisms. The analytic map which sends an operator to its negation is a symmetry about the point 0 of D. Therefore, to show that D is a bounded symmetric domain with respect to $Aut_a(D)$, it is enough to show that D is homogeneous. This will follow if for any operator $a \in D$, we construct an analytic automorphism φ_a of D such that $\varphi_a(0) = a$. For the construction of φ_a, we will use ideas from Transmission Line Theory. We will prove this result for type I domains first and then extend it to the unit ball D in an arbitrary JC^*-triple.

4.2.1 Mathematical formulation of Transmission Line Theory

One of the major engineering projects at the beginning of the 20th century was the laying of the transatlantic telephone line connecting Europe and America. During the final stages, when the entire line was finally in place, it had to be tested. The amount of noise produced by such a long line was so large that it was impossible to recognize the speech transmitted through the line. At first, all of the attempts to improve the quality of the transmission ended in failure. Due to the environmental conditions surrounding the cable, the characteristics of the noise were constantly changing. The engineers working on the project came up with the original solution of treating the transmission line as a linear black box. After the experimental study of the properties of this black box transformation, they were able to build devices which corrected the output signal to match the input signal.

The mathematical formulation of Transmission Line Theory is as follows. Refer to Figure 4.2. A signal sent into the line at port $\widetilde{H}_1$ on side 1 causes the appearance of two new signals: a signal at port $\widetilde{H}_2$ on side 2, called the *transmitted signal*, and a signal at port $\widetilde{K}_1$ on side 1, called the *reflected signal*. Likewise, a signal sent into the cable at port $\widetilde{K}_2$ on side 2 causes the appearance of two new signals: a transmitted signal at port $\widetilde{K}_1$ on side 1 and a reflected signal at port $\widetilde{H}_2$ on side 2. We assume that the input signal ξ_1 at port $\widetilde{H}_1$ is an element of a Hilbert space H_1 and that the input signal η_2 at port $\widetilde{K}_2$ is an element of a Hilbert space K_2. Similarly, we assume that the output signals η_1 at port $\widetilde{K}_1$ and ξ_2 at port $\widetilde{H}_2$ belong to the Hilbert spaces K_1 and H_2, respectively.

It is natural to model the set S of all signals on the same side of the line and in the same direction as a complex Hilbert space because, in fact, such signals are complex-valued functions $f(t)$ ($t =$ time) and have a natural linear structure. Any linear combination of the signals on the same side of the line and in the same direction is also a signal on this side and in this direction. The integral $\int |f(t)|^2 dt$ is the energy of the signal, which is assumed to be

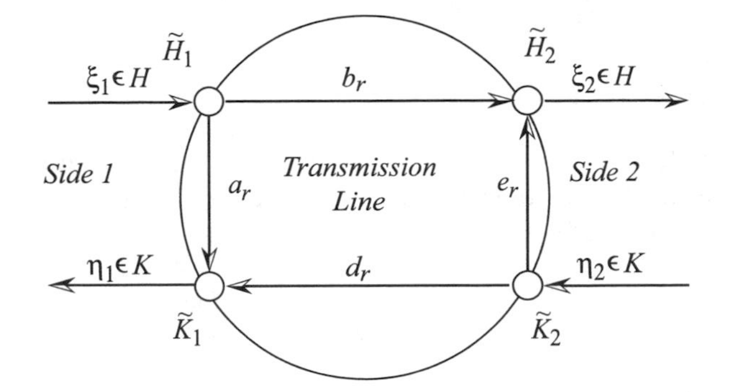

Fig. 4.2. Input and output signals of a transmission line. The circle represents a black box. An input signal ξ_1 sent into the line at port $\widetilde{H}_1$ of side 1 produces a transmitted signal at port $\widetilde{H}_2$ on side 2 and a reflected signal signal at port $\widetilde{K}_1$ on side 1. Similarly, a signal η_2 sent into the line at port $\widetilde{K}_2$ of side 2 produces a transmitted signal at port $\widetilde{K}_1$ on side 1 and a reflected signal at port $\widetilde{H}_2$ on side 2. The operators a_r and e_r describe the reflection of the two input signals, while the operators b_r and d_r describe the transmission of these signals.

finite. This allows one to define an inner product on S, making S a Hilbert space.

Signals going in different directions may have different physical characteristics, such as frequency. On the other hand, we assume that signals going in the same direction are of the same nature, even if they appear on opposite sides of the line. This means that the corresponding Hilbert spaces are identical, and we have $H_1 = H_2 = H$ and $K_1 = K_2 = K$. Hence, $\xi_1, \xi_2 \in H$, $\eta_1, \eta_2 \in K$, and the distinction between the sides of the line is expressed through the subscripts of the vectors ξ, η.

The total input signal into the line, which consists of the contributions of both ξ_1 and η_2, may now be represented by an element $\begin{pmatrix} \xi_1 \\ \eta_2 \end{pmatrix}$ of the direct sum $H \oplus K$. Similarly, the total output signal of the line may be represented by the vector $\begin{pmatrix} \xi_2 \\ \eta_1 \end{pmatrix}$ of $H \oplus K$. With this notation, the action of the line is described by a transformation $U : H \oplus K \to H \oplus K$, where

$$U \begin{pmatrix} \xi_1 \\ \eta_2 \end{pmatrix} = \begin{pmatrix} \xi_2 \\ \eta_1 \end{pmatrix}. \tag{4.39}$$

This combination of the input and output signals is called the *hybrid connection* and was introduced and studied in section 1.1.4 of Chapter 1.

TLT normally deals with *linear* transmission lines, that is, lines for which the corresponding transformation U is a linear operator. In this case, the operator U may be naturally decomposed as

$$U = a + b + e + d,$$

where

$$a = P_K U P_H,\ b = P_H U P_H,\ e = P_H U P_K,\ d = P_K U P_K, \tag{4.40}$$

and P_H and $P_K = 1 - P_H$ are the projections of the direct sum $H \oplus K$ onto H and K, respectively. Alternatively, we may write the operator U as a block matrix, and equation (4.39) becomes

$$U \begin{pmatrix} \xi_1 \\ \eta_2 \end{pmatrix} = \begin{pmatrix} b_r & e_r \\ a_r & d_r \end{pmatrix} \begin{pmatrix} \xi_1 \\ \eta_2 \end{pmatrix} = \begin{pmatrix} \xi_2 \\ \eta_1 \end{pmatrix}, \tag{4.41}$$

where the operators

$$a_r : H \to K,\ b_r : H \to H,\ e_r : K \to H,\ d_r : K \to K$$

are the natural restrictions of the operators a, b, e and d, respectively.

Later we shall see that the operators a, b, e and d are the ones which have a natural physical meaning, since only they may be determined through direct measurements on a real transmission line, rather than a_r, b_r, e_r and d_r, the ones usually used in TLT. Nevertheless, we suppress the subscript r, since it will always be clear from the context whether we are referring to a or a_r. The operator a is called the *reflection operator* of the transmission line.

4.2.2 The signal transformation in a lossless line

If the total energy of the output signals is equal to the total energy of the input signals, the transmission line is called *energy conserving*, or *lossless*. In our notation, this means that

$$|\xi_1|^2 + |\eta_2|^2 = |\xi_2|^2 + |\eta_1|^2, \tag{4.42}$$

or, equivalently,

$$|\xi_1|^2 - |\eta_1|^2 = |\xi_2|^2 - |\eta_2|^2. \tag{4.43}$$

Equation (4.42) implies that the transformation U is an isometry with respect to the natural norm in $H \oplus K$. Hence, both the operators a, b, e, d, defined by (4.40), and their natural restrictions have operator norm less than or equal to 1.

We show now that U is unitary with respect to the inner product of $H \oplus K$. First note that for any $\xi, \eta \in H \oplus K$, we have

$$|U(\xi + \eta)|^2 = |\xi + \eta|^2$$

since U is an isometry. Hence,

$$|U\xi|^2 + 2\mathrm{Re}\langle U\xi | U\eta \rangle + |U\eta|^2 = |\xi|^2 + 2\mathrm{Re}\langle \xi | \eta \rangle + |\eta|^2,$$

and so, by multiplying ξ by a unimodular λ, we obtain

$$\langle U\xi|U\eta\rangle = \langle \xi|\eta\rangle.$$

But from this, it follows that

$$\langle \xi|U^*U\eta\rangle = \langle \xi|\eta\rangle.$$

Therefore, $U^*U = I$, and U is unitary.

Since U is unitary, we have

$$\begin{pmatrix} b & e \\ a & d \end{pmatrix}^* \begin{pmatrix} b & e \\ a & d \end{pmatrix} = I, \tag{4.44}$$

or, equivalently,

$$\begin{pmatrix} b^* & a^* \\ e^* & d^* \end{pmatrix} \begin{pmatrix} b & e \\ a & d \end{pmatrix} = \begin{pmatrix} I_H & 0 \\ 0 & I_K \end{pmatrix}, \tag{4.45}$$

where I_H and I_K are identity matrices of size $\dim H$ and $\dim K$, respectively. From (4.45), we obtain the four equations

$$b^*b + a^*a = I_H, \tag{4.46}$$

$$b^*e + a^*d = 0, \tag{4.47}$$

$$e^*b + d^*a = 0, \tag{4.48}$$

$$e^*e + d^*d = I_K. \tag{4.49}$$

From (4.46), it follows that $|b| = \sqrt{1 - |a|^2}$, and by use of (4.26), we get $b = v\sqrt{1 - |a|^2}$ for some tripotent $v \in L(H)$. Similarly, equation (4.49) implies that $d = u\sqrt{1 - |e|^2}$ for some tripotent $u \in L(K)$. Substitute these expressions into (4.47) to obtain

$$\sqrt{1 - |a|^2}v^*e = -a^*u\sqrt{1 - |e|^2}. \tag{4.50}$$

Note that since $\|a\| \leq 1$ and $\|e\| \leq 1$, the operators $\sqrt{1 - |a|^2}$ and $\sqrt{1 - |e|^2}$ are positive. If they are not invertible, then, by replacing them with $\sqrt{1 + \varepsilon - |a|^2}$ and $\sqrt{1 + \varepsilon - |e|^2}$, with $\varepsilon > 0$, they become invertible, and we can perform all of the calculations and then take the limit as $\varepsilon \to 0$. Hence, we will assume without loss of generality that the above operators are invertible.

Equation (4.50) implies that

$$v^* e(1 - |e|^2)^{-1/2} = -(1 - |a|^2)^{-1/2} a^* u,$$

and, using (4.23), we get

$$v^* e(1 - |v^* e|^2)^{-1/2} = -(1 - |-u^* a|^2)^{-1/2} (u^* a)^*,$$

which, by (4.38), becomes

$$(1 - |(v^* e)^*|^2)^{-1/2} v^* e = -(1 - |-u^* a|^2)^{-1/2} (u^* a)^*.$$

Note that the function $f(x) = x(1 - x^2)^{-1/2}$ is invertible. In fact, its inverse is $f^{-1}(y) = \sqrt{\frac{y^2}{1+y^2}}$. Applying f^{-1} to both sides of the above equation, we get $v^* e = -(u^* a)^*$, implying that $e = -va^* u$ and

$$d = u\sqrt{1 - |a^* u|^2} = u\sqrt{1 - u^* |a^*|^2 u}.$$

By the definition of the operator function, we have $\sqrt{1 - u^* |a^*|^2 u} = u^* \sqrt{1 - |a^*|^2} u$. Thus,

$$d = uu^* \sqrt{1 - |a^*|^2} u = \sqrt{1 - |a^*|^2} u.$$

Hence, by substituting

$$b = v\sqrt{1 - |a|^2}, \; e = -va^* u, \; d = \sqrt{1 - |a^*|^2} u \tag{4.51}$$

into (4.41), the signal transformation for a lossless line becomes

$$U \begin{pmatrix} \xi_1 \\ \eta_2 \end{pmatrix} = \begin{pmatrix} v\sqrt{1 - |a|^2} & -va^* u \\ a & \sqrt{1 - |a^*|^2} u \end{pmatrix} \begin{pmatrix} \xi_1 \\ \eta_2 \end{pmatrix} = \begin{pmatrix} \xi_2 \\ \eta_1 \end{pmatrix}. \tag{4.52}$$

Note that by rotating the basis in H_2 (the Hilbert space H on side 2), we can take $v = I_H$. Similarly, by rotating the basis in K_2, we can take $u = I_K$.

4.2.3 Homogeneity of type I domains

We will now prove that type I domains are homogeneous. As mentioned at the start of this section, it is enough to construct, for each operator $a \in D$, a map $\varphi_a \in Aut_a(D)$ such that $\varphi_a(0) = a$. We connect a device called a *reflector* to side 2 of the line (see Figure 4.3). The reflector receives the output from port $\tilde{H}_2$ as input and sends its output to port $\tilde{G}_2$. The action of the reflector is described by a linear operator $z \in L(H, K)$. We assume that the reflector does not increase the energy of a signal, which means that the operator norm of z is less than or equal to 1. The "new" line (the line with the reflector) now has one input (at port $\tilde{H}_1$) and one output (at port $\tilde{G}_1$). Hence, the new line also behaves like a reflector itself. This implies that there is an operator $w \in L(H, K)$ which transforms the input signal $\xi_1 \in H$ at $\tilde{H}_1$ into the output signal $\eta_1 \in K$ at $\tilde{G}_1$. Thus, $w(\xi_1) = \eta_1$.

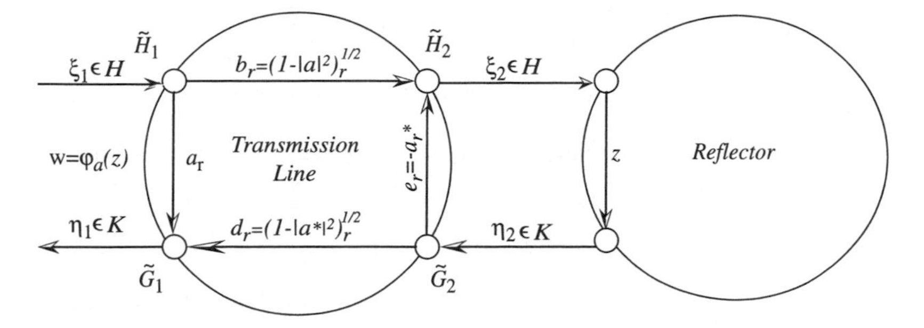

Fig. 4.3. Transmission line with a reflector. The added reflector receives as input the signal ξ_2 (an element of H) which is the output from port $\tilde{H}_2$. The reflector linearly transforms this signal and sends its output η_2 (an element of K) to port $\tilde{G}_2$ of the transmission line. The action of the reflector is described by a linear operator $z \in L(H,K)$.

We can calculate η_1 from ξ_1 directly via the infinite sum

$$\eta_1 = (a + dzb + dzezb + dzezezb + \cdots)\xi_1. \tag{4.53}$$

Here, the terms of the form $dzez \cdots ezb$ correspond to multiple "reflections" of the signal between the reflector and side 2 of the line. These terms are obtained from the following passes through the line:

$$\tilde{H}_1 \xrightarrow{a} \tilde{G}_1,$$

$$\tilde{H}_1 \xrightarrow{b} \tilde{H}_2 \xrightarrow{z} \tilde{G}_2 \xrightarrow{d} \tilde{G}_1,$$

$$\tilde{H}_1 \xrightarrow{b} \tilde{H}_2 \xrightarrow{z} \tilde{G}_2 \xrightarrow{e} \tilde{H}_2 \xrightarrow{z} \tilde{G}_2 \xrightarrow{d} \tilde{G}_1.$$

Hence, the linear operator $w : H \to K$ defined by $w\xi_1 = \eta_1$, is given by

$$w = a + dzb + dzezb + dzezezb + \cdots = a + dz(\sum (ez)^n)b. \tag{4.54}$$

Using the formula for the sum of a geometric series, we obtain

$$w(z) = a + dz(1 - ez)^{-1}b. \tag{4.55}$$

If the transmission line is lossless, we substitute (4.51) into (4.55) and obtain

$$w(z) = a + \sqrt{1 - |a^*|^2}\, uz(1 + va^*uz)^{-1}v\sqrt{1 - |a|^2}\,. \tag{4.56}$$

By adjusting the bases in the Hilbert spaces on side 2, we may take $u = I_K$ and $v = I_H$. Since w depends only on a and z, we rename it $\varphi_a(z)$. Hence,

$$\varphi_a(z) = a + \sqrt{1 - |a^*|^2}\, z(1 + a^*z)^{-1}\sqrt{1 - |a|^2}\,. \tag{4.57}$$

Clearly, $\varphi_a(0) = a$. We call $\varphi_a(z)$ a *translation* in D_1. Note that

$$uz(1 + va^*uz)^{-1}v = uzv - uzva^*uzv + \cdots = uzv(1 + a^*uzv)^{-1}.$$

Hence, the general transformation $w(z)$ of a lossless transmission line with a reflector is given by

$$w(z) = \varphi_a(uzv). \tag{4.58}$$

Note that the map $z \to uzv$, where u is a unitary operator on K and v is a unitary operator on H, is an isometry (or a *rotation*) on $L(H, K)$. Thus, the general transformation $w(z)$ of a lossless transmission line with a reflector is a composition of a rotation and a translation.

Fix a now and consider φ_a as a function of z. Since $z \in L(H, K)$ and has norm less than or equal to 1, the map φ_a is a map from the unit ball D_1 of $L(H, K)$ into $L(H, K)$. Since the line is lossless, the operator norm of $\varphi_a(z)$ is less than or equal to 1. So, in fact, φ_a is a map from D_1 to D_1. We claim that φ_a is an analytic automorphism of D_1. Equation (4.54) shows that $\varphi_a(z)$ is the sum of a converging series of homogeneous polynomials in z and, thus, is an analytic map of D_1 (by the definition of analyticity). See Chapter 5 for details. To show that the map φ_a is one-to-one, we will show later that its inverse φ_a^{-1} exists and equals φ_{-a}. For this, we will have to study the signal transformation in a composite transmission line.

4.2.4 Composition of two lossless lines

Consider now the signal transformation in a line composed of two lossless transmission lines. In Figure 4.4, we see a sequential composition of two transmission lines **TL1** and **TL2**. We choose the bases of the Hilbert spaces on side 2 so that the unitary operators u, v from (4.52) for the line **TL1** are identity operators. Similarly, we choose the bases of the Hilbert spaces on side 3 so that the unitary operators u, v from (4.52) for the line **TL2** are identity operators. The two lines can be considered as one line **TL12** with inputs $\xi_1 \in H$ and $\eta_3 \in K$ and outputs $\xi_3 \in H$ and $\eta_1 \in K$.

We will assume that the two transmission lines are lossless. Thus,

$$|\xi_1|^2 + |\eta_2|^2 = |\xi_2|^2 + |\eta_1|^2,$$

and

$$|\xi_2|^2 + |\eta_3|^2 = |\xi_3|^2 + |\eta_2|^2.$$

Adding these two equations, we get

$$|\xi_1|^2 + |\eta_3|^2 = |\xi_3|^2 + |\eta_1|^2,$$

implying that the line **TL12** is also lossless.

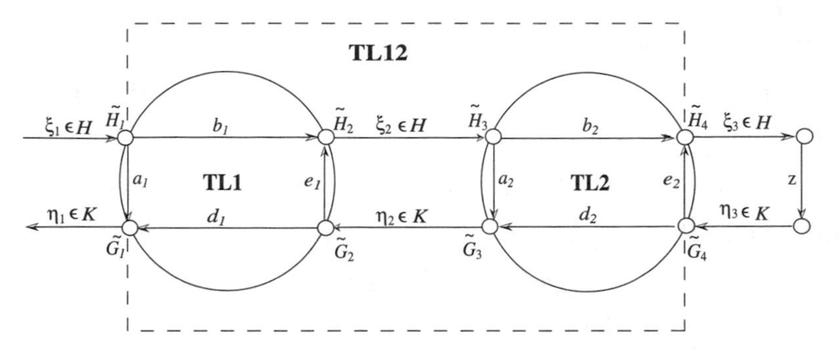

Fig. 4.4. A composite transmission line. The two transmission lines **TL1** and **TL2** can be considered as one line **TL12** with inputs $\xi_1 \in H$, $\eta_3 \in K$ and outputs $\xi_3 \in H$, $\eta_1 \in K$. A reflector z transforms $\xi_3 \in H$ to $\eta_3 \in K$.

The reflection operator a_{12} of the line **TL12** is, by definition, the operator which transforms the input signal $\xi_1 \in H$ into the output signal $\eta_1 \in K$, when the other input $\eta_3 = 0$. This is exactly the transformation by the line **TL1** with reflector a_2. Thus,

$$a_{12} = \varphi_{a_1}(a_2).$$

If we add now a reflector z, transforming $\xi_3 \in H$ to $\eta_3 \in K$, then the operator mapping ξ_1 to η_1, transformed by **TL12** and the reflector is, by (4.58),

$$w(z) = \varphi_{a_{12}}(uzv). \tag{4.59}$$

Note that we cannot assume now that u, v are identity operators, since we fixed the bases of the Hilbert spaces at an earlier stage.

We can regard the transformation $w(z)$ as the transformation of the line **TL1** with a reflector $\tilde{z}$ which maps ξ_2 to η_2. But by formula (4.57), applied to **TL2**, we get $\tilde{z} = \varphi_{a_2}(z)$. Therefore,

$$w(z) = \varphi_{a_1}(\tilde{z}) = \varphi_{a_1}(\varphi_{a_2}(z)) = \varphi_{\varphi_{a_1}(a_2)}(uzv). \tag{4.60}$$

We will use this formula to compute the inverse of φ_a.

4.2.5 The inverse of φ_a

We will show now that $\varphi_a^{-1} = \varphi_{-a}$. Actually, we will show that

$$\varphi_{-a}\varphi_a = I. \tag{4.61}$$

Consider now the composite line of the previous section with $a = a_2 = -a_1$. In this case, we have

$$a_{12} = \varphi_{-a}(a) = -a + \sqrt{1 - |a^*|^2}\, a(1 - a^*a)^{-1}\sqrt{1 - |a|^2}.$$

This can be written as

$$a_{12} = -a + \sqrt{1 - |a^*|^2}\, a(1 - |a|^2)^{-1/2}(1 - |a|^2)^{-1/2}\sqrt{1 - |a|^2}.$$

By use of (4.38), we get

$$a_{12} = -a + \sqrt{1 - |a^*|^2}(1 - |a^*|^2)^{-1/2}\, a = -a + a = 0.$$

Thus, from (4.59), we get

$$w(z) = \varphi_0(uzv) = uzv,$$

and from (4.60), it follows that

$$\varphi_{-a}(\varphi_a(z)) = uzv, \tag{4.62}$$

which is an isometry of $L(H, K)$. This already implies that φ_a is invertible. In order to show that $\varphi_{-a} = \varphi_a^{-1}$, we will show that u and v from (4.62) are, in fact, identity operators.

For this, we calculate the *derivative of the map* φ_a at an arbitrary point z_0 in the direction dz. From the definition (4.57) of φ_a, it follows that

$$\frac{d}{dz}\varphi_a(z_0)dz = \sqrt{1 - |a^*|^2}\, (\frac{d}{dz}z(1 + a^* z)^{-1}|_{z=z_0}dz)\sqrt{1 - |a|^2}.$$

But

$$\frac{d}{dz}z(1 + a^* z)^{-1}|_{z=z_0}dz = dz(1 + a^* z_0)^{-1} - z_0(1 + a^* z_0)^{-1}a^* dz(1 + a^* z_0)^{-1}.$$

Thus, using (4.38), we get

$$\frac{d}{dz}\varphi_a(z_0)dz = \sqrt{1 - |a^*|^2}(1 - z_0 a^*(1 + z_0 a^*)^{-1})dz(1 + a^* z_0)^{-1}\sqrt{1 - |a|^2}.$$

Finally, we arrive at the following formula for the derivative of φ_a:

$$\frac{d}{dz}\varphi_a(z_0)dz = \sqrt{1 - |a^*|^2}(1 + z_0 a^*)^{-1}dz(1 + a^* z_0)^{-1}\sqrt{1 - |a|^2}. \tag{4.63}$$

If $z_0 = 0$, we get

$$\frac{d}{dz}\varphi_a(0)dz = \sqrt{1 - |a^*|^2}dz\sqrt{1 - |a|^2},$$

and if $z_0 = -a$, we get

$$\frac{d}{dz}\varphi_a(-a)dz = (1 - |a^*|^2)^{-1/2}dz(1 - |a|^2)^{-1/2}.$$

Thus,

$$\frac{d}{dz}\varphi_{-a}\varphi_a(0)dz = \frac{d}{dz}\varphi_{-a}(a)(\frac{d}{dz}\varphi_a(0)dz)$$

$$= (1 - |a^*|^2)^{-1/2}\sqrt{1 - |a^*|^2}dz\sqrt{1 - |a|^2}(1 - |a|^2)^{-1/2} = dz.$$

But from (4.62), $\frac{d}{dz}\varphi_{-a}\varphi_a(0)dz = u\, dz\, v$. This can be only if u and v are equal to the identity. This completes the proof that $\varphi_{-a} = \varphi_a^{-1}$ and that the type I domain D_1 is a bounded symmetric domain with respect to $Aut_a(D)$.

4.2.6 Homogeneity of a unit ball in a JC^*-triple

We will now show that the domains D_2 and D_3 of type II and III are also homogeneous domains. Since both domains are unit balls of the JC^*-triples E_2 and E_3 defined by (4.7) and (4.8), we will prove the more general statement that the unit ball D of a JC^*-triple E is homogeneous. More precisely, we will show that for each $a \in D$, the map φ_a,

$$\varphi_a(z) = a + (1 - aa^*)^{1/2} z(1 + a^* z)^{-1}(1 - a^* a)^{1/2}, \tag{4.64}$$

defined above in (4.57), is an automorphism of D, implying that D is homogeneous with respect to the analytic bijections.

To show that the map φ_a maps D into D, we will write φ_a in terms of the triple product (4.9). This will imply that φ_a maps E into itself. From the results of the previous section, this will imply that φ_a is an automorphism of D.

In order to express φ_a in terms of the triple product, we now define, for each $a \in L(H, K)$, the following two operators on $L(H, K)$. The first operator is t_a, defined by $t_a(z) = a + z$, a translation by a. The second operator is $\tilde{t}_a$ and is defined by $\tilde{t}_a(z) = z(1 - a^* z)^{-1}$. Note that $\tilde{t}_a$ can be expressed in terms of the triple product as

$$\tilde{t}_a(z) = z + \{z, a, z\} + \{z, a, \{z, a, z\}\} + \{z, a, \{z, a\{z, a, z\}\}\} + \cdots$$

and, thus, if $a, z \in E$, then also $\tilde{t}_a(z) \in E$.

We also define, for each $a, b \in L(H, K)$, the *Bergman operator* $B(a, b)$ by

$$B(a, b)(z) = (1 - ab^*)z(1 - b^* a) = z - 2\{a, b, z\} + \{a, \{b, z, b\}, a\}. \tag{4.65}$$

Note that $B(a, b)(z)$ is a linear map in $L(H, K)$, and if $a, b, z \in E$, then also $B(a, b)(z) \in E$. If $\|a\| < 1$, then $B(a, a)z = (1 - |a^*|^2)z(1 - |a|^2)$, and, since both $1 - |a^*|^2$ and $1 - |a|^2$ are positive operators, the map

$$B(a, a)^{\frac{1}{2}} z = \sqrt{1 - |a^*|^2}\, z \sqrt{1 - |a|^2}$$

is well defined. Since $B(a, a)^{\frac{1}{2}}$ can be approximated by polynomials in $B(a, a)$ which map E into E, $B(a, a)^{\frac{1}{2}}$ also maps E into E. Now equation (4.64) becomes

$$\varphi_a(z) = t_a \circ B(a, a)^{\frac{1}{2}} \circ \tilde{t}_{-a}(z). \tag{4.66}$$

Thus, $\varphi_a(z)$ is a product of maps from E to E, which implies that φ_a also maps E to E.

The analytic automorphisms $\varphi_a(z)$ defined by (4.64) are a generalization of the Möbius transformations of the unit disk. This generalization was obtained by Potapov for a space of operators and by Harris for JC^*-triples.

The automorphisms $\varphi_a(z)$ are called *Möbius–Potapov–Harris transformations.* Note that by the use of the Bergman operator, we can now rewrite formula (4.63) for the derivative of φ_a as

$$\frac{d}{dz}\varphi_a(z_0)dz = B(a,a)^{\frac{1}{2}} \circ B(z_0,-a)^{-1}dz. \tag{4.67}$$

4.2.7 The Lie group $Aut_a(D)$ and its Lie algebra

Let D be the unit ball in a JC^*-triple E. Denote by $Aut_a(D)$ the group of analytic automorphisms of D. Define a set of maps

$$\Phi = \{\varphi_a(u \cdot v) : \ a \in D, \ v \in U(H), \ u \in U(K)\}, \tag{4.68}$$

where $\varphi_a(u \cdot v)$ maps $z \in E$ to $\varphi_a(uzv)$, φ_a is defined by (4.57) and $U(H), U(K)$ denote the set of unitary operators on H and K, respectively. By (4.58), Φ is a subset of $Aut_a(D)$. By (4.60), Φ is closed under composition, and by section 4.2.5, it is closed under inverses. Hence, Φ is a *subgroup* of $Aut_a(D)$.

Let $\psi \in Aut_a(D)$. Let $a = \psi(0)$ and consider the map $\phi = \varphi_{-a}\psi$. Clearly, $\phi \in Aut_a(D)$ and $\phi(0) = \varphi_{-a}\psi(0) = \varphi_{-a}a = 0$. By a theorem that we will prove in the next chapter, ϕ is a linear isometry of E. Thus, $Aut_a(D)$ is generated by the Möbius–Potapov–Harris transformations φ_a and the restrictions of linear isometries of E.

In [4], it was shown that an isometry on $L(H)$ has the form $z \to uzv$ or $z \to uz^tv$, with u, v unitary on H. A similar result holds for $L(H,K)$. For the connected component $Aut^0_a(D)$ of the identity of $Aut_a(D)$, the isometry must have the form $z \to uzv$. Thus, for the domain D_1 of type I defined in (4.2) as the unit ball of $L(H,K)$, we have

$$Aut^0_a(D_1) = \Phi.$$

For type II and III domains, the isometry must also preserve the subspaces E_2 and E_3 defined by (4.7) and (4.8). Thus, for these domains, we must satisfy

$$(uzv)^t = \pm uzv, \quad \text{for} \quad z^t = \pm z.$$

This implies that $uzv = v^t z u^t$ holds if and only if $v = u^t$. Thus, if D is a domain of type II or III, we have

$$Aut^0_a(D) = \{\varphi_a(u \cdot u^t) : \ a \in D, \ u \in U(H)\}.$$

The Lie algebra $aut_a(D)$ of the Lie group $Aut_a(D)$ consists of the generators of one-parameter groups of analytic automorphisms. These generators can be obtained by differentiating smooth curves $g(s)$ from a neighborhood I_0 of 0 into $Aut_a(D)$, with $g(0) = I$. For a domain D_1 of type I, such a curve is given by

$$g(s) = \varphi_{a(s)}(u(s) \cdot v(s)),$$

where $a(s) : I_0 \to D$ is a smooth map such that $a(0) = 0$, $u(s) : I_0 \to U(K)$ is a smooth map such that $u(0) = I_K$ and $v(s) : I_0 \to U(K)$ is a smooth map such that $v(0) = I_H$. Denote $b = a'(0)$, $A = u'(0), B = v'(0)$. Note that $b \in L(H, K)$ and $A^* = -A$, $B^* = -B$. By differentiating $g(s)$ at $s = 0$ and using (4.57), we get

$$\delta(z) = \frac{d}{ds}\varphi_{a(s)}(u(s) \cdot v(s))|_{s=0}$$

$$= b + Az + zB - zb^*z = b + Az + zB - \{z, b, z\},$$

for any $z \in E$. Thus,

$$aut_a(D_1) = \{\delta(z) = b + Az + zB - \{z, b, z\}\}, \tag{4.69}$$

where b ranges over $L(H, K)$, A ranges over $\{A \in L(K) : A^* = -A\}$ and B ranges over $\{B \in L(H) : B^* = -B\}$. Note that the generators are polynomial functions of z of order less than or equal to 2.

For a domain D of type II or III, we get

$$aut_a(D) = \{\delta(z) = b + Az + zA^t - \{z, b, z\}\}, \tag{4.70}$$

where b ranges over $L(H, K)$ and A ranges over $\{A \in L(H) : A^* = -A\}$. We have just characterized the Lie algebra $aut_a(D)$, where D is an arbitrary classical BSD. This establishes an intimate connection between Lie algebras and the triple product. And the connection goes both ways. Given a Lie algebra $aut_a(D)$ of a classical BSD, one can find the unique triple product $\{z, b, z\}$ such that $aut_a(D)$ is given by (4.70). Moreover, the triple product constructed from the Lie algebra will depend solely on the geometry of D and will not change if we perform an isometry on the JC^*-triple E of which D is the unit ball.

4.3 Peirce decomposition in JC^*-triples

The geometry of the state space of a quantum system must reflect the measuring process for such systems. Filtering projections which prepare states with a given definite value of some observable play a major role in the measuring process. In the category of bounded symmetric domains, the analog of a filtering projection is the *Peirce decomposition*, which is the subject of this section.

4.3.1 The operator $D(v)$

Let $E \subset L(H,K)$ be a JC^*-triple. Recall that E is equipped with the triple product defined by (4.9). Consider now the properties of the tripotents, the basic elements of this product. The element $v \in E$ is called a *tripotent* or a *partial isometry* if it satisfies equation (4.27);

$$v = \{v, v, v\} = vv^*v.$$

For any tripotent v, the *right support* of v, defined in section 4.1.4, is the operator $r(v) = v^*v$, because $v = vr(v)$. Note that $r(v)$ is a projection from H onto a subspace of H. Indeed,

$$(r(v))^2 = r(v)r(v) = (v^*v)(v^*v) = v^*(vv^*v) = v^*v = r(v).$$

Similarly, the *left support* of v is the operator $l(v) = vv^*$, because $v = l(v)v$. Like $r(v)$, the operator $l(v)$ is also a projection, but from K onto a subspace of K.

We will see that v is a *partial isometry* in the sense that it is an isometry between a subspace of H and a subspace of K. The projection $r(v)$ maps H onto a subspace H_r, and $l(v)$ maps K onto a subspace K_l. Let $\xi \in H_r$, so that $\xi = r(v)\xi$. Then

$$|v\xi|^2 = \langle v\xi|v\xi\rangle = \langle \xi|v^*v\xi\rangle = \langle \xi|r(v)\xi\rangle = \langle \xi|\xi\rangle = |\xi|^2,$$

so $|v\xi| = |\xi|$. Since $v = vv^*v$, we have $v\xi = (vv^*)v\xi = l(v)v\xi$, implying that $v\xi \in K_l$. Thus v is an isometry between H_r and K_l. Note that v vanishes on the orthogonal complement of H_r, since for any $\eta \in H_r^\perp$, we have $v\eta = v(1 - r(v))\eta = (v - vr(v))\eta = (v - v)\eta = 0$. Moreover, $\dim H_r = \dim K_l$ and is equal to $\operatorname{rank} v$.

Let v be a tripotent in a JC^*-triple E. Then, for any $z \in E$, the operator $D(v) = D(v, v)$ can be decomposed as

$$\begin{aligned} D(v)z = \{v, v, z\} &= \frac{vv^*z + zv^*v}{2} = \frac{1}{2}(l(v)z + zr(v)) \\ &= \frac{1}{2}l(v)z[r(v) + (1 - r(v))] + \frac{1}{2}[l(v) + (1 - l(v))]zr(v) \\ &= l(v)zr(v) + \frac{1}{2}[l(v)z(1 - r(v)) + (1 - l(v))zr(v)]. \end{aligned} \tag{4.71}$$

Write z in block matrix form as

$$\begin{pmatrix} z_{11} & z_{12} \\ z_{21} & z_{22} \end{pmatrix},$$

where $z_{11} = l(v)zr(v)$, $z_{12} = l(v)z(1 - r(v))$, $z_{21} = (1 - l(v))zr(v)$, and $z_{22} = (1 - l(v))z(1 - r(v))$. Then equation (4.71) implies that

$$D(v)z = \begin{pmatrix} z_{11} & \frac{1}{2}z_{12} \\ \frac{1}{2}z_{21} & 0 \end{pmatrix}, \quad D(v)^2 z = \begin{pmatrix} z_{11} & \frac{1}{4}z_{12} \\ \frac{1}{4}z_{21} & 0 \end{pmatrix}. \tag{4.72}$$

4.3.2 Peirce decomposition

Note that the eigenvalues of $D(v)$ are 0, $\frac{1}{2}$, and 1. Let $E_j(v)$ denote the j-eigenspace of $D(v)$, for $j \in \{0, \frac{1}{2}, 1.\}$ The natural projections $P_j(v)$ onto $E_j(v)$ are called the *Peirce projections* associated with the tripotent v. From (4.72), it follows that

$$P_1(v)(z) = l(v)zr(v), \tag{4.73}$$

$$P_{1/2}(v)(z) = l(v)z(1 - r(v)) + (1 - l(v))zr(v), \tag{4.74}$$

$$P_0(v)(z) = (1 - l(v))z(1 - r(v)). \tag{4.75}$$

From (4.72), we also have

$$P_1(v) = 2D(v)^2 - D(v),$$

$$P_{1/2}(v) = 4D(v) - 4D(v)^2,$$

and

$$P_0(v) = 2D(v)^2 - 3D(v) + I,$$

which implies that the Peirce projections map E into itself. It is obvious that $P_1(v)+P_{1/2}(v)+P_0(v) = I$. Thus, we have arrived at the *Peirce decomposition* of E with respect to a tripotent v:

$$E = P_1(v)E + P_{1/2}(v)E + P_0(v)E = E_1(v) + E_{1/2}(v) + E_0(v). \tag{4.76}$$

Since the norm of a projection on a Hilbert space equals 1, and the norm of the product of operators is less than or equal to the product of norms, the Peirce projections $P_1(v)$ and $P_0(v)$ are contractions, meaning that for any $z \in E$, we have $\|P_j(v)(z)\| \leq \|z\|$. To show that $P_{1/2}(v)$ is a contraction, we introduce a symmetry S_v which depends on v and is defined by

$$S_v(z) = (2l(v) - 1)z(2r(v) - 1).$$

Since the operators $2l(v) - 1$ and $2r(v) - 1$ are unitary maps, S_v is norm-preserving. But

$$P_{1/2}(v) = \frac{I - S_v}{2},$$

which implies that $P_{1/2}(v)$ is a contraction.

In quantum mechanics, the notion of *compatible observables* plays an important role. They represent physical quantities that can be measured simultaneously, meaning that measuring one quantity does not affect the result of measuring the other quantity. This property is equivalent to the commuting of the spectral projections of the operators representing the compatible observables and implies that the space generated by these observables is a commutative space. On the other hand, quantum mechanics is non-commutative. For JC^*-triples, we can build a full non-commutative basis with tripotents that generate a commuting Peirce decomposition. This follows from the following observation.

Let u, v be tripotents in a JC^*-triple E, and let $u \in E_j(v)$. Then, for any $k, n \in \{1, 1/2, 0\}$, we have

$$[P_k(v), P_n(u)] = P_k(v)\, P_n(u) - P_n(u)\, P_k(v) = 0. \tag{4.77}$$

Let us check (4.77) case by case. If $j = 1$, then $r(u)r(v) = r(v)r(u) = r(u)$, and, similarly $l(u)l(v) = l(v)l(u) = l(u)$, implying that $[P_1(v), P_1(u)] = 0$. We also have $r(u)(1 - r(v)) = r(u) - r(u)r(v) = r(u) - r(u) = 0$ and $(1 - r(v))r(u) = 0$. Similar identities hold for the left support, and so $[P_0(v), P_1(u)] = 0$. Next,

$$P_{1/2}(v)P_1(u) = (1 - P_1(v) - P_0(v))P_1(u)$$

$$= P_1(u)(1 - P_1(v) - P_0(v)) = P_1(u)P_{1/2}(v),$$

implying that $[P_{1/2}(v), P_1(u)] = 0$. Note that $r(v)(1 - r(u)) = r(v) - r(v)r(u) = r(v) - r(u)$ and $(1 - r(u))r(v) = r(v) - r(u)r(v) = r(v) - r(u)$. Thus, $[P_0(u), P_1(v)] = 0$. As above, this implies that $[P_{1/2}(u), P_1(v)] = 0$. The remaining identities are easily checked for this case. A similar proof holds for the case $j = 0$.

It remains to check the case $j = 1/2$. Decompose u as

$$u = l(v)u(1 - r(v)) + (1 - l(v))ur(v) = u_1 + u_2.$$

Note that $u_1 u_2^* = l(v)u(1-r(v))r(v)u^*(1-l(v)) = 0$ and, similarly, $u_2 u_1^* = 0$. Thus, $u = u^{(3)} = u_1^{(3)} + u_2^{(3)} = u_1 + u_2$, and both u_1 and u_2 are tripotents. Note that $l(u)l(v) = l(u_1)l(v) = l(u_1)$ and $l(v)l(u) = l(u_1)$. Similarly, $r(u)r(v) = r(u_2)r(v) = r(u_2)$ and $r(v)r(u) = r(u_2)$. Hence, the Peirce decompositions commute in this case also.

4.3.3 Peirce calculus

Let v be a tripotent in $E \subset L(H, K)$. Define a one-parameter family of unitary operators on H by $u(t) = e^{it/2}r(v) + (1 - r(v))$. Similarly, define a one-parameter family of unitary operators on K by $w(t) = e^{it/2}l(v) + (1 - l(v))$. Then, define a map $g(t) : H \to K$ by

$$g(t)z = w(t)zu(t). \tag{4.78}$$

By applying the Peirce decomposition to z, we obtain

$$g(t)z = w(t)(P_1(v)z + P_{1/2}(v)z + P_0(v)z)u(t)$$

$$= e^{it}P_1(v)z + e^{it/2}P_{1/2}(v)z + P_0(v)z, \tag{4.79}$$

implying that $g(t)$ maps E to E. Thus, $g(t)$ is a one-parameter family of isometries on E.

Note also that from (4.78), it follows that $g(t)$ is a triple product automorphism. Indeed,

$$g(t)\{a,b,c\} = w(t)\{a,b,c\}u(t) = \frac{w(t)ab^*cu(t) + w(t)cb^*au(t)}{2}$$

$$= \frac{w(t)au(t)u(t)^*b^*w(t)^*w(t)cu(t) + w(t)cu(t)u(t)^*b^*w(t)^*w(t)au(t)}{2}$$

$$= \{w(t)au(t), w(t)bu(t), w(t)cu(t)\} = \{g(t)a, g(t)b, g(t)c\}. \tag{4.80}$$

The generator of the family $g(t)$ is

$$\delta = \frac{d}{dt}g(t)|_{t=0} = iP_1(v) + i\frac{1}{2}P_{1/2}(v) = iD(v). \tag{4.81}$$

Differentiating equation (4.80) at $t = 0$, we get

$$iD(v)\{a,b,c\} = \{iD(v)a,b,c\} + \{a,iD(v)b,c\} + \{a,b,iD(v)c\}. \tag{4.82}$$

Using the linearity and the conjugate linearity of the triple product, we have

$$D(v)\{a,b,c\} = \{D(v)a,b,c\} - \{a,D(v)b,c\} + \{a,b,D(v)c\}, \tag{4.83}$$

for any $a, b, c \in E$.

Assume now that $a \in E_j(v)$, $b \in E_k(v)$, and $c \in E_n(v)$, where j, k, and n all belong to $\{1, 1/2, 0\}$. Then, from (4.83), we get

$$D(v)\{a,b,c\} = \{ja,b,c\} - \{a,kb,c\} + \{a,b,nc\} = (j-k+n)\{a,b,c\},$$

implying that the triple product $\{a,b,c\}$ belongs to $E_{j-k+n}(v)$. Thus, we have shown that

$$\{E_j(v), E_k(v), E_n(v)\} \subseteq E_{j-k+n}(v). \tag{4.84}$$

This relation is called the *Peirce calculus*, and it shows how to compute the triple product for elements of various parts of the Peirce decomposition.

Note that if $j - k + n$ does not belong to the set $\{1, 1/2, 0\}$, then $\{E_j(v), E_k(v), E_n(v)\} = 0$. If $j = k = n$, then (4.84) yields

$$\{E_j(v), E_j(v), E_j(v)\} \subseteq E_j(v),$$

implying that $E_j(v)$ is a JC^*-triple. If $j = 1$ and $k = 0$, then

$$ab^* = ar(v)(b(1 - r(v)))^* = ar(v)(1 - r(v))b^* = 0$$

and

$$b^*a = ((1 - l(v))b)^* l(v)a = b^*(1 - l(v))l(v)a = 0,$$

for a, b as above. Thus $\{a, b, c\} = 0$ and so

$$\{E_1(v), E_0(v), E\} = 0. \tag{4.85}$$

Similarly,

$$\{E_0(v), E_1(v), E\} = 0. \tag{4.86}$$

Equations (4.84),(4.85) and (4.86) provide a complete set of rules for computing the triple product of elements of various parts of the Peirce decomposition.

4.3.4 The main identity of the triple product

We now derive the main identity of the triple product, which is a generalization of (4.83). Let $E \subset L(H, K)$ be a JC^*-triple. Suppose $d \in E$ has discrete singular decomposition (4.32) $d = \sum s_j v_j$, where the v_j's are algebraically orthogonal tripotents. Then, from (4.85) and (4.86), it follows that

$$D(d)z = \left\{\sum s_j v_j, \sum s_k v_k, z\right\} = \sum s_j^2 D(v_j)z.$$

Thus, equation (4.83) implies that

$$D(d)\{a, b, c\} = \{D(d)a, b, c\} - \{a, D(d)b, c\} + \{a, b, D(d)c\}, \tag{4.87}$$

for any $a, b, c \in E$. If d has non-discrete spectrum, then it is the limit of elements with discrete spectrum, for which equation (4.87) holds. Thus, by passing to the limit, this equation will hold also for d.

For arbitrary $x, y \in E$ and any complex number λ, let $d = x + \lambda y$. Then

$$D(d) = D(x + \lambda y, x + \lambda y) = D(x) + D(y) + \lambda D(y, x) + \overline{\lambda} D(x, y).$$

Substituting this into (4.87) and considering the coefficient of $\overline{\lambda}$, we get

$$D(x, y)\{a, b, c\} = \{D(x, y)a, b, c\} - \{a, D(y, x)b, c\} + \{a, b, D(x, y)c\}, \tag{4.88}$$

which is called the *main identity* of the triple product.

4.4 Non-commutative perturbation

Any quantum theory has to handle *perturbations*, changes of certain quantities caused by small changes in the state. The state space of a quantum system may be described by a trace class operator a. If the considered quantity is an operator function of the state, the perturbation will be the derivative of the operator function at a in direction of the change. This derivative is described by use of the joint Peirce decomposition about a, which we introduce now.

4.4.1 Joint Peirce decomposition

In this section, we define the joint Peirce decomposition with respect to a fixed element a with a discrete spectrum in a JC^*-triple E. By use of the discrete singular decomposition (4.32), we decompose a as $a = \sum s_j v_j$, where the v_j's are a family of algebraically orthogonal tripotents, with $j \in J = \{1, 2, 3, ...\}$. Denote by v the support tripotent of a from the polar decomposition (4.26). The algebraically orthogonal family of tripotents v_j generates a partition of the Hilbert space H via the family of mutually orthogonal projections $r(v_1), r(v_2), ...$ and $1 - r(v) = 1 - \sum_{j \in J} r(v_j)$. Similarly, the Hilbert space K is partitioned by the family $l(v_1), l(v_2), ...$ and $1 - l(v) = 1 - \sum_{j \in J} l(v_j)$.

Define now two maps $P_+(v)$ and $P_-(v)$ by

$$P_\pm(v)(z) = \frac{z \pm Q(v)z}{2} = \frac{z \pm vz^*v}{2} = z_\pm. \tag{4.89}$$

Note that since $P_1(v) = Q(v)^2$, the real linear (not complex linear) maps $P_\pm(v)$ are projections from the subspace $E_1(v)$ into itself. The images of these projections are the real subspaces $\{z \in E_1(v) : \ z = \pm Q(v)z\}$, corresponding to the spaces of self-adjoint and anti-self-adjoint operators with respect to v. Each element $z \in E$ can be decomposed as

$$z = P_+(v)z + P_-(v)z = z_+ + z_-.$$

As a result, we can decompose each element $z \in E$ using the partitions of the spaces H and K induced by the families $r(v_j)$ and $l(v_j)$, respectively, and by the operators $P_\pm(v)$ (see Figure 4.5).

More precisely, for $|j|, k \in J, \ |j| < k$, define a projection

$$P_{j,k} = P_\pm(v) P_{1/2}(v_j) P_{1/2}(v_k)$$

and denote its range by

$$E_{j,k} = P_{j,k}E = \{z : \ z = l(v_j)zr(v_k) + l(v_k)zr(v_j) = \pm Q(v)z\}, \tag{4.90}$$

with the sign $+$ for positive j and $-$ for negative j. Similarly, for $|j|, k \in J, \ |j| = k$, define a projection

	$r(v_1)$	$r(v_2)$	$r(v_3)$	$1-r(v)$
$l(v_1)$	$E_{11}+E_{-11}$	$E_{12}+E_{-12}$	$E_{13}+E_{-13}$	E_{01}
$l(v_2)$	$E_{12}+E_{-12}$	$E_{22}+E_{-22}$	$E_{23}+E_{-23}$	E_{02}
$l(v_3)$	$E_{12}+E_{-12}$	$E_{23}+E_{-23}$	$E_{33}+E_{-33}$	E_{03}
$1-l(v)$	E_{01}	E_{02}	E_{03}	E_{00}

Fig. 4.5. Partition of the joint Peirce decomposition with respect to an element a with singular decomposition $a = s_1v_1 + s_2v_2 + s_3v_3$ and polar decomposition $a = v|a|$. The Hilbert space H is partitioned by the family of mutually orthogonal projections $r(v_1), r(v_2), r(v_3)$ and $1-r(v)$. The Hilbert space K is partitioned by the family $l(v_1), l(v_2), l(v_3)$ and $1-l(v)$. In addition, the subspace $E_1(v)$ is partitioned by $P_{\pm}(v)$.

$$P_{\pm k,k} = P_{\pm}(v)P_1(v)P_1(v_k)$$

and denote its range by

$$E_{\pm k,k} = P_{\pm k,k}E = \{z : z = l(v_k)zr(v_k) = \pm Q(v)z\}. \tag{4.91}$$

For $j = 0$ and $k \in J$, define

$$P_{0,k} = P_{1/2}(v_k)P_{1/2}(v)$$

and denote its range by

$$E_{0,k} = P_{0,k}E = \{z : z = l(v_k)z(1 - r(v)) + (1 - l(v))zr(v_k)\}. \tag{4.92}$$

Finally, define $P_{0,0} = P_0(v)$ and

$$E_{0,0} = P_0(v)E = \{z : z = (1 - l(v))z(1 - r(v))\}. \tag{4.93}$$

With the above definitions, we have

$$\sum P_{j,k} = I, \tag{4.94}$$

and the JC^*-triple E can be decomposed with respect to a as

$$E = \bigoplus_{|j|\leq k} E_{j,k}, \quad j \in (-J) \cup \{0\} \cup J, \quad k \in J. \tag{4.95}$$

This is called the *joint Peirce decomposition* of E with respect to a.

4.4.2 Differentiation of operator functions

For any real-valued smooth function $f(x)$, defined on R^+ (the positive semi-axis), we defined, in section 4.1.6, the meaning of an operator function $f(z) \in$

E for $z \in E$. By use of the joint Peirce decomposition, we can now define the derivative $\frac{d}{dz}f(a)dz$ of f at any given point $a \in E$ in the direction $dz \in E$.

Let a be a fixed element with a discrete spectrum in a JC^*-triple E. By use of the discrete singular decomposition (4.32), decompose a as $a = \sum s_j v_j$. Let $s_0 = 0$. Decompose E with respect to a by the joint Peirce decomposition (4.95). Note that the subspaces $E_{j,k}$ of the joint Peirce decomposition are invariant subspaces for both operators $D(a)$ and $Q(a)$. More precisely, for any $z \in E_{j,k}$ we have

$$D(a)z = \frac{s_j^2 + s_k^2}{2} z \tag{4.96}$$

and

$$Q(a)z = \begin{cases} s_j s_k z & j > 0, \\ -s_j s_k z & j < 0. \end{cases} \tag{4.97}$$

Consider first the case $f(x) = x^3$. By definition, the derivative $\frac{d}{dz}f(a)dz$ is the linear term in dz in the expansion of $f(a + dz)$. But

$$f(a + dz) = (a + dz)^{(3)} = \{a + dz, a + dz, a + dz\}$$

$$= a^{(3)} + 2D(a)dz + Q(a)dz + o(dz),$$

implying that

$$\frac{d}{dz}f(a)dz = 2D(a)dz + Q(a)dz.$$

Use (4.94) to decompose dz as

$$dz = \sum P_{j,k}dz,$$

where the sum is over all possible j, k, and use the abbreviation $dz_{j,k} = P_{j,k}dz$. Since $z_{j,k} \in E_{j,k}$ from (4.96) and (4.97) and a formula from elementary algebra, we get, for $j \geq 0, j \neq k$,

$$\frac{d}{dz}f(a)dz_{j,k} = (s_j^2 + s_k^2 + s_j s_k)dz_{j,k} = \frac{s_j^3 - s_k^3}{s_j - s_k}dz_{j,k} = \frac{f(s_j) - f(s_k)}{s_j - s_k}dz_{j,k},$$

and for $j = k$, we get

$$\frac{d}{dz}f(a)dz_{k,k} = 3s_k^2 dz_{k,k} = f'(s_k)dz_{k,k}.$$

Similarly, for $j < 0$ we get

$$\frac{d}{dz}f(a)dz_{j,k} = (s_j^2 + s_k^2 - s_j s_k)dz_{j,k} = \frac{s_j^3 + s_k^3}{s_j + s_k}dz_{j,k} = \frac{f(s_j) + f(s_k)}{s_j + s_k}dz_{j,k}.$$

Thus, we can write the above derivative as

$$\frac{d}{dz}f(a)dz = \sum_{j\neq k} \frac{f(s_j) \mp f(s_k)}{s_j \mp s_k} dz_{\pm|j|,k} + \sum_k f'(s_k)dz_{k,k}, \tag{4.98}$$

showing that the derivative acts on the components of the joint Peirce decomposition about a by multiplication by the divided differences $\frac{f(s_j)-f(s_k)}{s_j-s_k}$ (or divided sums $\frac{f(s_j)+f(s_k)}{s_j+s_k}$) outside the diagonal (when $j \neq k$) and by multiplication by the derivative $f'(s_k)$ on the diagonal (when $j = k$).

We will show that a similar formula holds for the derivative of any function f having a smooth derivative, $f'(x)$, as a real function. We will show this first for the case in which f is an odd polynomial. Note that if (4.98) holds for functions f and g, it will also hold for any linear combination of f and g. Thus, in this case, it is enough to prove it for $f(x) = x^{2n+1}$. This can be done by induction as follows.

Let $f(x) = x^{2n+1}$, $g(x) = x^{2n-1}$, and let a be as above. Then

$$f(a+dz) = (a+dz)^{(2n+1)} = \{a+dz, a+dz, g(a+dz)\},$$

implying that

$$\frac{d}{dz}f(a)dz = \{dz, a, g(a)\} + \{a, dz, g(a)\} + \{a, a, \frac{d}{dz}g(a)dz\}.$$

By the induction hypothesis, formula (4.98) holds for the function g, and from the definition of the operator function (4.20), we have $g(a) = \sum s_j^{2n-1}v_j$. Thus, for $j \geq 0, j \neq k$, we get

$$\frac{d}{dz}f(a)dz_{j,k}$$

$$= \frac{1}{2}(s_j^{2n} + s_k^{2n} + s_j^{2n-1}s_k + s_j s_k^{2n-1} + (s_j^2 + s_k^2)\frac{s_j^{2n-1} - s_k^{2n-1}}{s_j - s_k})dz_{j,k}$$

$$= \frac{s_j^{2n+1} - s_k^{2n+1}}{s_j - s_k}dz_{j,k} = \frac{f(s_j) - f(s_k)}{s_j - s_k}dz_{j,k},$$

and, for $j = k$, we get

$$\frac{d}{dz}f(a)dz_{k,k} = (s_k^{2n} + s_k^{2n} + s_k^2(2n-1)s_k^{2n-2})dz_{k,k}$$

$$= (2n+1)s_k^{2n}dz_{k,k} = f'(s_k)dz_{k,k}.$$

Similarly, for $j < 0$ we get

$$\frac{d}{dz}f(a)dz_{j,k}$$

$$= \frac{1}{2}(s_j^{2n} + s_k^{2n} - s_j^{2n-1}s_k - s_j s_k^{2n-1} + (s_j^2 + s_k^2)\frac{s_j^{2n-1} + s_k^{2n-1}}{s_j + s_k})dz_{j,k}$$

$$= \frac{s_j^{2n+1} + s_k^{2n+1}}{s_j + s_k}dz_{j,k} = \frac{f(s_j) + f(s_k)}{s_j + s_k}dz_{j,k}.$$

Thus, (4.98) holds also for $f(x) = x^{2n+1}$ and for any odd polynomial in x.

The above formula can be extended to a point a with non-discrete spectrum by approximating it with elements with discrete spectrum constructed from the singular decomposition (4.31). The formula (4.98) can also be extended to functions f with a smooth derivative. It is not enough for the derivative to be continuous, but if f is twice differentiable, then the formula holds. For more precise conditions for operator differentiability, see [6].

4.5 The dual space to a JC^*-triple

In this section, we will present the structure of the dual space - the space of complex linear functionals on a JC^*-triple.

4.5.1 The description of the dual space

We assume now that $E = L(H, K)$ has finite dimension, namely $\dim H = n$ and $\dim K = m$. Choose an orthonormal basis $e_1, e_2, ..., e_n$ in H and an orthonormal basis $h_1, h_2, ..., h_m$ in K. This defines a basis in E consisting of elements $e_{j,k}$, for $j = 1, \dots, m,\ k = 1, ..., n$, defined by

$$e_{j,k}\xi = \langle \xi | e_k \rangle h_j. \tag{4.99}$$

Any $z \in E$ can be expanded as

$$z = \sum_{j,k} z_{j,k}e_{j,k}, \text{ with } z_{j,k} = \langle z e_j | h_k \rangle \tag{4.100}$$

and represented by an $m \times n$ matrix $M(z)$ with complex entries $z_{j,k}$. Thus, E can be regarded as the set of $m \times n$ matrices. We denote the unit ball of E in the operator norm by D.

The space of linear functionals, called the dual space, will be denoted by E^*. We define a norm for an element f of E^* by

$$\|f\|_* = \sup\{|f(z)| : \ z \in D\}. \tag{4.101}$$

The unit ball S of E^* plays the role of a state space. Let f be any linear functional in E^*. Let $f_{j,k} = f(e_{j,k})$ and let $M(f)$ be the $m \times n$ matrix with entries $\overline{f}_{j,k}$. Thus, f can also be considered as a linear map from H to K. From the linearity of f, it follows that

$$f(z) = \sum_{j,k} z_{j,k} f(e_{j,k}) = \sum_{j,k} z_{j,k} f_{j,k} = \sum_{j} (\sum_{k} z_{j,k} f_{j,k}) = \sum_{j} d_{j,j},$$

where the $m \times m$ matrix $d = M(z)M(f)^*$. So we can write

$$f(z) = \operatorname{tr}(M(z)M(f)^*), \tag{4.102}$$

which is the usual definition for functionals on operators.

The last formula can be viewed from a different perspective. Define a Hilbert space of dimension mn with $e_{j,k}$ as an orthonormal basis. This Hilbert space is called the *tensor product* of H and K and is denoted by $H \otimes K$. As a linear space, it is isomorphic to E. Thus, any linear functional f on E is also a linear functional on the Hilbert space $H \otimes K$. Denote the dual basis to $e_{j,k} \in H \otimes K$ by $h_{j,k}$, meaning that $h_{j,k}(e_{p,q}) = \delta_j^p \delta_k^q$. By Riesz representation, we can represent any linear functional f on the Hilbert space $H \otimes K$ by

$$f(z) = \langle z | \check{f} \rangle_{H \otimes K},$$

where the inner product $\langle \cdot | \cdot \rangle_{H \otimes K}$ is that of the Hilbert space $H \otimes K$ and $\check{f}$ is the image of f under the Riesz representation, defined as a conjugate linear map $E^* \to E$ which maps the basis elements $h_{j,k}$ of E^* to $\check{h}_{j,k} = e_{j,k}$, implying that $\check{f} = \sum \overline{f(e_{j,k})} e_{j,k}$. Hence, formula (4.102) can be rewritten as

$$f(z) = \operatorname{tr}(M(z)M(f)^*) = \langle z | \check{f} \rangle_{H \otimes K}. \tag{4.103}$$

This shows that the duality in operator spaces and Hilbert spaces is similar. The difference lies only in the norm and the group of isometries of these spaces.

Consider now the case in which the dimension of H or K is infinite. Define an increasing sequence of Hilbert spaces H_n converging to H and a sequence K_m converging to K. The restriction of a functional f to the subspace $L(H_n, K_m)$ of $L(H, K)$ must satisfy (4.103). Thus, also in this case f can be considered as a linear operator from H to K, but it will have to satisfy some norm condition (described below) that will insure that the sum in the trace and in the inner product of (4.103) converges.

4.5.2 Decomposition of a functional into pure states

For any pair of fixed elements $\xi \in H$ and $\eta \in K$, we define an element $g = \eta \otimes \xi$ of E^* by

$$g(z) = \eta \otimes \xi(z) = \langle \eta | z\xi \rangle, \tag{4.104}$$

for $z \in E$. Since $|\langle\eta|z\xi\rangle| \leq |\eta|\,|z|\,|\xi|$, the norm $\|\eta\otimes\xi\|_* \leq |\eta|\,|\xi|$. If the vectors η, ξ are of norm 1, we may choose the orthonormal basis in H such that $\xi = e_1$ and a basis in K such that $\eta = h_1$. In this basis, for any element z with decomposition (4.100), we have $g(z) = z_{1,1}$, implying that $\|g(z)\|_* = \|\eta\otimes\xi\|_* = 1$. We will call such a g a *pure state*. Note that with this notation, the basis of E^*, defined above, consists of pure states $h_{j,k} = h_j \otimes e_k$.

For any functional $f \in E^*$, we can define a linear map $\check{f}$ from H to K, as above. If the Hilbert spaces H and K are infinite dimensional, we have to apply the following arguments first to the restriction of f to $L(H_n, K_m)$, as above, and then pass to the limit. We use the singular decomposition (4.32) of $\check{f}$ to decompose it as

$$\check{f} = \sum s_l v_l, \tag{4.105}$$

where the v_l's form an algebraically orthogonal family of tripotents. If the rank of some v_l (which is equal to the dimension of the projection $r(v_l)$) is larger than 1, we decompose this tripotent into a sum of rank one algebraically orthogonal tripotents. Hence, we will assume that all tripotents in the decomposition (4.105) are of rank 1. Choose $\xi_l \in H$ such that $\xi_l = r(v_l)\xi_l$ and $|\xi_l| = 1$. If we denote $\eta_l = v_l\xi_l$, then $\eta_l = l(v_l)\eta_l$ and $|\eta_l| = 1$. From (4.105), we get

$$f = \sum_l s_l(\eta_l \otimes \xi_l), \tag{4.106}$$

which is a decomposition of a functional f into mutually orthogonal pure states. Obviously,

$$\|f\|_* \leq \sum s_l\|\eta_l \otimes \xi_l\|_* = \sum s_l. \tag{4.107}$$

But for the support tripotent $v = \sum v_l$ of $\check{f}$, we get

$$f(v) = \sum s_l(\eta_l \otimes \xi_l)v = \sum s_l\langle\eta_l|v\xi_l\rangle = \sum s_l\langle\eta_l|\eta_l\rangle = \sum s_l. \tag{4.108}$$

Therefore,

$$\|f\|_* = \sum s_l, \tag{4.109}$$

where s_l are the singular numbers of $\check{f}$. Thus, any norm 1 functional f in E^* is a convex combination of mutually orthogonal pure states.

Note that from (4.109), it follows that a linear map f from H to K defines a linear functional on E if the sum of its singular values $\sum s_l$ is finite. Such an operator is called a *trace class operator* and the norm $\|f\|_*$, defined by (4.109), is called the *trace norm*. This is equivalent to

$$\mathrm{tr}|f| < \infty.$$

To see this we note the following: if f has the decomposition (4.106), then $|f| = \sum_l s_l(\xi_l \otimes \xi_l)$, and by choosing the orthonormal family ξ_l as a part of the orthonormal basis in H, we get

$$\mathrm{tr}|f| = \sum_l s_l = \|f\|_*. \tag{4.110}$$

4.5.3 Facial structure of the state space

In general, the building blocks of the affine geometric structure of any convex subset S of a Banach space are the extreme points of S, or more generally, the extreme subsets, or faces, of S. We denote by S the unit ball of E^* with the norm defined by (4.110). This convex set can play the role of a state space for quantum systems.

E. Effros in [19] and R. Prosser in [60] studied the geometry of the state space $\widetilde{S}$, consisting of positive norm less than or equal to 1 elements in the dual to the space of compact operators on a Hilbert space H or in the predual of the space of bounded operators on H, and even more generally, in the predual of a von Neumann algebra M. They showed that the norm-closed faces of $\widetilde{S}$ are in one-to-one correspondence with the self-adjoint projections in M. Moreover, orthogonality of faces corresponds to orthogonality of projections.

To describe the *extreme points* of S, let $g = \eta \otimes \xi$, with $|\eta| = |\xi| = 1$, be a pure state. The support tripotent u of g is defined by $u\rho = \eta\langle\rho|\xi\rangle$ for $\rho \in H$. If $f \in S$ is exposed by u, from (4.108) we get

$$1 = f(u) = \sum s_l(\eta_l \otimes \xi_l)u = \sum s_l\langle\eta_l|\eta\rangle\langle\xi|\xi_l\rangle \le \sum s_l \le 1. \tag{4.111}$$

Thus, $\eta_l = \eta$, $\xi_l = \xi$, implying that $f = g$. So g is the only state exposed by u. This implies that any pure state is a norm-exposed face and an extreme point of S. Conversely, any extreme point of S cannot be a non-trivial convex combination of elements of S. Thus, from the decomposition (4.106), it follows that only the pure states are extreme points.

From (4.108), it follows that for any functional f, there exists a tripotent v, which is the support tripotent of $\check{f}$, on which f assumes its norm, *i.e.*,

$$f(v) = \|f\|_*. \tag{4.112}$$

Thus, f belongs to a norm-exposed face generated by a tripotent v. Conversely, if a face F_x is norm exposed by a norm 1 element x of E, it can be verified directly that F_x is also exposed by the tripotent u corresponding to the singular value 1 in the singular decomposition (4.31). Moreover, for any $f \in F_x$, its support tripotent v is dominated by u, which we denote by $v \preceq u$. Thus, there is a one-to-one correspondence between the norm-exposed faces of S and the tripotents of E.

To any tripotent u of E, we can associate not only a norm-exposed face on S, but also a decomposition defined by the action of the Peirce projections

$P_j(u)$. This corresponds to the decomposition coming from the measuring process. For any tripotent u of E and $j = 1, 1/2, 0$, we define projections P_j^* on E^* as follows: for any $f \in E^*$ and any $z \in E$, we define

$$(P_j^*(u)f)z = f(P_j(u)z).$$

Obviously, $P_j^*(u)f \in E^*$.

We show now that the projection $P_1^*(u)$ is *neutral*, meaning that the equality $\|P_1^*(u)f\|_* = \|f\|_*$ implies $P_1(u)f = f$. As explained in section 3.3.5 of Chapter 3, the filtering projection on the state space from the measuring process corresponding to the Peirce projection P_1^* is neutral. Let u be a tripotent in E and let $f^* \in E^*$ be such that $\|P_1^*(u)f\|_* = \|f\|_*$. Denote by z a norm-one element in E such that the functional $P_j^*(u)f$ assumes its norm on it, *i.e.*, $|(P_j^*(u)f)z| = \|(P_j(u)f\|_*$. By using the decomposition (4.106), we get

$$\|(P_1^*(u)f\|_* = |(P_1^*(u)f)z| = |f(P_j(u)z| \leq \sum s_l|(\eta_l \otimes \xi_l)P_1(u)z|$$

$$= \sum s_l|\langle \eta_l|l(u)zr(u)\xi_l\rangle| \leq \sum s_l|l(u)\eta_l|\,|r(u)\xi_l|\,\|z\| \leq \sum s_l = \|f\|_*.$$

Since we must now have equality in each inequality, for any l we have $|l(u)\eta_l| = |\eta_l|$, implying that $l(u)\eta_l = \eta_l$ and, similarly, $r(u)\xi_l = \xi_l$. Thus,

$$\|P_1^*(u)f\|_* = \|f\|_* \Rightarrow P_1(u)f = f, \tag{4.113}$$

implying that the projection $P_1^*(u)$ is neutral.

4.6 The infinite-dimensional classical domains

The *infinite-dimensional domains* consist of bounded operators. The natural basis in the spaces in which these domains reside is constructed from minimal rank operators. The norm closure of the span of the basis consists of compact operators, which form a subspace of the space of bounded operators. However, for our applications, it is not enough to use compact operators. For example, the support tripotent of a compact operator may not be compact. Hence, we need to define a new convergence, called w^*-convergence, and with respect to this convergence, the w^*-closure of the span of the basis elements will lead to the space of bounded operators for which the classical domain is the unit ball.

4.6.1 The natural basis of classical domains

Let D be a classical domain of type I, which is a unit ball in $E = L(H, K)$. Choose an orthonormal basis $\{e_j : j \in J\}$ in H and an orthonormal basis

$\{f_i : i \in I\}$ in K. For any pair of indexes $i, j \in I \times J$, we define $\mathbf{u}_{i,j} \in L(H, K)$ by

$$\mathbf{u}_{i,j}\xi = (f_i \otimes e_j)\xi = \langle \xi | e_j \rangle f_i, \tag{4.114}$$

for $\xi \in H$. Since $\mathbf{u}_{i,j}^* = e_j \otimes f_i$, we have

$$\mathbf{u}_{i,j}\mathbf{u}_{i,j}^*\mathbf{u}_{i,j}\xi = \langle \xi | e_j \rangle \langle f_i | f_i \rangle \langle e_j | e_j \rangle f_i = \mathbf{u}_{i,j}\xi,$$

implying that $\mathbf{u}_{i,j}$ is a tripotent. Moreover, from section 4.3.2, it follows that these tripotents are compatible.

The family of tripotents $\{\mathbf{u}_{i,j}\}_{i \in I, j \in J}$ in a JC^*-triple E is said to form a *rectangular grid* if

(i) any distinct $\mathbf{u}_{j,k}, \mathbf{u}_{i,l}$ are *co-orthogonal*, meaning that $\mathbf{u}_{j,k} \in E_{1/2}(\mathbf{u}_{i,l})$ and $\mathbf{u}_{i,l} \in E_{1/2}(\mathbf{u}_{j,k})$, when they share a common row index ($j = i$) or column index ($k = l$), and otherwise they are algebraically orthogonal, meaning that $\mathbf{u}_{j,k} \in E_0(\mathbf{u}_{i,l})$ and $\mathbf{u}_{i,l} \in E_0(\mathbf{u}_{j,k})$,

(ii) for any $j \neq i, l \neq k$, we have

$$\mathbf{u}_{j,k} = \frac{1}{2}\{\mathbf{u}_{j,l}, \mathbf{u}_{i,l}, \mathbf{u}_{i,k}\},$$

(iii) all triple products of the basis elements vanish unless they are of the form $\{\mathbf{u}_{j,l}, \mathbf{u}_{j,l}, \mathbf{u}_{i,k}\}$ for $\{j, l\} \cap \{i, k\} \neq \varnothing$ or $\{\mathbf{u}_{j,l}, \mathbf{u}_{i,l}, \mathbf{u}_{i,k}\}$.

By direct verification, we see that $\mathbf{u}_{i,j}$ defined by (4.114) form a rectangular grid. In case E is finite dimensional, the span of the grid is equal to E, and the grid is a natural basis in E. In case E is infinite dimensional, we will see later how to define convergence in E so that the closure of the span of the grid will be equal to E.

Consider now a domain D of type III in $E \subset L(H)$, with $E = \{\mathbf{a} \in L(H) : \mathbf{a}^t = \mathbf{a}\}$. Choose an orthonormal basis $\{e_j : j \in J\}$ in H. Define for any pair of indexes $i, j \in J \times J$ the element $\mathbf{u}_{i,j} \in E$ by

$$\mathbf{u}_{i,j} = \frac{e_i \otimes e_j + e_j \otimes e_i}{2}, \tag{4.115}$$

where $e_i \otimes e_j$ is defined as in (4.114). Also in this case, it is easy to verify that the family $\mathbf{u}_{i,j}$ consists of compatible tripotents.

Let $F = \{\mathbf{u}_{ij} | i, j \in I\}$ for some index set I be a family of tripotents in a JC^*-triple E. We will say that F is a *Hermitian grid* if for every $i, j, k, l \in I$, we have $\mathbf{u}_{i,j} = \mathbf{u}_{j,i}$ and the tripotents satisfy the following relations:

(i)

$$\mathbf{u}_{i,i} \in E_1(\mathbf{u}_{i,j}),\ \mathbf{u}_{i,j} \in E_{1/2}(\mathbf{u}_{i,i}),\ \mathbf{u}_{i,j} \in E_{1/2}(\mathbf{u}_{j,k}),\ \mathbf{u}_{j,k} \in E_{1/2}(\mathbf{u}_{i,j}),$$

when i, j, k are different, and

$$\mathbf{u}_{i,j} \in E_0(\mathbf{u}_{k,l}),\ \mathbf{u}_{k,l} \in E_0(\mathbf{u}_{i,j}),$$

when $\{i,j\} \cap \{k,l\} = \varnothing$.

(ii) Every triple product among elements of F which cannot be brought to the form $\{\mathbf{u}_{i,j}, \mathbf{u}_{j,k}, \mathbf{u}_{k,l}\}$ vanishes, and for arbitrary i,j,k,l, the triple products involving at least two different elements satisfy

$$\{\mathbf{u}_{i,j}, \mathbf{u}_{j,k}, \mathbf{u}_{k,l}\} = \frac{1}{2}\mathbf{u}_{i,l}, \text{ for } i \neq l,$$

and

$$\{\mathbf{u}_{i,j}, \mathbf{u}_{j,k}, \mathbf{u}_{k,i}\} = \mathbf{u}_{i,i}.$$

By direct verification, we see that $\mathbf{u}_{i,j}$ defined by (4.115) form a Hermitian grid.

Consider now a domain D of type II in $E \subset L(H)$, with $E = \{\mathbf{a} \in L(H) : \mathbf{a}^t = -\mathbf{a}\}$. Choose an orthonormal basis $\{e_j : j \in J\}$ in H. Define for any pair of indexes $i,j \in J \times J$, $i \neq j$, the element $\mathbf{u}_{i,j} \in E$ by

$$\mathbf{u}_{i,j} = \frac{e_i \otimes e_j - e_j \otimes e_i}{2}, \tag{4.116}$$

where $e_i \otimes e_j$ is defined as in (4.114). Also in this case, it is easy to verify that the family $\mathbf{u}_{i,j}$ consists of compatible tripotents.

A family of tripotents $F = \{\mathbf{u}_{i,j} : i,j \in I\}$ in a JC^*-triple E is called a *symplectic grid* if $\mathbf{u}_{i,j} = -\mathbf{u}_{i,j}$ for all $i,j \in I$, implying that $\mathbf{u}_{i,i} = 0$, and

(i) if $j \neq k$, then

$$\mathbf{u}_{i,j} \in E_{1/2}(\mathbf{u}_{k,j}),\ \mathbf{u}_{k,j} \in E_{1/2}(\mathbf{u}_{i,j}),$$

and if all indexes i,j,k,l are distinct,

$$\mathbf{u}_{i,j} \in E_0(\mathbf{u}_{k,l}),\ \mathbf{u}_{k,l} \in E_0(\mathbf{u}_{i,j}),$$

(ii) if all indexes i,j,k,l are distinct,

$$\mathbf{u}_{i,j} = \frac{1}{2}\{\mathbf{u}_{i,l}, \mathbf{u}_{k,l}, \mathbf{u}_{k,j}\}$$

and

$$\mathbf{u}_{k,j} = \frac{1}{2}\{\mathbf{u}_{i,j}, \mathbf{u}_{i,j}, \mathbf{u}_{k,j}\},$$

(iii) all triple products among elements of the family that are not of the form in (ii) vanish.

By direct verification, we see that $\mathbf{u}_{i,j}$ defined by (4.116) form a symplectic grid.

4.6.2 w^*-convergence and JCW^*-triples

Let D be a classical domain of infinite rank, meaning that $\dim H = \infty$, and, for type I, also $\dim K = \infty$. In this case, we can choose a basis $\mathbf{u}_{i,j}$ for some index sets I, J ($I = J$ for type II and III domains) consisting of compatible tripotents, as described above. We will denote the linear subspace consisting of all finite linear combinations of elements of the basis by E_0 and write $E_0 = \operatorname{span}\{\mathbf{u}_{i,j}\}_{i,j \in I \times J}$. We can use the operator norm for elements of E_0. This space will consist of finite rank operators.

The space E_0 is not complete, meaning that there are sequences of operators in E_0 converging in norm to an operator which is not in E_0. A typical example is an operator $\mathbf{a}$ with singular decomposition

$$\mathbf{a} = \sum_{j=1}^{\infty} s_j \mathbf{v}_j,$$

with orthogonal tripotents $\mathbf{v}_j$ of finite rank and strictly positive singular numbers s_j monotonically decreasing to 0. Such an operator is called a *compact operator*. Obviously, for any integer m, the operator $\mathbf{a}_m = \sum_{j=1}^{m} s_j \mathbf{v}_j$ is of finite rank and is an element of E_0. It is also obvious that

$$\|\mathbf{a} - \mathbf{a}_m\| = \|\sum_{j=m+1}^{\infty} s_j \mathbf{v}_j\| = s_{m+1},$$

and goes to 0 as $m \to \infty$, implying that $\mathbf{a} = \lim \mathbf{a}_m$, where the limit is taken in the operator norm. But $\mathbf{a}$ is of infinite rank and thus is not in E_0.

We can complete the space E_0 by adding to it operators which are the limit of a norm-converging sequence from E_0. We will denote this space by E_c and write

$$E_c = \overline{\operatorname{span}}\{\mathbf{u}_{i,j}\}_{i,j \in I \times J}. \tag{4.117}$$

It can be shown that for each type of classical domain, the space E_c consists of compact operators. However, for physical applications, we have to consider also bounded non-compact operators and even sometimes unbounded ones. So we need a completion of E_0 that will consist of all bounded operators in D. This could be done by embedding E_0 into E_c^{**} and completing it there with respect to w^*-convergence, defined below.

We will describe now the space E_c^*, the dual to the space of compact operators. From the previous section, the dual E_0^* of E_0 consists of operators with pure state decomposition (4.106) for which $\|f\|_* = \sum s_l < \infty$. Such operators are called trace class operators. They are used to represent states of quantum systems. It can be shown that for type I domains, E_c^* consists of all trace class operators from H to K, and for type II and III domains, E_c^* consists of all anti-symmetric and symmetric trace class operators on H, respectively.

The dual to the space of trace class operators is the space of bounded operators which we considered in the definition of the classical domains. To obtain this space from E_0, we have to consider w^*-convergence, which is defined as follows. Let E be a Banach (complete normed) space, which is the dual of a Banach space E_*, called its *predual.* Denote the norm of an element $f \in E_*$ by $\|f\|_*$. We will say that a sequence $\mathbf{a}_m$ of elements of E *converges* w^* to $\mathbf{a} \in E$ if for any $\mathbf{f} \in E_*$, we have

$$\lim_{m\to\infty} \mathbf{a}_m(\mathbf{f}) = \mathbf{a}(\mathbf{f}). \tag{4.118}$$

This will be denoted by $\mathbf{a}_m \xrightarrow{w^*} \mathbf{a}$. It can be shown that the Banach space $E := E_c^{**}$, which is the dual of the Banach space E_c^*, the space of all trace class operators, is the w^*-closure of E_0. We will write this as

$$E = \overline{\operatorname{span}}^{w^*} \{\mathbf{u}_{i,j}\}_{i,j\in I\times J}. \tag{4.119}$$

This means that for any bounded operator $\mathbf{a}$, we can find a sequence $\mathbf{a}_m \in E_0$ such that (4.118) will hold for any trace class operator $\mathbf{f} \in E_* = E_c^*$. Moreover, the finite rank operators can be chosen as $\mathbf{a}_m = P_m a Q_m$, where P_m, Q_m are families of finite-dimensional projections of the Hilbert spaces K, H respectively, converging to the identity.

We will say that the triple product is w^*-continuous in a JC^*-triple E, which is the dual of E_*, if

$$\mathbf{a}_m \xrightarrow{w^*} \mathbf{a}, \mathbf{b}_m \xrightarrow{w^*} \mathbf{b}, \mathbf{c}_m \xrightarrow{w^*} \mathbf{c} \Rightarrow \{\mathbf{a}_m, \mathbf{b}_m, \mathbf{c}_m\} \xrightarrow{w^*} \{\mathbf{a}, \mathbf{b}, \mathbf{c}\}. \tag{4.120}$$

In this case, we call E a JCW^*-triple. It can be shown that $E = E_c^{**}$ is a JCW^*-triple.

4.7 Notes

The theory of non–self-adjoint operators on a Hilbert space is well presented in [38]. The classical bounded symmetric domains and JC^*-triples were studied by L. Harris [40]. The connection between the Möbius–Potapov–Harris transformations and Transmission Line Theory was explained in [30] and [25]. Operator differentiability for self-adjoint operators was studied in [14], and, for elements of JC^*-triples, in [6]. For the general theory of operator algebras, see [44] and [61].

5 The algebraic structure of homogeneous balls

In this chapter we will present the main ideas of the theory of homogeneous balls and bounded symmetric domains and the algebraic structure associated with them. The domains will be domains in complex Banach spaces and their homogeneity and symmetry will be with respect to analytic (called also holomorphic) maps. Thus we will start with the definition and study of the analytic mappings on Banach spaces. Next we will consider the group of analytic automorphisms $Aut_a(D)$ of a bounded domain D and show that the elements of this group are uniquely defined by their value and the value of their derivative at some point.

The Lie algebra $aut_a(D)$ can be defined as the tangent of the group at the identity and as the *generators of global* (for any time t) *analytic flows* of the domain. If the domain D is homogeneous, the elements of $aut_a(D)$ are polynomials of degree 2. This allows us to define a triple product structure for these domains. We will show that this triple product is a hermitian Jordan triple product. We will describe the geometry of these domains and their duals in terms of the triple product.

5.1 Analytic mappings on Banach spaces

In the previous chapter we have seen that the classical domains are homogeneous under the analytic maps. Also the unit ball of the complex spin factor of Chapter 3 is homogeneous under the analytic maps. We saw in Chapters 1 and 2 that real domains, which are homogeneous under the analytic maps, can be embedded in a complex domain. Thus, we will now define the notion of an analytic map of a Banach space and study the properties of such maps.

5.1.1 Homogeneous polynomials

A Banach space over the complex numbers is a complete normed vector space. Let X,Y be Banach spaces over the complex numbers. Denote the unit ball of X by

$$B = \{\mathbf{x} \in X : \ \ \|\mathbf{x}\| < 1\}. \tag{5.1}$$

A mapping $\widetilde{f} : \underbrace{X \times X \times \cdots \times X}_{m} \to Y$ such that

$$\widetilde{f}(\mathbf{x}_1, \mathbf{x}_2, \ldots, \mathbf{x}_m) \text{ is linear in each } \mathbf{x}_j \tag{5.2}$$

is called an *m-linear map.* An m-linear map $\widetilde{f}$ is continuous if the norm of $\widetilde{f}$ defined by

$$\| \widetilde{f} \| = \sup_{\mathbf{x}_1, \mathbf{x}_2, \ldots, \mathbf{x}_m \in B} \| \widetilde{f}(\mathbf{x}_1, \mathbf{x}_2, \ldots, \mathbf{x}_m) \| \tag{5.3}$$

is finite. We denote the space of all m-linear and continuous maps from X to Y by $L^m(X, Y)$. We denote $L^m(X) = L^m(X, X)$.

Define $L^0(X) = X$. If the dimension n of X is finite, let $\mathbf{e}_1, \mathbf{e}_1, \ldots, \mathbf{e}_n$ be a basis in X. In order to analyze $L^1(X)$ (consisting of 1-linear maps), decompose any element in X as

$$\mathbf{x} = \sum_{j=1}^{n} x^j \mathbf{e}_j = x^j \mathbf{e}_j. \tag{5.4}$$

We will use the Einstein summation convention, where there is an assumed summation for repeated lower and upper indexes, in order to simplify formulas involving summation. According to the linearity property, we have

$$\widetilde{f}(\mathbf{x}) = x^j \widetilde{f}(\mathbf{e}_j). \tag{5.5}$$

In order to define $\widetilde{f}$, we denote

$$\widetilde{f}(\mathbf{e}_j) = a_j^k \mathbf{e}_k \tag{5.6}$$

and then combine (5.4) and (5.5) together:

$$\widetilde{f}(\mathbf{x}) = x^j a_j^k \mathbf{e}_k = a_j^k x^j \mathbf{e}_k \tag{5.7}$$

This expression can also expressed by $\widetilde{f}(\mathbf{x}) = A\overline{x}$, where

$$A = \begin{pmatrix} a_1^1 & a_2^1 & \cdots & a_n^1 \\ a_1^2 & a_2^2 & \cdots & a_n^2 \\ \vdots & \vdots & \vdots & \vdots \\ a_1^n & a_2^n & \cdots & a_n^n \end{pmatrix}, \quad \overline{x} = \begin{pmatrix} x^1 \\ x^2 \\ \vdots \\ x^n \end{pmatrix}.$$

To define $L^2(X)$, note that every $\widetilde{f} \in L^2(X)$ operates on pairs of vectors $\mathbf{x} = x^j \mathbf{e}_j$ and $\mathbf{y} = y^e \mathbf{e}_e$. Using the linearity property, we obtain

$$\widetilde{f}(\mathbf{x}, \mathbf{y}) = \widetilde{f}(x^j \mathbf{e}_j, y^e \mathbf{e}_e) = x^j y^e \widetilde{f}(\mathbf{e}_j, \mathbf{e}_e). \tag{5.8}$$

We will now denote

$$\widetilde{f}(\mathbf{e}_j, \mathbf{e}_e) = a^k_{je}\mathbf{e}_k \tag{5.9}$$

and then combine (5.8) and (5.9) together in order to produce the following formula:

$$\widetilde{f}(\mathbf{x}, \mathbf{y}) = a^k_{je}x^j y^e \mathbf{e}_k. \tag{5.10}$$

This representation creates a three-dimensional matrix.

For every $\widetilde{f} \in L^m(X)$ that operates on m vectors, we have the following formulas:

$$\widetilde{f}(\mathbf{x_1}, \mathbf{x_2}, \dots, \mathbf{x_m}) = a^k_{j1,j2,\dots,jm} x_1^{j1} x_2^{j2} \cdots x_m^{jm} \mathbf{e}_k \tag{5.11}$$

$$\mathbf{x}_e = x_e^k \mathbf{e}_k \quad 1 \leq e \leq m \tag{5.12}$$

$$\widetilde{f}(\mathbf{e}_{j1}, \mathbf{e}_{j2}, \dots, \mathbf{e}_{jm}) = a^k_{j1,j2,\dots,jm}\mathbf{e}_k. \tag{5.13}$$

Let m be a natural number. We say that the map $f : X \to Y$ is an *m-homogeneous continuous polynomial* if there is an m-linear, continuous $\widetilde{f} \in L^m(X, Y)$ such that

$$f(\mathbf{x}) = \widetilde{f}(\mathbf{x}, \mathbf{x}, \dots, \mathbf{x}). \tag{5.14}$$

From this definition, it follows that

$$f(s\mathbf{x}) = \widetilde{f}(s\mathbf{x}, s\mathbf{x}, \dots, s\mathbf{x}) = s^m f(\mathbf{x}) \tag{5.15}$$

for any number s. Define the *norm* of f by

$$|| f || = \sup_{\mathbf{x} \in B} || f(\mathbf{x}) ||. \tag{5.16}$$

Denote by $P_m(X, Y)$ the set of all m-homogeneous, continuous polynomials from X to Y.

An m-linear map is called *symmetric* if

$$\widetilde{f}(\mathbf{x}_1, \mathbf{x}_2, \dots, \mathbf{x}_m) = \widetilde{f}(\mathbf{x}_{\sigma 1}, \mathbf{x}_{\sigma 2}, \dots, \mathbf{x}_{\sigma m})$$

for every permutation σ of $\{1, 2, \dots, m\}$. If $f \in P_m(X, Y)$ and the corresponding $\widetilde{f} \in L^m(X, Y)$ is not symmetric, we can replace $\widetilde{f}(\mathbf{x}_1, \mathbf{x}_2, \dots, \mathbf{x}_m)$ with

$$\widetilde{f}_s(\mathbf{x}_1, \mathbf{x}_2, \dots, \mathbf{x}_m) = \frac{1}{m!} \sum_{\sigma} \widetilde{f}(\mathbf{x}_{\sigma 1}, \mathbf{x}_{\sigma 2}, \dots, \mathbf{x}_{\sigma m}),$$

where the sum is taken over the $m!$ permutations σ of $\{1, 2, ..., m\}$. It is obvious that $\widetilde{f}_s$ is a *symmetric* m-linear map and that $f(\mathbf{x}) = \widetilde{f}_s(\mathbf{x}, \mathbf{x}, ..., \mathbf{x}) = \widetilde{f}(\mathbf{x}, \mathbf{x}, ..., \mathbf{x})$.

Denote by $\frac{df}{d\mathbf{x}}(\mathbf{x})d\mathbf{x}$ the derivative of the function f at $\mathbf{x} \in X$ in the direction $d\mathbf{x} \in X$, which is defined by

$$\frac{df}{d\mathbf{x}}(\mathbf{x})d\mathbf{x} = \frac{d}{ds}f(\mathbf{x} + sd\mathbf{x})\big|_{s=0} = \lim_{s \to 0} \frac{f(\mathbf{x} + sd\mathbf{x}) - f(\mathbf{x})}{s}, \tag{5.17}$$

where s is a complex number. Suppose $\widetilde{f} \in L_m(X, Y)$. Then

$$\frac{d\widetilde{f}}{d\mathbf{x}}(\mathbf{x}, ..., \mathbf{x})d\mathbf{x} = \lim_{s \to 0} \frac{\widetilde{f}(\mathbf{x} + sd\mathbf{x}, ..., \mathbf{x} + sd\mathbf{x}) - \widetilde{f}(\mathbf{x}, ..., \mathbf{x})}{s}$$

$$= \lim_{s \to 0} \frac{\widetilde{f}(\mathbf{x}, ..., \mathbf{x}) + s\widetilde{f}(d\mathbf{x}, \mathbf{x}, ..., \mathbf{x}) + \cdots + s\widetilde{f}(\mathbf{x}, ..., \mathbf{x}, d\mathbf{x}) + O(s^2) - \widetilde{f}(\mathbf{x}, ..., \mathbf{x})}{s}$$

$$= \widetilde{f}(d\mathbf{x}, \mathbf{x}, ..., \mathbf{x}) + \widetilde{f}(\mathbf{x}, d\mathbf{x}, ..., \mathbf{x}) + \cdots + \widetilde{f}(\mathbf{x}, ..., \mathbf{x}, d\mathbf{x}). \tag{5.18}$$

Thus, if $f^{(m)}(\mathbf{x})$ is an m-homogeneous polynomial defined by a symmetric m-linear map $\widetilde{f}$, then

$$\frac{df^{(m)}}{d\mathbf{x}}(\mathbf{x})d\mathbf{x} = m\widetilde{f}(d\mathbf{x}, \mathbf{x}, ..., \mathbf{x}) = mg^{(m-1)}(\mathbf{x})d\mathbf{x}, \tag{5.19}$$

where $g^{(m-1)}$ is an $m - 1$ homogeneous polynomial from X to $L^1(X, Y)$. In particular, when $\mathbf{x} = d\mathbf{x}$ we obtain:

$$\frac{df^{(m)}}{d\mathbf{x}}(\mathbf{x})\mathbf{x} = mf^{(m)}(\mathbf{x}). \tag{5.20}$$

5.1.2 Analytic maps on Banach spaces

A *power series* from X to Y is an infinite sum of the form

$$\sum_{m=0}^{\infty} f^{(m)}(\mathbf{x}), \tag{5.21}$$

where $f^{(m)} \in P_m(X, Y)$. We say that the power series *converges uniformly on a set* S if for any $\varepsilon > 0$, there is a number N such that for any $N < k < l$ we have

$$\|\sum_{m=k}^{l} f^{(m)}(\mathbf{x})\| < \varepsilon \tag{5.22}$$

for any $\mathbf{x} \in S$.

A mapping $f : X \to Y$ is called *analytic at the point* $\mathbf{a}$ if

$$f(\mathbf{x}) = \sum_{m=0}^{\infty} f^{(m)}(\mathbf{x} - \mathbf{a}), \tag{5.23}$$

where $f^{(m)}(\mathbf{x}-\mathbf{a}) \in P_m(X,Y)$, and there is $r > 0$ such that the power series converges uniformly for $\| \mathbf{x} - \mathbf{a} \| < r$. A mapping f from an open set $D \subset X$ into Y is said to be *analytic* on D if it is analytic at each point of D.

If we denote $\mathbf{h} = \mathbf{x} - \mathbf{a}$, then we can rewrite (5.23) as

$$f(\mathbf{a}+\mathbf{h}) = f(\mathbf{a}) + \sum_{m=1}^{\infty} f^{(m)}(\mathbf{h}). \tag{5.24}$$

Then, for any number s, from (5.15) we obtain

$$f(\mathbf{a}+s\mathbf{h}) = f(\mathbf{a}) + \sum_{m=1}^{\infty} f^{(m)}(s\mathbf{h}) = f(\mathbf{a}) + \sum_{m=1}^{\infty} s^m f^{(m)}(\mathbf{h}). \tag{5.25}$$

Thus,

$$\frac{d^k}{ds^k} f(\mathbf{a}+s\mathbf{h})\bigg|_{s=0} = k! f^{(k)}(\mathbf{h}), \tag{5.26}$$

from which we obtain the well-known Taylor formula

$$f^{(k)}(\mathbf{h}) = \frac{1}{k!}\frac{d^k}{ds^k} f(\mathbf{a}+s\mathbf{h})\bigg|_{s=0}. \tag{5.27}$$

This formula not only gives us an explicit definition of $f^{(k)}(\mathbf{h})$ but also proves that the power series expansion (5.23) of an analytic function is *unique.*

5.1.3 Examples of analytic mappings

As a first example of an analytic function, take $X = \mathcal{S}^n$, the complex spin factor, whose underlying set is C^n for n finite or infinite. The norm is defined by use of a triple product. For a fixed $\mathbf{a} \in \mathcal{S}^n$, define a map $q_{\mathbf{a}} : \mathcal{S}^n \to \mathcal{S}^n$ by

$$q_{\mathbf{a}}(\mathbf{z}) = \{\mathbf{z}, \mathbf{a}, \mathbf{z}\} = 2\langle \mathbf{z}|\mathbf{a}\rangle \mathbf{z} - \langle \mathbf{z}|\overline{\mathbf{z}}\rangle \overline{\mathbf{a}}.$$

Then $q_{\mathbf{a}}(\mathbf{z}) = \tilde{q}_{\mathbf{a}}(\mathbf{z},\mathbf{z})$, where $\tilde{q}_{\mathbf{a}}(\mathbf{x},\mathbf{y}) = \{\mathbf{x},\mathbf{a},\mathbf{y}\}$. Clearly $\tilde{q}_{\mathbf{a}}$ is bilinear and continuous, and so $q_{\mathbf{a}} \in P_2(X,X)$. Thus $q_{\mathbf{a}}$ is analytic on $\mathcal{S}^n$.

Another example of an analytic function was considered in the previous chapter. There, the Banach space is $X = L(H,K)$, the space of all bounded linear operators on a Hilbert space H, with the operator norm. The unit ball B in X is defined by (5.1). Let $\mathbf{a} \in B$. We define a mapping $\tilde{t}_{\mathbf{a}} : B \to B$ by

$$\tilde{t}_{\mathbf{a}}(\mathbf{b}) = \mathbf{b}(1-\mathbf{a}^*\mathbf{b})^{-1} = \mathbf{b} + \mathbf{b}\mathbf{a}^*\mathbf{b} + \mathbf{b}\mathbf{a}^*\mathbf{b}\mathbf{a}^*\mathbf{b} + \cdots \ . \tag{5.28}$$

We show that $f(\mathbf{b}) = \tilde{t}_{\mathbf{a}}(\mathbf{b})$ is analytic on B.

Note that, $f^{(0)} = 0, f^{(1)}(\mathbf{b}) = \mathbf{b},\ f^{(2)}(\mathbf{b}) = \mathbf{b}\mathbf{a}^*\mathbf{b},\ f^{(3)}(\mathbf{b}) = \mathbf{b}\mathbf{a}^*\mathbf{b}\mathbf{a}^*\mathbf{b}$, and so on. The norm of $f^{(1)}$ is

$$||f^{(1)}|| = \sup_{\mathbf{b}\in B} ||f^{(1)}(\mathbf{b})|| = \sup_{\mathbf{b}\in B} ||\mathbf{b}|| = 1. \tag{5.29}$$

The norm of $f^{(2)}$ is

$$||f^{(2)}|| = \sup_{\mathbf{b}\in B} ||\mathbf{b}\mathbf{a}^*\mathbf{b}|| \leq \sup_{\mathbf{b}\in B} ||\mathbf{b}|| \cdot ||\mathbf{a}|| \cdot ||\mathbf{b}|| \leq ||\mathbf{a}||. \tag{5.30}$$

Similarly, $||f^{(3)}|| \leq ||\mathbf{a}||^2$, and for any m, we have $||f^{(m)}|| \leq ||\mathbf{a}||^{m-1}$. Thus, using the formula for the sum of a geometric series, we obtain

$$||f|| \leq \sum ||f^{(m)}|| \leq 1 + ||\mathbf{a}|| + ||\mathbf{a}||^2 + ||\mathbf{a}||^3 + \cdots = \frac{1}{1-||\mathbf{a}||} < \infty, \tag{5.31}$$

implying that the power series (5.28) converges uniformly on B.

For a third example, we take $X = L(H)$. Define a function $\exp : X \to X$ by

$$f(\mathbf{a}) = \exp(\mathbf{a}) = \sum_{m=0}^{\infty} \frac{\mathbf{a}^m}{m!}. \tag{5.32}$$

In this case $f^{(0)} = I,\ f^{(1)}(\mathbf{a}) = \mathbf{a},\ f^{(2)}(\mathbf{a}) = \frac{\mathbf{a}^2}{2!}$, and so on. The norm of $f^{(m)}$ is

$$||f^{(m)}|| = \frac{1}{m!} \sup_{\mathbf{a}\in X} ||\mathbf{a}^m|| \leq \frac{||\mathbf{a}||^m}{m!}. \tag{5.33}$$

Thus, for any r we have

$$\sum ||f^{(m)}|| r^m \leq \sum \frac{r^m}{m!} = e^r < \infty,$$

implying that the power series (5.32) converges for all $\mathbf{a} \in X$ and the function exp is analytic on X.

5.1.4 Cauchy estimates for analytic functions

Let f be an analytic map from an open set $D \subset X$ to Y. We define

$$||f||_D = \sup_{\mathbf{x}\in D} ||f(\mathbf{x})||.$$

For an interior point $\mathbf{a} \in D$, we want to estimate the norm of $f^{(m)}$ (defined by (5.16)) from the decomposition (5.24) and the norm $||\widetilde{f}_s^{(m)}||$ (defined by (5.3)) of its symmetric m-linear map. These estimates depend on $||f||_D$ and $\delta = dist(\mathbf{a}, \partial D)$, the distance from the point $\mathbf{a}$ to the boundary of D, which is a positive number. We will show that

$$||\widetilde{f}_s^{(m)}|| \leq (e/\delta)^m ||f||_D. \tag{5.34}$$

Let $\mathbf{h}^1, \mathbf{h}^2, ..., \mathbf{h}^m$ be elements of $B \subset X$, and let $z_1, z_2, ..., z_m$ be complex numbers. Define $\varepsilon = \frac{\delta}{m}$. If $|z_1| = |z_2| = \cdots = |z_m| = \varepsilon$, then $||z_j \mathbf{h}^j|| < \delta$, implying that $\mathbf{a} + z_j \mathbf{h}^j \in D$. Furthermore, note that

$$f^{(k)}(z_j \mathbf{h}^j) = \widetilde{f}_s^{(k)}(z_{j_1}\mathbf{h}^{j_1}, z_{j_2}\mathbf{h}^{j_2}, ..., z_{j_k}\mathbf{h}^{j_k})$$

$$= z_{j_1} z_{j_2} ... z_{j_k} \widetilde{f}_s^{(k)}(\mathbf{h}^{j_1}, \mathbf{h}^{j_2}, ..., \mathbf{h}^{j_k}),$$

where the summation is over all $1 \leq j_n \leq m$. Since for any $\varepsilon > 0$ and any integer n, we have

$$\int_{|z|=\varepsilon} \frac{dz}{z^n} = \int_0^{2\pi} \frac{\varepsilon e^{i\varphi} i d\varphi}{\varepsilon^n e^{in\varphi}} = \frac{i}{\varepsilon^{n-1}} \int_0^{2\pi} e^{i(1-n)\varphi} d\varphi = \begin{cases} 2\pi i, & \text{if n=1,} \\ 0, & \text{otherwise,} \end{cases}$$

the expression

$$\int_{|z_m|=\varepsilon} \cdots \int_{|z_1|=\varepsilon} \frac{\widetilde{f}_s^{(k)}(z_{j_1}\mathbf{h}^{j_1}, z_{j_2}\mathbf{h}^{j_2}, ..., z_{j_k}\mathbf{h}^{j_k})}{z_1^2 z_2^2 ... z_m^2} dz_1 ... dz_m \tag{5.35}$$

is not equal to zero only if $k = m$ and $(j_1, j_2, ..., j_k)$ is a permutation of $(1, 2, ..., m)$. Since the number of such permutations is $m!$, we have

$$\int_{|z_m|=\varepsilon} \cdots \int_{|z_1|=\varepsilon} \frac{f(\mathbf{a} + z_j \mathbf{h}^j)}{z_1^2 ... z_m^2} dz_1 ... dz_m$$

$$= \sum_{k=1}^{\infty} \int_{|z_m|=\varepsilon} \cdots \int_{|z_1|=\varepsilon} \frac{\widetilde{f}_s^{(k)}(z_{j_1}\mathbf{h}^{j_1}, z_{j_2}\mathbf{h}^{j_2}, ..., z_{j_k}\mathbf{h}^{j_k})}{z_1^2 z_2^2 ... z_m^2} dz_1 ... dz_m$$

$$= m!(2\pi i)^m \widetilde{f}_s^{(m)}(\mathbf{h}^1, \mathbf{h}^2, ..., \mathbf{h}^m).$$

Therefore,

$$\widetilde{f}_s^{(m)}(\mathbf{h}^1, \mathbf{h}^2, ..., \mathbf{h}^m) = \frac{1}{m!}\left(\frac{1}{2\pi i}\right)^m \int_{|z_m|=\varepsilon} \cdots \int_{|z_1|=\varepsilon} \frac{f(\mathbf{a} + z_j \mathbf{h}^j)}{z_1^2 ... z_m^2} dz_1 ... dz_m.$$

This gives the following estimate:

$$||\widetilde{f}_s^{(m)}(\mathbf{h}^1, \mathbf{h}^2, ..., \mathbf{h}^m)|| \leq \frac{1}{m!}(\frac{1}{2\pi})^m ||f||_D \frac{1}{\varepsilon^{2m}}(2\pi\varepsilon)^m$$

$$= \frac{1}{m!\varepsilon^m}||f||_D = \frac{m^m}{m!\delta^m}||f||_D. \tag{5.36}$$

Using the Stirling inequality $\frac{m^m}{m!} \leq e^m$, we obtain

$$||\widetilde{f}_s^{(m)}(\mathbf{h}^1, \mathbf{h}^2, ..., \mathbf{h}^m)|| \leq (\frac{e}{\delta})^m ||f||_D. \tag{5.37}$$

This establishes (5.34). Since $f^{(m)}(\mathbf{h}) = \widetilde{f}_s^{(m)}(\mathbf{h}, \mathbf{h}, ..., \mathbf{h})$, we get

$$||f^{(m)}|| \leq (\frac{e}{\delta})^m ||f||_D. \tag{5.38}$$

It can even be shown that

$$||f^{(m)}|| \leq (\frac{1}{\delta})^m ||f||_D. \tag{5.39}$$

As a consequence of this formula, we get the following result, known as Liouville's Theorem: if f is analytic on the entire space X and bounded, then it must be constant. The proof is as follows: since f is bounded and analytic everywhere, it follows that $||f^{(m)}|| \leq (\frac{1}{\delta})^m ||f||$ for all δ. Hence, $||f^{(m)}|| = 0$ for all $m \geq 1$, and, thus, $f = f^{(0)}$ is constant.

5.2 The group $Aut_a(D)$

Let D be a domain in a Banach space X. We will say that the domain is *bounded* if there is a positive number b_0 such that for any $\mathbf{x} \in D$, we have $||\mathbf{x}|| < b_0$. We denote by $Aut_a(D)$ the set of all analytic maps f from D to D for which there is an analytic map g such that $f \circ g = g \circ f = I$, where I denotes the identity map on D. These maps are called automorphisms of D, and $Aut_a(D)$ is called *the automorphism group of* D. To show that $Aut_a(D)$ is a group, we have to show that it is closed under composition.

5.2.1 Composition of analytic functions

We want to show that a composition of two analytic functions is also analytic. Let X, Y, Z be Banach spaces. Let f be an analytic function from some domain $D \subset X$ to a domain $G \subset Y$, and let g be an analytic function from the domain G in Y to Z, as expressed in the diagram

$$X \xrightarrow{f} Y \xrightarrow{g} Z.$$

Let $\mathbf{a} \in D$, and let $f(\mathbf{a}) = \mathbf{b} \in G$. We will show that $g \circ f$ is analytic at $\mathbf{a}$ (see Figure 5.1).

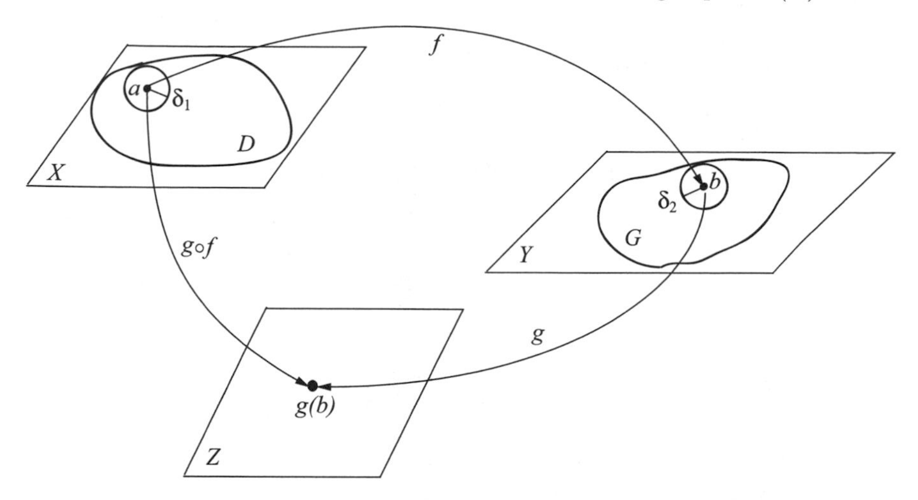

Fig. 5.1. Composition of analytic functions

Expand the analytic functions f and g into homogeneous polynomials by (5.24) as

$$f(\mathbf{a}+\mathbf{h}) \;=\; f(\mathbf{a})+\sum_{m=1}^{\infty} f^{(m)}(\mathbf{h}), \quad g(\mathbf{b}+\mathbf{y}) \;=\; g(\mathbf{b})+\sum_{k=1}^{\infty} g^{(k)}(\mathbf{y}).$$

Then the composition $f \circ g$ is decomposed as

$$(g\circ f)(\mathbf{a}+\mathbf{h}) \;=\; g(\mathbf{b})+\sum_{k=1}^{\infty}\sum_{m_1\ldots m_k=1}^{\infty} \widetilde{g_s}^{(k)}(f^{(m_1)}(\mathbf{h}), f^{(m_2)}(\mathbf{h}), ..., f^{(m_k)}(\mathbf{h})).$$

We introduce a multi-index $\mathbf{m}=(m_1,m_2,...,m_k)$, where $m_j \in N$ are natural numbers, and denote $|\mathbf{m}| = m_1+m_2+\cdots+m_k$. By combining terms of the same order, we can rewrite the previous expression as

$$(g\circ f)(\mathbf{a}+\mathbf{h}) \;=\; g(\mathbf{b})+\sum_{m=1}^{\infty}(g\circ f)^{(m)}(\mathbf{h}) \tag{5.40}$$

with

$$(g\circ f)^{(m)}(\mathbf{h}) := \sum_{k=1}^{m}\sum_{|\mathbf{m}|=m} \widetilde{g}_s^{(k)}(f^{(m_1)}(\mathbf{h}), f^{(m_2)}(\mathbf{h}), ..., f^{(m_k)}(\mathbf{h})). \tag{5.41}$$

To show that $g\circ f$ is analytic, we need to show that the power series (5.40) converges uniformly.

Denote $\delta_1 \;=\; dist(\mathbf{a}, \partial D)$ and $\delta_2 \;=\; dist(\mathbf{b}, \partial G)$. Using (5.34) and (5.39), we have

$$||\widetilde{g}_s^{(k)}|| \le (\frac{e}{\delta_2})^k \, ||g||_G, \;\; ||f^{(m)}(\mathbf{h})|| \le (\frac{||\mathbf{h}||}{\delta_1})^m ||f||_D.$$

This gives the following estimate:

$$||\widetilde{g}_s^{(k)}(f^{(m_1)}(\mathbf{h}), f^{(m_2)}(\mathbf{h}), ..., f^{(m_k)}(\mathbf{h}))||$$

$$\le (\frac{e}{\delta_2})^k \, ||g||_G (\frac{||\mathbf{h}||}{\delta_1})^{m_1} \, ||f||_D (\frac{||\mathbf{h}||}{\delta_1})^{m_2} \, ||f||_D \cdots (\frac{||\mathbf{h}||}{\delta_1})^{m_k} \, ||f||_D$$

$$= ||g||_G (\frac{e\,||f||_D}{\delta_2})^k \, (\frac{||\mathbf{h}||}{\delta_1})^m.$$

Note that the number of summands in (5.41) is at most m^3, since k has at most m values, m_j has at most m values and the number of such m_j is at most m. Thus,

$$||(g \circ f)^{(m)}(\mathbf{h})|| \le ||g||_G m^3 (\frac{e\,||f||_D}{\delta_2})^m (\frac{||\mathbf{h}||}{\delta_1})^m. \tag{5.42}$$

If we take a positive $\alpha < 1$, then for sufficiently small $\mathbf{h}$ we obtain

$$m^3 (\frac{e\,||f||_D}{\delta_2})^m (\frac{||\mathbf{h}||}{\delta_1})^m \le \alpha^m.$$

More precisely, using that $m^{3/m} < 3$, it is sufficient to take

$$||\mathbf{h}|| \le \frac{\alpha\,\delta_1\delta_2}{3e\,||f||_D}.$$

Thus

$$||(g \circ f)^{(m)}(\mathbf{h})|| \le \alpha^m.$$

Thus, the power series (5.40) converges uniformly. This completes the proof that the composition of two analytic functions is analytic.

From this it follows that $Aut_a(D)$ is closed under composition, meaning that if f and g belong to $Aut_a(D)$, then also $g \circ f$ belongs to $Aut_a(D)$. Since $f^{-1} \circ g^{-1} \circ g \circ f = f^{-1} \circ f = I$, the map $f^{-1} \circ g^{-1}$ is the inverse of $g \circ f$ and is also analytic. Thus, $Aut_a(D)$ is a group.

5.2.2 Generalized Classical Schwartz Lemma

The elements of $Aut_a(D)$ have several properties similar to those of the Möbius transformations of the unit disk in the complex plane. One of them is given by the *Generalized Classical Schwartz Lemma* for bounded domains. This lemma states: Let D be a bounded domain containing $\mathbf{0}$. If $f \in Aut_a(D)$

satisfies $f(\mathbf{0}) = \mathbf{0}$ and $f^{(1)}(\mathbf{0}) = I$ (the identity map), then $f(\mathbf{x}) = \mathbf{x}$ for all $\mathbf{x} \in D$.

To show this, assume that $f(\mathbf{x})$ is not identically equal to $\mathbf{x}$. Let k be the smallest integer ≥ 2 such that the k-homogeneous polynomial $f^{(k)}$ in decomposition of f is not equal identically to $\mathbf{0}$. Since $f(\mathbf{0}) = \mathbf{0}$ and $f^{(1)}(\mathbf{x}) = \mathbf{x}$, we have

$$f(\mathbf{x}) = \mathbf{x} + f^{(k)}(\mathbf{x}) + \cdots \qquad \text{when} \quad f^{(k)} \neq 0 \tag{5.43}$$

where $f^{(k)}$ is a k-homogeneous polynomial with values in X. Since the composition of analytic functions is analytic, $f^2 \in Aut_a(D)$, and we can decompose it by use of (5.14) as follows:

$$f^2(\mathbf{x}) = (f \circ f)(\mathbf{x}) = f(\mathbf{x} + f^{(k)}(\mathbf{x}) + \cdots) = f(\mathbf{x} + \widetilde{f}^{(k)}(\mathbf{x}, ..., \mathbf{x}) + \cdots)$$

$$= \mathbf{x} + \widetilde{f}^{(k)}(\mathbf{x}, ..., \mathbf{x}) + \widetilde{f}^{(k)}(\mathbf{x} + f^{(k)}(\mathbf{x}) + ..., ..., \mathbf{x} + f^{(k)}(\mathbf{x}) + \cdots) + \cdots$$

$$= \mathbf{x} + 2\widetilde{f}^{(k)}(\mathbf{x}, ..., \mathbf{x}) + \cdots = \mathbf{x} + 2f^{(k)}(\mathbf{x}) + \cdots$$

Repeating this argument, we obtain

$$f^m(\mathbf{x}) = \underbrace{(f \circ f \circ \cdots \circ f)}_{m}(\mathbf{x}) = \mathbf{x} + mf^{(k)}(\mathbf{x}) + \cdots \tag{5.44}$$

Since $f^m \in Aut_a(D)$ and D is bounded, we have $||f^m(\mathbf{x})|| \leq b_0$. From (5.39) we get

$$||mf^{(k)}|| \leq \frac{b_0}{\delta^k}, \tag{5.45}$$

where $\delta = dist(0, \partial D)$. But this inequality cannot hold as $m \to \infty$, so the assumption that there is $f^{(k)} \neq \mathbf{0}$ is wrong. Therefore, $f(\mathbf{x}) = \mathbf{x}$, proving the Generalized Classical Schwartz Lemma.

5.2.3 The Cartan Linearity Theorem and the Cartan Uniqueness Theorem

A domain D is called *circular* if $e^{i\theta}\mathbf{x} \in D$ for any $\theta \in R$ and $\mathbf{x} \in D$. From now on, the domain D will be a bounded circular domain in a Banach space X. All of the domains considered in the previous chapters were unit balls in some norm in a Banach space. Since $\|e^{i\theta}\mathbf{x}\| = \|\mathbf{x}\|$, any unit ball is a bounded circular domain.

We will now prove the *Cartan Linearity Theorem,* which states: Let D be a bounded circular domain in a Banach space X and let $f \in Aut_a(D)$ with $f(\mathbf{0}) = \mathbf{0}$. Then f is *linear.*

From the definition of $Aut_a(D)$, if $f \in Aut_a(D)$, then also $f^{-1} \in Aut_a(D)$. Since D is circular, it follows that for any θ, the map $g_\theta : D \to D$ defined by

$$g_\theta(\mathbf{x}) = e^{-i\theta} f^{-1}(e^{i\theta} f(\mathbf{x})) \tag{5.46}$$

is in $Aut_a(D)$. But

$$g_\theta(\mathbf{0}) = e^{-i\theta} f^{-1}(e^{i\theta} f(\mathbf{0})) = e^{-i\theta} f^{-1}(\mathbf{0}) = \mathbf{0}$$

and, by use of the chain rule, its derivative at $\mathbf{0}$ is

$$\frac{d}{d\mathbf{x}} g_\theta(\mathbf{0}) = e^{-i\theta} \frac{df^{-1}}{d\mathbf{x}}(\mathbf{0})(e^{i\theta} \frac{df}{d\mathbf{x}}(\mathbf{0}))$$

$$= e^{-i\theta} e^{i\theta} \frac{df^{-1}}{d\mathbf{x}}(\mathbf{0}) \frac{df}{d\mathbf{x}}(\mathbf{0}) = \frac{d}{d\mathbf{x}}(f^{-1} \circ f)(\mathbf{0}) = I.$$

By the Generalized Classical Schwartz Lemma, we have $g_\theta(\mathbf{x}) = \mathbf{x}$ for all $\mathbf{x} \in D$ and from (5.46), we obtain

$$e^{i\theta} f(\mathbf{x}) = f(e^{i\theta}\mathbf{x}). \tag{5.47}$$

Since f is analytic, we have $f(\mathbf{x}) = \sum f^{(m)}(\mathbf{x})$. If we substitute this into (5.47), we get

$$e^{i\theta} \sum f^{(m)}(\mathbf{x}) = \sum f^{(m)}(e^{i\theta}\mathbf{x}) = \sum e^{im\theta} f^{(m)}(\mathbf{x}). \tag{5.48}$$

In order to fulfill (5.48), m must be equal to 1. Hence, for all $m > 1$ we obtain $f^{(m)}(\mathbf{x}) = \mathbf{0}$. So $f(\mathbf{x}) = f^{(1)}(\mathbf{x})$, which means that f is linear. This proves the Cartan Linearity Theorem.

We will now show that an element of $Aut_a(D)$, where D is a bounded circular domain, is uniquely determined by its value and the value of its derivative at the point $\mathbf{x} = \mathbf{0}$. This is called the *Cartan Uniqueness Theorem,* which states: Let D be a bounded and circular domain and let $f, g \in Aut_a(D)$, such that $f(\mathbf{0}) = g(\mathbf{0})$, and $\frac{df}{d\mathbf{x}}(\mathbf{0}) = \frac{dg}{d\mathbf{x}}(\mathbf{0})$. Then $f = g$.

Define $h = f^{-1} \circ g$. Then $h \in Aut_a(D)$, $h(\mathbf{0}) = f^{-1}(g(\mathbf{0})) = f^{-1}f(\mathbf{0}) = \mathbf{0}$ and

$$\frac{dh}{d\mathbf{x}}(\mathbf{0}) = \frac{df^{-1}}{d\mathbf{x}}(g(\mathbf{0})) \frac{dg}{d\mathbf{x}}(\mathbf{0}) = \frac{df^{-1}}{d\mathbf{x}}(f(\mathbf{0})) \frac{df}{d\mathbf{x}}(\mathbf{0}) = I.$$

Thus, from the Generalized Classical Schwartz Lemma, we have $h(\mathbf{x}) = \mathbf{x}$. This implies that $(f^{-1} \circ g)(\mathbf{x}) = \mathbf{x}$, and so $g(\mathbf{x}) = f(\mathbf{x})$ for all $\mathbf{x} \in D$. This proves the Cartan Uniqueness Theorem.

Define a subgroup K of $Aut_a(D)$ by

$$K = \{f \in Aut_a(D) : f(\mathbf{0}) = \mathbf{0}\}. \tag{5.49}$$

From the Cartan Linearity Theorem, it follows that K consists of all linear maps in $Aut_a(D)$. Elements of K are called *rotations* of D. A domain D is called a *homogeneous domain* if for any $\mathbf{a} \in D$ there is $f \in Aut_a(D)$ such that $f(\mathbf{0}) = \mathbf{a}$. For a homogeneous domain we can identify the factor space $Aut_a(D)/K$ with the domain D by identifying any element $f \in Aut_a(D)$ with its value $f(\mathbf{0})$.

5.3 The Lie Algebra of $Aut_a(D)$

In this section we will describe the Lie algebra $aut_a(D)$ of the group $Aut_a(D)$ and define the triple product associated with the domain D.

5.3.1 The tangent space at the identity of $Aut_a(D)$

The Lie algebra of $Aut_a(D)$ consists of the tangent vectors to the identity I of the group. To define a tangent vector at the point I, we start with a smooth curve in the set $Aut_a(D)$, *i.e.*, a real analytic function $g : (-\varepsilon, \varepsilon) \to Aut_a(D)$ with $g(0) = I$. Then $\delta_g = \frac{d}{dt} g(t)|_{t=0}$, the derivative of $g(t)$ at $t = 0$, is a tangent vector at I. Since the elements $g(t)$ are analytic functions from D to $D \subset X$, and X is a linear vector space, δ_g is a function from D to X. We will consider only smooth curves $g(t)$ for which the tangent δ_g at I is an analytic function on D. Denote by $Hol(D, X)$ the set of *analytic* (called also *holomorphic*) maps from D to X.

Thus, the *tangent space* of $Aut_a(D)$ at the identity I is defined by

$$T_I(Aut_a(D)) = \{\delta_g \in Hol(D, X) : \delta_g = \frac{d}{d_t} g(t)|_{t=0}\}, \tag{5.50}$$

$g : (-\varepsilon, \varepsilon) \to Aut_a(D)$ is analytic and $g(0) = I$. Any element δ_g of the tangent space can be considered as a vector field on the domain D, since for any point $\mathbf{x} \in D$, it defines a vector $\delta_g(\mathbf{x}) \in X$ and X can be considered as a tangent space to D at $\mathbf{x}$. The vector fields play the role of infinitesimal generators of the group. Examples of such vector fields were given in earlier sections (see Figures 1.30, 1.31, 2.7, 2.8 and 3.6). The vector $\delta_g(\mathbf{x})$ is a tangent vector to the curve $g(t)\mathbf{x}$ in D. This means that for small $\Delta t > 0$, we have

$$g(\Delta t)\mathbf{x} \approx \mathbf{x} + \delta_g(\mathbf{x})\Delta t. \tag{5.51}$$

Even though both $Aut_a(D)$ and $T_I(Aut_a(D))$ consist of analytic functions on D, they differ significantly in their mathematical structure. The set $Aut_a(D)$ is a group, implying that it is closed under composition and inverses. However, there is no linear structure on $Aut_a(D)$. On the other hand, composition and inverses are not defined for $T_I(Aut_a(D))$, but this set is a linear space. This follows from the fact that if $\delta_g \in T_I(Aut_a(D))$ and α is a constant, then $g(\alpha t)$ is a smooth curve in $Aut_a(D)$, with the tangent vector $\alpha\delta_g \in T_I(Aut_a(D))$. Also for any two vector fields $\delta_g, \delta_h \in T_I(Aut_a(D))$ we can define a smooth curve $\gamma(t) = g(t) \circ h(t)$ in $Aut_a(D)$. To calculate the tangent vector of this curve, notice that $g(t)\mathbf{x} = \mathbf{x} + t\delta_g(\mathbf{x}) + o(t)$. Therefore, $g(t) \circ h(t)\mathbf{x} = h(t)\mathbf{x} + t\delta_g(h(t)\mathbf{x}) + o(t)$. So

$$\lim_{t\to 0} \frac{\gamma(t)\mathbf{x} - \mathbf{x}}{t} = \lim_{t\to 0} \left(\frac{g(t) \circ h(t)\mathbf{x} - g(t)\mathbf{x}}{t} + \frac{g(t)\mathbf{x} - \mathbf{x}}{t} \right)$$

$$= \lim_{t\to 0} \frac{h(t)\mathbf{x} + t\delta_g(h(t)\mathbf{x}) - \mathbf{x} - t\delta_g(\mathbf{x}) + o(t)}{t} + \delta_g(\mathbf{x}) = \delta_h(\mathbf{x}) + \delta_g(\mathbf{x}).$$

Thus $\gamma(t)$ generates a tangent vector $\delta_g + \delta_h \in T_I(Aut_a(D))$. This shows that $T_I(Aut_a(D))$ is a linear space.

In addition to the linear structure, the set $T_I(Aut_a(D))$ is also closed under the Lie bracket, defined as follows. For any two smooth curves $g(t), h(t)$ in $Aut_a(D)$, with tangent vectors $\delta_g, \delta_h \in T_I(Aut_a(D))$, define a smooth curve $\gamma(t) = g(t) \circ h(t) \circ g(t)^{-1} \circ h(t)^{-1}$. It is obvious that $\gamma(0) = I$ and $\gamma(t) \in Aut_a(D)$ for some t in a neighborhood of 0. Moreover, if we introduce a new parameter $s = t^2$, then $\gamma(\sqrt{s})$, which is a smooth curve in $Aut_a(D)$, is differentiable at $s = 0$, and for any $\mathbf{x} \in D$ we have

$$\frac{d}{ds}\gamma(\sqrt{s})|_{s=0}(\mathbf{x}) = \delta'_g(\mathbf{x})\delta_h(\mathbf{x}) - \delta'_h(\mathbf{x})\delta_g(\mathbf{x}), \tag{5.52}$$

where $\delta'_g(\mathbf{x})\delta_h(\mathbf{x})$ denotes the derivative of δ_g at the point $\mathbf{x}$ in the direction $\delta_h(\mathbf{x})$. This implies that $T_I(Aut_a(D))$ is closed under the *Lie bracket*, defined as

$$[\delta_g, \delta_h](\mathbf{x}) = \delta'_g(\mathbf{x})\delta_h(\mathbf{x}) - \delta'_h(\mathbf{x})\delta_g(\mathbf{x}), \tag{5.53}$$

for any $\delta_g, \delta_h \in T_I(Aut_a(D))$ and any $\mathbf{x} \in D$. It is known that this bracket satisfies the so-called Jacobi identity

$$[\delta_f, [\delta_g, \delta_h]] = [[\delta_f, \delta_g], \delta_h] + [\delta_g, [\delta_f, \delta_h]], \tag{5.54}$$

for any $\delta_f, \delta_g, \delta_h \in T_I(Aut_a(D))$.

5.3.2 Complete analytic vector fields and $aut_a(D)$

For any fixed point $\mathbf{x}_0 \in D$ and vector field $\delta \in Hol(D, X)$, define a *local flow* $\varphi_\delta(t)$ on D to be the solution of the first-order differential equation

$$\frac{d}{dt}\varphi_\delta(t) = \delta(\varphi_\delta(t)) \tag{5.55}$$

satisfying the initial condition $\varphi_\delta(0) = \mathbf{x}_0$. From the general theory of differential equations, this solution is defined and unique for t in some real interval $J_{\mathbf{x}_0}$ containing the point $t = 0$. Moreover, the solution will depend analytically on the initial condition $\mathbf{x}_0$.

We will say that a vector field $\delta \in Hol(D, X)$ is a *complete vector field* if the solution $\varphi_\delta(t)$ of (5.55) exists for any $\mathbf{x}_0 \in D$ and any real t. If the vector field $\delta \in Hol(D, X)$ is complete, then it defines a *global* flow $\varphi_\delta(t, \mathbf{x})$ on D, defined for any $t \in R$ and any fixed $\mathbf{x} \in D$, which is the solution of the initial-value problem

$$\frac{\partial}{\partial t}\varphi_\delta(t, \mathbf{x}) = \delta(\varphi_\delta(t, \mathbf{x})), \quad \varphi_\delta(0, \mathbf{x}) = \mathbf{x}. \tag{5.56}$$

The set of *complete analytic vector fields* on a domain D is denoted by $aut_a(D)$. It was proved ([70] and [69]) that

$$aut_a(D) = T_I(Aut_a(D)) \tag{5.57}$$

and is a real Lie algebra.

The property of completeness of a vector field is connected with the property of being tangent on the boundary. We will say that a real linear functional $f : X \to R$ exposes a boundary point $\mathbf{x}_0 \in \partial D$, if $f(\mathbf{x}) \leq f(\mathbf{x}_0)$ for any $\mathbf{x} \in D$. A vector field δ is said to be tangent on the boundary of D if for any boundary point $\mathbf{x}_0 \in \partial D$ and any functional f exposing $\mathbf{x}_0$, we have $f(\delta(\mathbf{x}_0)) = 0$. In [64] it was shown that an analytic vector field on a bounded domain D is complete if it is tangent on the boundary ∂D of this domain, see also [69]. The meaning of this condition is the following. Since a local flow for an analytic vector field always exists inside the domain, in order that the flow will exist for any t, we have to take care that the flow will not leave the interior of the domain. The vector field's being tangent on the boundary prevents the flow from leaving the interior of the domain. See the examples of the vector fields in sections 1.5.1, 2.4.3 and 3.4.1.

Fix now some $t > 0$ and $\delta \in aut_a(D)$. Then an approximate solution $\tilde{\varphi}_\delta(t, \mathbf{x})$ of (5.56) can be obtained as follows. Divide the interval $[0, t]$ into n intervals by introducing intermediate points $t_j = jt/n$ with $j = 0, 1, ..., n$. Then, for any $1 \leq k \leq n$, we have

$$\varphi_\delta(t_k, \mathbf{x}) \approx \varphi_\delta(t_{k-1}, \mathbf{x}) + \delta(\varphi_\delta(t_{k-1}, \mathbf{x}))t/n$$

and, by using (5.51) with $\delta_g = \delta$ we have

$$\varphi_\delta(t_k, \mathbf{x}) \approx g(t/n)(\varphi_\delta(t_{k-1}, \mathbf{x})).$$

Thus

$$\varphi_\delta(t, \mathbf{x}) \approx g(t/n)^n(\mathbf{x}).$$

By taking the limit as n goes to infinity, we obtain

$$\varphi_\delta(t, \mathbf{x}) = \lim_{n\to\infty} g(t/n)^n(\mathbf{x}). \tag{5.58}$$

Since for each fixed t and n, the map $g(t/n)^n$ belongs to $Aut_a(D)$, the map $\varphi_\delta(t,\mathbf{x})$ also belongs to $Aut_a(D)$.

If δ in the initial value problem (5.56) is a linear map, then this initial-value problem has the solution $\varphi_\delta(t,\mathbf{x}) = \exp(t\delta)\mathbf{x}$. Moreover, by substituting (5.51) into (5.58), we have in this case

$$\varphi_\delta(t,\mathbf{x}) = \lim_{n\to\infty}(I + \frac{\delta t}{n})^n\mathbf{x},$$

justifying the notation $\varphi_\delta(t,\mathbf{x}) = \exp(t\delta)\mathbf{x}$. From (5.58) one can show that $\varphi_\delta(t,\mathbf{x})$ generates a commuting *one-parameter subgroup* of $Aut_a(D)$ satisfying

$$\varphi_\delta(t_1+t_2,\mathbf{x}) = \varphi_\delta(t_1,\mathbf{x})\circ\varphi_\delta(t_2,\mathbf{x}) = \varphi_\delta(t_2,\mathbf{x})\circ\varphi_\delta(t_1,\mathbf{x}). \tag{5.59}$$

This equation clearly holds for $\frac{t_1}{t_1+t_2} = \frac{m}{n}$ (and then $\frac{t_2}{t_1+t_2} = \frac{n-m}{n}$) and since such t_1 are dense, this will hold for any t_1.

The Lie algebra $aut_a(D)$ is *purely real*, meaning that if both δ and $i\delta$ belong to $aut_a(D)$, then $\delta \equiv 0$. This results from the following argument. Let $\mathbf{x}$ be any fixed point in D. For any complex number $w = t + is$, define a map $f : C \to X$ by $f(w) = \exp(w\delta)\mathbf{x}$. Since $[\delta, i\delta] = 0$, the one-parameter groups $\exp(t\delta)$ and $\exp(s\, i\delta)$ commute. Thus,

$$f(w)\mathbf{x} = \exp(w\delta)\mathbf{x} = \exp(t\delta)\exp(s\, i\delta)\mathbf{x}$$

is an analytic map on C with image in D, which is bounded. By Liouville's Theorem, $f(w)$ is constant, implying that $\delta = 0$.

5.3.3 The Cartan Uniqueness Theorem for $aut_a(D)$

The elements of $Aut_a(D)$, are determined by their value and their derivative at $\mathbf{x} = \mathbf{0}$. The complete vector fields of $aut_a(D)$ have a similar uniqueness. The *Cartan Uniqueness Theorem for* $aut_a(D)$ states: Let D be a bounded domain in a Banach space and let δ, ξ be compete holomorphic vector fields in $aut_a(D)$ such that

$$\delta(\mathbf{0}) = \xi(\mathbf{0}) \text{ and } \frac{d\delta(\mathbf{x})}{d\mathbf{x}}(\mathbf{0}) = \frac{d\xi(\mathbf{x})}{d\mathbf{x}}(\mathbf{0}).$$

Then $\delta(\mathbf{x}) = \xi(\mathbf{x})$ for all $\mathbf{x} \in D$.

Indeed, let $\eta = \delta - \xi$. Then $\eta \in aut_a(D)$ satisfies $\eta(\mathbf{0}) = \mathbf{0}$ and $\frac{d\eta}{d\mathbf{x}}(\mathbf{0}) = \mathbf{0}$. To prove the theorem, it is enough to show that $\eta(\mathbf{x}) = \mathbf{0}$ for any $\mathbf{x} \in D$.

Denote by $\varphi_\eta(t,\mathbf{x})$ the solution of the initial-value problem

$$\frac{\partial}{\partial t}\varphi_\eta(t,\mathbf{x}) = \eta(\varphi_\eta(t,\mathbf{x})),\ \varphi_\eta(0,\mathbf{x}) = \mathbf{x}. \tag{5.60}$$

Note that since $\eta(\mathbf{0}) = (\mathbf{0})$, the initial-value problem (5.60) has a solution $\varphi_\eta(t,\mathbf{0}) \equiv \mathbf{0}$ for $\mathbf{x} = \mathbf{0}$.

Let us calculate now $\frac{\partial \varphi_\eta(t,\mathbf{x})}{\partial \mathbf{x}}(\mathbf{0})$. Note that

$$\frac{\partial}{\partial t}\frac{\partial \varphi_\eta(t,\mathbf{x})}{\partial \mathbf{x}} = \frac{\partial}{\partial \mathbf{x}}\frac{\partial \varphi_\eta(t,\mathbf{x})}{\partial t} = \frac{\partial}{\partial \mathbf{x}}\eta(\varphi_\eta(t,\mathbf{x})) = \frac{d\eta}{d\mathbf{x}}(\varphi_\eta(t,\mathbf{x}))\frac{\partial}{\partial \mathbf{x}}(\varphi_\eta(t,\mathbf{x})).$$

Substituting $\mathbf{x} = \mathbf{0}$ and using that $\frac{d\eta}{d\mathbf{x}}(\mathbf{0}) = \mathbf{0}$, we obtain

$$\frac{\partial}{\partial t}\frac{\partial \varphi_\eta(t,\mathbf{x})}{\partial \mathbf{x}}\Big|_{\mathbf{x}=\mathbf{0}} = \mathbf{0},$$

implying that

$$\frac{\partial \varphi_\eta(t,\mathbf{x})}{\partial \mathbf{x}}\Big|_{\mathbf{x}=\mathbf{0}} = \frac{\partial \varphi_\eta(0,\mathbf{x})}{\partial \mathbf{x}}\Big|_{\mathbf{x}=\mathbf{0}} = \frac{\partial \mathbf{x}}{\partial \mathbf{x}}\Big|_{\mathbf{x}=\mathbf{0}} = I.$$

Thus, for any fixed t, the map $\varphi_\eta(t,\mathbf{x}) \in Aut_a(D)$ satisfies $\varphi_\eta(t,\mathbf{0}) = \mathbf{0}$ and $\frac{\partial \varphi_\eta(t,\mathbf{x})}{\partial \mathbf{x}}(\mathbf{0}) = I$, implying by the Cartan Uniqueness Theorem for $Aut_a(D)$ that $\varphi_\eta(t,\mathbf{x}) = \mathbf{x}$. By (5.60) from this follows that $\eta \equiv \mathbf{0}$. This proves the Cartan Uniqueness Theorem for $aut_a(D)$.

5.3.4 The structure of $aut_a(D)$

We will now derive the key result which is needed in order to construct a triple product on a bounded symmetric domain (which will be defined in the next section). It is known [12] that any bounded symmetric domain is analytically equivalent to a bounded circular domain. So we will assume that the domain D is a bounded circular domain in a Banach space X.

Let $\xi \in aut_a(D)$. We will show that the power series expansion (5.24) of ξ at $\mathbf{x} = \mathbf{0}$ is

$$\xi(\mathbf{x}) = \mathbf{b} + \xi^{(1)}(\mathbf{x}) + \xi^{(2)}(\mathbf{x}), \tag{5.61}$$

where $\mathbf{b} = \xi(\mathbf{0})$. Moreover, we will show that

$$\xi^{(1)}(\mathbf{x}) \in aut_a(D), \quad \text{and} \quad \mathbf{b} + \xi^{(2)}(\mathbf{x}) \in aut_a(D), \tag{5.62}$$

and that $\xi^{(2)}(\mathbf{x})$ is uniquely determined by $\mathbf{b}$.

To verify this result, we introduce a one-parameter family of maps $\psi(t), t \in R$, on X defined by

$$\psi(t)(\mathbf{x}) = e^{it}\mathbf{x}.$$

Since the domain D is assumed to be circular, $\psi(t) : D \to D$ and $\psi(t) \in Aut_a(D)$. The tangent to this family at $\psi(0) = I$ is given by

$$\eta(\mathbf{x}) = \frac{d}{dt}\psi(t)(\mathbf{x})\big|_{t=0} = i\mathbf{x}.$$

From (5.57) follows that $\eta \in aut_a(D)$.

Using this η and the Lie bracket, defined by (5.53), we define a map $T : aut_a(D) \to aut_a(D)$ by

$$T\xi = [\eta, \xi].$$

The action of T on the m-homogeneous polynomials of $\xi(x)$ is, by use of (5.20), given by

$$T\xi^{(m)}(\mathbf{x}) = [\eta, \xi^{(m)}] = \frac{d\eta}{d\mathbf{x}}(\mathbf{x})\xi^{(m)}(\mathbf{x}) - \frac{d\xi^{(m)}}{d\mathbf{x}}(\mathbf{x})\eta(\mathbf{x})$$

$$= i\xi^{(m)}(\mathbf{x}) - im\xi^{(m)}(\mathbf{x})\mathbf{x} = i(1-m)\xi^{(m)}(\mathbf{x}).$$

By decomposing $\xi(\mathbf{x}) = \sum_{m=0}^{\infty} \xi^{(m)}(\mathbf{x})$, we have

$$T\xi(\mathbf{x}) = \sum_{m=0}^{\infty} i(1-m)\xi^{(m)}(\mathbf{x}) \in aut_a(D). \tag{5.63}$$

Since $aut_a(D)$ is a real linear space, any polynomial $p(T)$ with real coefficients maps $aut_a(D)$ into $aut_a(D)$. Since T acts on homogeneous polynomials by multiplication by a constant, the polynomial $p(T)$ acts on $\xi^{(m)}$ by

$$p(T)\xi = \sum_{m=0}^{\infty} p(i(1-m))\xi^{(m)}.$$

Let $p(x) = x(1+x^2)$. Then

$$p(i(1-m)) = i(1-m)(1-(1-m)^2) = i(1-m)m(2-m)$$

and thus

$$p(T)\xi = i\sum_{m=0}^{\infty} m(m-1)(m-2)\xi^{(m)} \in aut_a(D). \tag{5.64}$$

But from this equation, we have

$$p(T)\xi(\mathbf{0}) = \mathbf{0} \text{ and } \frac{d\,p(T)\xi(\mathbf{x})}{d\mathbf{x}}(\mathbf{0}) = \mathbf{0},$$

which, by the Cartan Uniqueness Theorem for $aut_a(D)$, implies that $p(T)\xi \equiv 0$. Since for any $m > 2$, the coefficient of $\xi^{(m)}$ in the power series of (5.64) is non-zero, while the sum equals zero, we conclude that $\xi^{(m)} = 0$ for such m, proving (5.61).

The same argument for the polynomial $\tilde{p}(x) = (1+x^2)$ yields

$$\tilde{p}(T)\xi = \sum_{m=0}^{2} m(2-m)\xi^{(m)} = -\xi^{(1)} \in aut_a(D).$$

Thus, also $\xi^{(0)} + \xi^{(2)} = \xi - \xi^{(1)} \in aut_a(D)$, proving (5.62).

Finally, to show that $\xi^{(2)}(\mathbf{x})$ is uniquely defined by $\mathbf{b} = \xi(\mathbf{0})$, let $\mathbf{b} + \xi^{(2)}(\mathbf{x}) \in aut_a(D)$ and also $\mathbf{b} + \tilde{\xi}^{(2)}(\mathbf{x}) \in aut_a(D)$; then their difference $\zeta = (\xi^{(2)}(\mathbf{x}) - \tilde{\xi}^{(2)}(\mathbf{x})) \in aut_a(D)$. But, since $\zeta(\mathbf{0}) = \mathbf{0}$ and $\frac{d\zeta(\mathbf{x})}{d\mathbf{x}}(\mathbf{0}) = \mathbf{0}$, the Cartan Uniqueness Theorem for $aut_a(D)$ implies that $\zeta \equiv 0$, and, thus, $\xi^{(2)}(\mathbf{x}) = \tilde{\xi}^{(2)}(\mathbf{x})$. Hence, the quadratic polynomial in the power series expansion of ξ of $aut_a(D)$ is uniquely defined by $\mathbf{b} = \xi(\mathbf{0})$. We denote this polynomial by

$$q_{\mathbf{b}}(\mathbf{x}) = -\xi^{(2)}(\mathbf{x}). \tag{5.65}$$

Thus,

$$aut_a(D) = \{\xi(\mathbf{x}) = \mathbf{b} + \kappa\mathbf{x} - q_{\mathbf{b}}(\mathbf{x})\}, \tag{5.66}$$

where $\mathbf{b} \in X$, a map $\kappa \in L(X)$ is a linear map such that $\exp\kappa$ is a linear automorphism of D, and $q_{\mathbf{b}}$ is the quadratic polynomial uniquely defined by $\mathbf{b}$.

We denote by $\xi_{\mathbf{b}}(\mathbf{x}) = \mathbf{b} - q_{\mathbf{b}}(\mathbf{x})$ the element ξ of $aut_a(D)$ such that $\xi(\mathbf{0}) = \mathbf{b}$ and $\xi'(\mathbf{0}) = 0$. We will call such elements the *generators of translations* and denote the set of all such generators by

$$\underline{\mathbf{p}} = \{\xi_{\mathbf{b}} = \mathbf{b} - q_{\mathbf{b}} \in aut_a(D) : \ \mathbf{b} \in X\}. \tag{5.67}$$

Similarly, we can define the generators of rotations by

$$\underline{\mathbf{k}} = \{\kappa\mathbf{x} \in aut_a(D) : \ \kappa \in L(X), \ \exp(\kappa) \in Aut_a(D)\}. \tag{5.68}$$

This defines the *Cartan decomposition* of $aut_a(D)$:

$$aut_a(D) = \underline{\mathbf{p}} + \underline{\mathbf{k}},$$

where the elements of $\underline{\mathbf{p}}$ are symmetric with respect to the symmetry $s(\mathbf{x}) = -\mathbf{x}$, and the elements of $\underline{\mathbf{k}}$ are antisymmetric with respect to s.

5.3.5 The triple product defined by $aut_a(D)$

A bounded domain D is called *homogeneous* if for any $\mathbf{a} \in D$ there is a map $\varphi \in Aut_a(D)$ such that $\varphi(\mathbf{0}) = \mathbf{a}$. Homogeneity of the domain D implies that for any $\mathbf{b} \in X$, there is an element $\xi_{\mathbf{b}}$, defined by (5.67), in $aut_a(D)$. A domain D is called a *symmetric* domain if for any $\mathbf{a} \in D$ there is a symmetry $s_{\mathbf{a}} \in Aut_a(D)$ (i.e., $s_{\mathbf{b}}^2 = I$) fixing only $\mathbf{a}$. It is known that any bounded symmetric domain is analytically equivalent to a bounded homogeneous circular domain. From now on we will assume that a bounded symmetric domain D is a bounded homogeneous circular domain in some Banach space X.

The triple product of a a bounded symmetric domain D is defined by use of the quadratic map $q_{\mathbf{b}}(\mathbf{x})$ from $aut_a(D)$. Since $aut_a(D)$ is real linear, for any two elements $\xi_{\mathbf{b}}, \xi_{\mathbf{d}} \in aut_a(D)$, their real linear combination $\alpha\xi_{\mathbf{b}} + \beta\xi_{\mathbf{d}}$ also belongs to $aut_a(D)$. But

$$\alpha\xi_{\mathbf{b}} + \beta\xi_{\mathbf{d}} = \alpha\mathbf{b} + \beta\mathbf{d} - \alpha q_{\mathbf{b}} - \beta q_{\mathbf{d}} = \xi_{\alpha\mathbf{b}+\beta\mathbf{d}},$$

implying that

$$q_{\alpha\mathbf{b}+\beta\mathbf{d}} = \alpha q_{\mathbf{b}} + \beta q_{\mathbf{d}}$$

and that the map $\mathbf{b} \to q_{\mathbf{b}}$ is real linear. From (5.63), it follows that

$$T\xi_{\mathbf{b}} = i\mathbf{b} - i\xi_{\mathbf{b}}^{(2)} = i\mathbf{b} + iq_{\mathbf{b}} \in aut_a(D),$$

implying that $\xi_{i\mathbf{b}} = i\mathbf{b} + iq_{\mathbf{b}}$ and that $q_{i\mathbf{b}} = -iq_{\mathbf{b}}$. Thus, the map $\mathbf{b} \to q_{\mathbf{b}}$ is a conjugate linear map.

Apply now the definition of an m-homogeneous polynomial (5.14) to the (quadratic) 2-homogeneous polynomial $q_{\mathbf{b}}(\mathbf{x})$ and write this polynomial as

$$q_{\mathbf{b}}(\mathbf{x}) = \tilde{q}_{\mathbf{b}}(\mathbf{x}, \mathbf{x}),$$

where $\tilde{q}_{\mathbf{b}}(\mathbf{x}, \mathbf{x})$ is a symmetric bilinear map on X. Since this map is also conjugate linear in $\mathbf{b}$, we will write it as a real trilinear map, called the *triple product,* defined on elements of X by

$$q_{\mathbf{b}}(\mathbf{x}) = \tilde{q}_{\mathbf{b}}(\mathbf{x}, \mathbf{x}) = \{\mathbf{x}, \mathbf{b}, \mathbf{x}\}. \tag{5.69}$$

Note that this triple product is linear in the first and last components and conjugate linear in the middle one. This is consistent with the triple product (4.9) in Chapter 4 for JC^*-triples consisting of spaces of operators.

If we write $\mathbf{x} = \mathbf{a} + \mathbf{c}$ for arbitrary $\mathbf{a}, \mathbf{c} \in X$, then we can rewrite (5.69) as

$$\tilde{q}_{\mathbf{b}}(\mathbf{a} + \mathbf{c}, \mathbf{a} + \mathbf{c}) = \{(\mathbf{a} + \mathbf{c}), \mathbf{b}, (\mathbf{a} + \mathbf{c})\}$$

$$= \{\mathbf{a}, \mathbf{b}, \mathbf{a}\} + 2\{\mathbf{a}, \mathbf{b}, \mathbf{c}\} + \{\mathbf{c}, \mathbf{b}, \mathbf{c}\} = q_{\mathbf{b}}(\mathbf{a}) + 2\{\mathbf{a}, \mathbf{b}, \mathbf{c}\} + q_{\mathbf{b}}(\mathbf{c}).$$

Thus, we can now define the triple product for any $\mathbf{a}, \mathbf{b}, \mathbf{c} \in X$ by

$$\{\mathbf{a}, \mathbf{b}, \mathbf{c}\} = \frac{1}{2}(q_{\mathbf{b}}(\mathbf{a} + \mathbf{c}) - q_{\mathbf{b}}(\mathbf{a}) - q_{\mathbf{b}}(\mathbf{c})). \tag{5.70}$$

For any pair of elements $\mathbf{a}, \mathbf{b} \in X$, we define two real linear maps $X \to X$. The first one, denoted $D(\mathbf{a}, \mathbf{b})$, is also a complex linear map:

$$D(\mathbf{a}, \mathbf{b})\mathbf{x} = \{\mathbf{a}, \mathbf{b}, \mathbf{x}\}. \tag{5.71}$$

We write $D(\mathbf{a})$ instead of $D(\mathbf{a}, \mathbf{a})$. The second map is a complex conjugate map defined by

$$Q(\mathbf{a}, \mathbf{b})\mathbf{x} = \{\mathbf{a}, \mathbf{x}, \mathbf{b}\}. \tag{5.72}$$

We write $Q(\mathbf{a})$ instead of $Q(\mathbf{a}, \mathbf{a})$.

5.4 Algebraic properties of the triple product

In the previous section, we introduced a triple product for a bounded symmetric domain D. The product was derived from the quadratic term elements of the Lie algebra $aut_a(D)$. We will continue to study this algebra and derive consequences of its properties for the triple product.

5.4.1 The main identity

By use of the triple product, defined by (5.70), we can now rewrite the formula (5.66) of $aut_a(D)$ as

$$aut_a(D) = \{\xi(\mathbf{x}) = \mathbf{b} + \kappa\mathbf{x} - Q(\mathbf{x})(\mathbf{b}) = \mathbf{b} + \kappa\mathbf{x} - \{\mathbf{x}, \mathbf{b}, \mathbf{x}\}\}, \tag{5.73}$$

where $\mathbf{b} \in X$ and the map $\kappa \in L(X)$ is a linear map such that $\exp\kappa$ is a linear automorphism of D. Let us calculate now the Lie bracket (5.53) of two elements $\zeta = \mathbf{d} + \nu\mathbf{x} - \{\mathbf{x}, \mathbf{d}, \mathbf{x}\}$ and $\xi = \mathbf{b} + \kappa\mathbf{x} - \{\mathbf{x}, \mathbf{b}, \mathbf{x}\}$ in $aut_a(D)$. Since the derivative of ξ at $\mathbf{x}$ in the direction $\mathbf{y}$ is

$$\frac{d\xi}{d\mathbf{x}}(\mathbf{x})\mathbf{y} = \kappa\mathbf{y} - 2\{\mathbf{x}, \mathbf{b}, \mathbf{y}\} = (\kappa - 2D(\mathbf{x}, \mathbf{b}))\mathbf{y},$$

we have

$$[\zeta, \xi](\mathbf{x}) = (\nu - 2D(\mathbf{x}, \mathbf{d}))(\mathbf{b} + \kappa\mathbf{x} - \{\mathbf{x}, \mathbf{b}, \mathbf{x}\})$$

$$-(\kappa - 2D(\mathbf{x}, \mathbf{b}))(\mathbf{d} + \nu\mathbf{x} - \{\mathbf{x}, \mathbf{d}, \mathbf{x}\}), \tag{5.74}$$

which is an element of $aut_a(D)$.

Consider now the case $\kappa = \nu = 0$. Then, using the notation from (5.67), we have

$$[\xi_{i\mathbf{d}}, \xi_{\mathbf{d}}](\mathbf{x}) = 4i\{\mathbf{x}, \mathbf{d}, \mathbf{d}\} = 4iD(\mathbf{d})\mathbf{x} \in aut_a(D), \tag{5.75}$$

implying that $iD(\mathbf{d})$ is a linear map in $aut_a(D)$. Hence, the exponent $\exp(itD(\mathbf{d}))$ is a linear automorphism of D for any real t. If D is the unit ball in X, then $\exp(itD(\mathbf{d}))$ is a linear isometry of X. Denote by $\sigma D(\mathbf{d})$ the spectrum of the operator $D(\mathbf{d})$. If $\lambda \in \sigma D(\mathbf{d})$, then we can find a non-zero $\mathbf{x} \in D$ such that $D(\mathbf{d})\mathbf{x} \approx \lambda\mathbf{x}$. But if λ is not real, then $\exp(itD(\mathbf{d}))\mathbf{x} \approx e^{it\lambda}\mathbf{x}$ will not be bounded for all t, contradicting the fact that this exponent generates a flow in a bounded domain. Thus, the spectrum of the operator $D(\mathbf{d})$ is real for any $\mathbf{d} \in X$.

Since $iD(\mathbf{d})$ is a linear map in $aut_a(D)$, by use of (5.74), the bracket of this map with $\xi_{\mathbf{b}}$ for any $\mathbf{z} \in D$ is

$$[iD(\mathbf{d}), \xi_{\mathbf{b}}](\mathbf{z}) = iD(\mathbf{d})(\mathbf{b} - Q(\mathbf{z})\mathbf{b}) + 2D(\mathbf{z}, \mathbf{b})iD(\mathbf{d})\mathbf{z},$$

which is another element of $aut_a(D)$ with no linear term, hence equal to $\xi_{iD(\mathbf{d})\mathbf{b}}$. Thus,

$$Q(\mathbf{z})iD(\mathbf{d})\mathbf{b} = -q_{iD(\mathbf{d})\mathbf{b}}(\mathbf{z}) + iD(\mathbf{d})\mathbf{b} = iD(\mathbf{d})Q(\mathbf{z})\mathbf{b} - 2D(\mathbf{z},\mathbf{b})iD(\mathbf{d})\mathbf{z},$$

or

$$Q(\mathbf{z})D(\mathbf{d})\mathbf{b} = 2D(\mathbf{z},\mathbf{b})D(\mathbf{d})\mathbf{z} - D(\mathbf{d})Q(\mathbf{z})\mathbf{b}.$$

Now, by substituting $\mathbf{x}+\lambda\mathbf{y}$ for $\mathbf{d}$ and $\mathbf{a}+\mu\mathbf{c}$ for $\mathbf{z}$, and using the linearity and conjugate linearity of the triple product, we obtain

$$\{\mathbf{a},\{\mathbf{y},\mathbf{x},\mathbf{b}\},\mathbf{c}\}=\{\mathbf{a},\mathbf{b},\{\mathbf{x},\mathbf{y},\mathbf{c}\}\}+\{\mathbf{c},\mathbf{b},\{\mathbf{x},\mathbf{y},\mathbf{a}\}\}-\{\mathbf{x},\mathbf{y},\{\mathbf{a},\mathbf{b},\mathbf{c}\}\}.$$

By moving around the terms we get the usual form of the *main identity* of the triple product:

$$\{\mathbf{x},\mathbf{y},\{\mathbf{a},\mathbf{b},\mathbf{c}\}\}=\{\{\mathbf{x},\mathbf{y},\mathbf{a}\},\mathbf{b},\mathbf{c}\}-\{\mathbf{a},\{\mathbf{y},\mathbf{x},\mathbf{b}\},\mathbf{c}\}+\{\mathbf{a},\mathbf{b},\{\mathbf{x},\mathbf{y},\mathbf{c}\}\}. \tag{5.76}$$

5.4.2 Hermitian Jordan triple system

A complex Banach space X is called a *Hermitian Jordan triple system* if there is a triple product on X mapping any triple of elements $\mathbf{a},\mathbf{b},\mathbf{c}\in X$ to an element of X, denoted by $\{\mathbf{a},\mathbf{b},\mathbf{c}\}$ such that:

1. the product is linear in the outer variables $\mathbf{a},\mathbf{c}$ and conjugate linear in the middle one $\mathbf{b}$,
2. the product is symmetric in the outer variables, meaning that

$$\{\mathbf{a},\mathbf{b},\mathbf{c}\} = \{\mathbf{c},\mathbf{b},\mathbf{a}\}, \tag{5.77}$$

3. the product satisfies the main identity (5.76),
4. the operator $D(\mathbf{a}) = \{\mathbf{a},\mathbf{a},\cdot\}$ is Hermitian, meaning that $\exp(iD(\mathbf{a}))$ is an isometry.

We have already shown that if the unit ball D in a Banach space X is a bounded symmetric domain, then the triple product on X defined from the group $aut_a(D)$ satisfies the above properties. Thus, X with the triple product is a Hermitian Jordan triple system.

Let $g\in K\subset Aut_a(D)$ be a linear automorphism (one-to-one and onto analytic map) of the domain D. We will show now that g preserves the triple product, and, thus is a triple product automorphism of D. To see this, let $\mathbf{a}$ be an arbitrary element of X. Denote by $\delta_{\mathbf{a}} = \mathbf{a} - \{\mathbf{x},\mathbf{a},\mathbf{x}\}$ the complete holomorphic vector field in $aut_a(D)$. The vector field $\delta_{\mathbf{a}}$ generates a flow $\varphi_\delta(t,\mathbf{x})$ which is a solution of the initial-value problem (5.56). We define now the adjoint map $Ad(g)$, for $g\in Aut_a(D)$, which acts on elements

$f \in Aut_a(D)$ or $f \in aut_a(D)$ by $Ad(g)f(\mathbf{x}) = g^{-1}f(g\mathbf{x})$. Then the flow $Ad(g)\varphi_\delta(t,\mathbf{x}) := g^{-1}\varphi_\delta(t, g\mathbf{x})$ satisfies the equation

$$\frac{\partial}{\partial t}Ad(g)\varphi_\delta(t,\mathbf{x}) = g^{-1}\frac{\partial}{\partial t}\varphi_\delta(t,g\mathbf{x}) = g^{-1}\delta_{\mathbf{a}}(\varphi_\delta(t,g\mathbf{x}))$$

$$= g^{-1}\delta_{\mathbf{a}}(g\,g^{-1}\varphi_\delta(t,g\mathbf{x})) = Ad(g)\delta_{\mathbf{a}}(Ad(g)\varphi_\delta(t,\mathbf{x})),$$

implying that $Ad(g)\delta_{\mathbf{a}}$ is a complete vector field on D hence belongs to $aut_a(D)$. But

$$Ad(g)\delta_{\mathbf{a}}(\mathbf{x}) = g^{-1}(\mathbf{a} - \{g\mathbf{x},\mathbf{a},g\mathbf{x}\}) = g^{-1}\mathbf{a} - g^{-1}\{g\mathbf{x},\mathbf{a},g\mathbf{x}\}).$$

Since $Ad(g)(aut_a(D)) \subseteq aut_a(D)$, we have $Ad(g)\delta_{\mathbf{a}} = \delta_{g^{-1}(\mathbf{a})}$ and thus

$$g^{-1}\{g\mathbf{x},\mathbf{a},g\mathbf{x}\} = \{\mathbf{x}, g^{-1}\mathbf{a}, \mathbf{x}\}$$

and

$$\{g\mathbf{x},\mathbf{a},g\mathbf{x}\} = g\{\mathbf{x}, g^{-1}\mathbf{a}, \mathbf{x}\}.$$

If we write $\mathbf{b} = g^{-1}\mathbf{a}$, then

$$g\{\mathbf{x},\mathbf{b},\mathbf{x}\} = \{g\mathbf{x}, g\mathbf{b}, g\mathbf{x}\},$$

and by polarization of $\mathbf{x}$ we get

$$g\{\mathbf{x},\mathbf{b},\mathbf{y}\} = \{g\mathbf{x}, g\mathbf{b}, g\mathbf{y}\},$$

for any $\mathbf{x},\mathbf{b},\mathbf{y} \in X$. Thus g is a triple product automorphism.

5.4.3 Peirce decomposition

The basic elements of any binary algebraic operation are the idempotents. For a triple structure, the corresponding elements are tripotents $\mathbf{v}$ satisfying $\mathbf{v} = \{\mathbf{v},\mathbf{v},\mathbf{v}\}$. We will derive first some identities for the operators $D(\mathbf{v})$ and $Q(\mathbf{v})$ associated with a tripotent and defined by (5.71) and (5.72).

Let $\mathbf{v}$ be a tripotent in a Hermitian Jordan triple system X. By substituting $\mathbf{y} = \mathbf{a} = \mathbf{b} = \mathbf{c} = \mathbf{v}$ in (5.76), we have

$$D(\mathbf{v}) = D(\mathbf{v})^2 - Q(\mathbf{v})^2 + D(\mathbf{v})^2,$$

implying that the operators $Q(\mathbf{v})$ and $D(\mathbf{v})$ are related by

$$Q(\mathbf{v})^2 = 2D(\mathbf{v})^2 - D(\mathbf{v}). \tag{5.78}$$

Similarly, by taking $\mathbf{x} = \mathbf{a} = \mathbf{b} = \mathbf{c} = \mathbf{v}$, we have

$$Q(\mathbf{v}) = D(\mathbf{v})Q(\mathbf{v}) - Q(\mathbf{v})D(\mathbf{v}) + D(\mathbf{v})Q(\mathbf{v}),$$

and by taking $\mathbf{x} = \mathbf{y} = \mathbf{a} = \mathbf{c} = \mathbf{v}$, we have

$$D(\mathbf{v})Q(\mathbf{v}) = Q(\mathbf{v}) - Q(\mathbf{v})D(\mathbf{v}) + Q(\mathbf{v}).$$

Adding the last two equations, we obtain

$$D(\mathbf{v})Q(\mathbf{v}) = Q(\mathbf{v}), \tag{5.79}$$

showing that the image of $Q(\mathbf{v})$ is in the 1-eigenspace of $D(\mathbf{v})$. Multiply both sides of (5.78) by $D(\mathbf{v})$ and use (5.79) to obtain

$$2D(\mathbf{v})^3 - D(\mathbf{v})^2 = D(\mathbf{v})Q(\mathbf{v})^2 = (D(\mathbf{v})Q(\mathbf{v}))Q(\mathbf{v}) = Q(\mathbf{v})^2.$$

Finally, substituting from (5.78) the expression for $Q(\mathbf{v})^2$, we have

$$2D(\mathbf{v})^3 - 3D(\mathbf{v})^2 + D(\mathbf{v}) = 0, \tag{5.80}$$

which is an algebraic identity for the operator $D(\mathbf{v})$ for any tripotent $\mathbf{v}$.

This identity implies that

$$D(\mathbf{v})(D(\mathbf{v}) - 1)(2D(\mathbf{v}) - 1) = 0. \tag{5.81}$$

Thus, if we define a map

$$P_1(\mathbf{v}) = D(\mathbf{v})(2D(\mathbf{v}) - 1) = 2D(\mathbf{v})^2 - D(\mathbf{v}), \tag{5.82}$$

then, for any $\mathbf{x} \in X$, from (5.81) we have

$$D(\mathbf{v})P_1(\mathbf{v})\mathbf{x} = D(\mathbf{v})(2D(\mathbf{v}) - 1)\mathbf{x} = P_1(\mathbf{v})\mathbf{x},$$

implying that $P_1(\mathbf{v})$ maps the space X into a subspace corresponding to the 1-eigenvalue of the operator $D(\mathbf{v})$. Since $D(\mathbf{v})P_1(\mathbf{v}) = P_1(\mathbf{v})$, we have

$$P_1(\mathbf{v})^2 = (2D(\mathbf{v})^2 - D(\mathbf{v}))P_1(\mathbf{v}) = P_1(\mathbf{v}),$$

so that $P_1(\mathbf{v})$ is a projection. Moreover, from (5.78) we have

$$P_1(\mathbf{v}) = Q^2(\mathbf{v}). \tag{5.83}$$

Similarly, the operators

$$P_{\frac{1}{2}}(\mathbf{v}) = 4D(\mathbf{v}) - 4D(\mathbf{v})^2 \text{ and } P_0(\mathbf{v}) = 2D(\mathbf{v})^2 - 3D(\mathbf{v}) + I \tag{5.84}$$

are projections from the space X to the subspaces corresponding to the 1/2- and 0-eigenvalues of operator $D(\mathbf{v})$, respectively. Note that

$$P_1(\mathbf{v}) + P_{\frac{1}{2}}(\mathbf{v}) + P_0(\mathbf{v}) = I. \tag{5.85}$$

This implies that the spectral values of the operator $D(\mathbf{v})$ belong to the set $\{1, \frac{1}{2}, 0\}$ and that the spectral decomposition of $D(\mathbf{v})$ is

$$D(\mathbf{v}) = P_1(\mathbf{v}) + \frac{1}{2} P_{\frac{1}{2}}(\mathbf{v}). \tag{5.86}$$

If we denote by $X_j(\mathbf{v})$ the j-eigenspace of $D(\mathbf{v})$, for $j \in \{1, 1/2, 0\}$, then $X_j(\mathbf{v}) = P_j(\mathbf{v})X$, and, by use of (5.85), the space X can be decomposed as a direct sum

$$X = X_1(\mathbf{v}) \oplus X_{1/2}(\mathbf{v}) \oplus X_0(\mathbf{v}). \tag{5.87}$$

This decomposition is called the *Peirce decomposition* of X with respect to a tripotent $\mathbf{v}$.

As mentioned in section 4.3.2 for quantum mechanics applications it is important to know which tripotents generate commuting Peirce decompositions. We will show now that for any two tripotents $\mathbf{u}, \mathbf{v}$ in a Hermitian Jordan triple system X, we have

$$\mathbf{u} \in X_j(\mathbf{v}) \;\Rightarrow\; [P_k(\mathbf{v}), P_n(\mathbf{u})] = P_k(\mathbf{v})\, P_n(\mathbf{u}) - P_n(\mathbf{u})\, P_k(\mathbf{v}) = 0, \tag{5.88}$$

implying that such tripotents generate commuting Peirce decompositions. Note that from (5.76), we get

$$D(\mathbf{x}, \mathbf{y}) D(\mathbf{a}, \mathbf{b})\mathbf{c} = D(D(\mathbf{x}, \mathbf{y})\mathbf{a}, \mathbf{b})\mathbf{c} - D(\mathbf{a}, D(\mathbf{y}, \mathbf{x})\mathbf{b})\mathbf{c} + D(\mathbf{a}, \mathbf{b}) D(\mathbf{x}, \mathbf{y})\mathbf{c},$$

which can be rewritten as

$$[D(\mathbf{x}, \mathbf{y}), D(\mathbf{a}, \mathbf{b})] = D(D(\mathbf{x}, \mathbf{y})\mathbf{a}, \mathbf{b}) - D(\mathbf{a}, D(\mathbf{y}, \mathbf{x})\mathbf{b}). \tag{5.89}$$

Applying this formula with $\mathbf{x} = \mathbf{y} = \mathbf{v}$ and $\mathbf{a} = \mathbf{b} = \mathbf{u}$ and using the fact that $D(\mathbf{v})\mathbf{u} = j\mathbf{u}$ with real j, we have

$$[D(\mathbf{v}), D(\mathbf{u})] = D(D(\mathbf{v})\mathbf{u}, \mathbf{u}) - D(\mathbf{u}, D(\mathbf{v})\mathbf{u}) = jD(\mathbf{u}) - jD(\mathbf{u}) = 0.$$

Thus, the operators $D(\mathbf{v})$ and $D(\mathbf{u})$ commute. Consequently, their spectral projections $P_k(\mathbf{v})$ and $P_n(\mathbf{u})$ also commute and (5.88) follows.

5.4.4 Peirce calculus

We will show now that each component of the Peirce decomposition is a Hermitian Jordan triple. We will also derive the rules for the product of different parts of this decomposition, which is called the Peirce calculus. These rules provide an important tool in working with the Jordan triple system.

Let $\mathbf{v}$ be a tripotent in a Hermitian Jordan triple system X. By substituting in (5.76) $\mathbf{x} = \mathbf{y} = \mathbf{v}$, we have

$$D(\mathbf{v})\{\mathbf{a}, \mathbf{b}, \mathbf{c}\} = \{D(\mathbf{v})\mathbf{a}, \mathbf{b}, \mathbf{c}\} - \{\mathbf{a}, D(\mathbf{v})\mathbf{b}, \mathbf{c}\} + \{\mathbf{a}, \mathbf{b}, D(\mathbf{v})\mathbf{c}\}, \tag{5.90}$$

for any $\mathbf{a}, \mathbf{b}, \mathbf{c} \in X$.

Assume now that $\mathbf{a} \in X_j(\mathbf{v})$, $\mathbf{b} \in X_k(\mathbf{v})$ and $\mathbf{c} \in X_n(\mathbf{v})$ where j, k, and n all belong to $\{0, \frac{1}{2}, 1\}$. Then, from (5.90) we have

$$D(\mathbf{v})\{\mathbf{a}, \mathbf{b}, \mathbf{c}\} = \{j\mathbf{a}, \mathbf{b}, \mathbf{c}\} - \{\mathbf{a}, k\mathbf{b}, \mathbf{c}\} + \{\mathbf{a}, \mathbf{b}, n\mathbf{c}\} = (j - k + n)\{\mathbf{a}, \mathbf{b}, \mathbf{c}\},$$

implying that the triple product $\{\mathbf{a}, \mathbf{b}, \mathbf{c}\}$ belongs to $X_{j-k+n}(\mathbf{v})$. Thus, we have shown that

$$\{X_j(\mathbf{v}), X_k(\mathbf{v}), X_n(\mathbf{v})\} \subseteq X_{j-k+n}(\mathbf{v}). \tag{5.91}$$

This relation is called the *Peirce calculus* and it shows how to compute the triple product on different parts of the Peirce decomposition.

Note that if $j - k + n$ does not belong to the set $\{1, 1/2, 0\}$, then $\{X_j(\mathbf{v}), X_k(\mathbf{v}), X_n(\mathbf{v})\} = \mathbf{0}$. For example,

$$\{X_0(\mathbf{v}), X_1(\mathbf{v}), X_0(\mathbf{v})\} = \mathbf{0}, \ \{X_0(\mathbf{v}), X_1(\mathbf{v}), X_{1/2}(\mathbf{v})\} = \mathbf{0}, \tag{5.92}$$

implying that

$$\{X_0(\mathbf{v}), X_1(\mathbf{v}), X\} = \{X_0(\mathbf{v}), X_1(\mathbf{v}), P_1(\mathbf{v})X\}. \tag{5.93}$$

If $j = k = n$, then (5.91) yields

$$\{X_j(\mathbf{v}), X_j(\mathbf{v}), X_j(\mathbf{v})\} \subseteq X_j(\mathbf{v}),$$

implying that $X_j(\mathbf{v})$ is a Hermitian Jordan triple system.

We will show now that

$$\{X_0(\mathbf{v}), X_1(\mathbf{v}), X\} = \mathbf{0}. \tag{5.94}$$

Let $\mathbf{x} \in X_0(\mathbf{v})$, $\mathbf{y} \in X_1(\mathbf{v})$ and $\mathbf{c} \in X$. From (5.93) we have

$$\{\mathbf{x}, \mathbf{y}, \mathbf{c}\} = \{\mathbf{x}, \mathbf{y}, P_1(\mathbf{v})\mathbf{c}\}.$$

Thus, in order to prove that $\{\mathbf{x}, \mathbf{y}, \mathbf{c}\} = 0$, we may assume without loss of generality that $\mathbf{c} \in X_1(\mathbf{v})$. Let us substitute into the main identity (5.76) $\mathbf{a} = \mathbf{b} = \mathbf{v}$ and assume that $\mathbf{x} \in X_0(\mathbf{v})$ and $\mathbf{y}, \mathbf{c} \in X_1(\mathbf{v})$. Then this identity becomes

$$\{\mathbf{x}, \mathbf{y}, \mathbf{c}\} = \{\mathbf{x}, \mathbf{y}, \{\mathbf{v}, \mathbf{v}, \mathbf{c}\}\}$$

$$= \{\{\mathbf{x}, \mathbf{y}, \mathbf{v}\}, \mathbf{v}, \mathbf{c}\} - \{\mathbf{v}, \{\mathbf{y}, \mathbf{x}, \mathbf{v}\}, \mathbf{c}\} + \{\mathbf{v}, \mathbf{v}, \{\mathbf{x}, \mathbf{y}, \mathbf{c}\}\}. \tag{5.95}$$

From (5.91), $\{\mathbf{y}, \mathbf{x}, \mathbf{v}\} \subseteq X_{1-0+1}(\mathbf{v}) = \{\mathbf{0}\}$ and $\{\mathbf{x}, \mathbf{y}, \mathbf{c}\} \subseteq X_{0-1+1}(\mathbf{v}) = X_0(\mathbf{v})$, implying that $\{\mathbf{v}, \mathbf{v}, \{\mathbf{x}, \mathbf{y}, \mathbf{c}\}\} = \mathbf{0}$. Thus, if we assume for a moment $\mathbf{c} = \mathbf{v}$ in (5.95), we have $\{\mathbf{x}, \mathbf{y}, \mathbf{v}\} = \mathbf{0}$. Then, for any c we have $\{\mathbf{x}, \mathbf{y}, \mathbf{c}\} = \mathbf{0}$, proving (5.94).

In a similar way, one may also show that

$$\{X_1(\mathbf{v}), X_0(\mathbf{v}), X\} = \mathbf{0}. \tag{5.96}$$

5.4.5 Orthogonality in Hermitian Jordan triple systems

Let X be a Hermitian Jordan triple system. We will say that two non-zero elements $\mathbf{a}$ and $\mathbf{b}$ are *algebraically orthogonal*, written $\mathbf{a} \perp \mathbf{b}$, if there is a tripotent $\mathbf{v}$ such that $\mathbf{a} \in X_1(\mathbf{v})$ and $\mathbf{b} \in X_0(\mathbf{v})$ and denote it as $\mathbf{a} \perp \mathbf{b}$. From (5.96) and (5.94), it follows that if $\mathbf{a}$ and $\mathbf{b}$ are algebraically orthogonal, then $D(\mathbf{a}, \mathbf{b}) = D(\mathbf{b}, \mathbf{a}) = 0$. In this case,

$$(\mathbf{a} + \mathbf{b})^{(3)} = D(\mathbf{a} + \mathbf{b})(\mathbf{a} + \mathbf{b}) = D(\mathbf{a})\mathbf{a} + D(\mathbf{a})\mathbf{b} + D(\mathbf{b})\mathbf{a} + D(\mathbf{b})\mathbf{b}.$$

Since $D(\mathbf{a})\mathbf{b} = D(\mathbf{b}, \mathbf{a})\mathbf{a} = 0$, we have

$$\mathbf{a} \perp \mathbf{b} \Rightarrow (\mathbf{a} + \mathbf{b})^{(3)} = \mathbf{a}^{(3)} + \mathbf{b}^{(3)}. \tag{5.97}$$

We can now define the notion of an *algebraically orthogonal decomposition.* We say that X is decomposed as a sum of two algebraically orthogonal subspaces Y and Z if $X = Y \oplus Z$ and every element $\mathbf{y} \in Y$ is algebraically orthogonal to every element $\mathbf{z} \in Z$. In this case, each subspace Y and Z is a Hermitian Jordan triple system. Moreover, they are ideals in X in the following sense. We say that a subspace Y of a Hermitian Jordan triple system X is an *ideal* if $\{\mathbf{a}, \mathbf{b}, \mathbf{c}\} \in Y$ whenever one of $\mathbf{a}, \mathbf{b}, \mathbf{c} \in Y$. To see that if X is decomposed as a sum of two algebraically orthogonal subspaces Y and Z, then Y is an ideal; choose arbitrary $\mathbf{a}, \mathbf{b}, \mathbf{c} \in X$. Decompose $\mathbf{a}$ as $\mathbf{a} = \mathbf{a}_1 + \mathbf{a}_2$ with $\mathbf{a}_1 \in Y$ and $\mathbf{a}_2 \in Z$. Do the same for $\mathbf{b}, \mathbf{c}$. Using the fact that the D operator of algebraically orthogonal tripotents is zero, if $\mathbf{a} \in Y$, then

$$\{\mathbf{a}, \mathbf{b}, \mathbf{c}\} = \{\mathbf{a}_1, \mathbf{b}_1 + \mathbf{b}_2, \mathbf{c}_1 + \mathbf{c}_2\} = \{\mathbf{a}_1, \mathbf{b}_1, \mathbf{c}_1 + \mathbf{c}_2\} = \{\mathbf{a}_1, \mathbf{b}_1, \mathbf{c}_1\},$$

which is an element of Y.

For algebraically orthogonal tripotents $\mathbf{u}$ and $\mathbf{v}$, the equation (5.97) implies that $\mathbf{u} \pm \mathbf{v}$ are tripotents. Conversely, if for two tripotents $\mathbf{u}$ and $\mathbf{v}$, both $\mathbf{u} \pm \mathbf{v}$ are tripotents, then

$$(\mathbf{u} \pm \mathbf{v})^{(3)} = \mathbf{u} \pm \mathbf{v},$$

implying that

$$2D(\mathbf{v})\mathbf{u} + Q(\mathbf{v})\mathbf{u} \pm 2D(\mathbf{u})\mathbf{v} \pm Q(\mathbf{u})\mathbf{v} = 0,$$

and thus

$$2D(\mathbf{v})\mathbf{u} + Q(\mathbf{v})\mathbf{u} = 0. \tag{5.98}$$

We decompose $\mathbf{u}$ by the Peirce decomposition with respect to $\mathbf{v}$ as $\mathbf{u} = \mathbf{u}_1 + \mathbf{u}_{1/2} + \mathbf{u}_0$. Then (5.98) implies that

$$2\mathbf{u}_1 + \mathbf{u}_{1/2} + Q(\mathbf{v})\mathbf{u}_1 = 0.$$

Thus, $\mathbf{u}_{1/2} = 0$, and, multiplying this equation by $Q(\mathbf{v})$ and using (5.83), we have $\mathbf{u}_1 + 2Q(\mathbf{v})\mathbf{u}_1 = 0$, implying that $P_1(\mathbf{v})\mathbf{u} = 0$. Therefore, $\mathbf{u} = P_0(\mathbf{v})\mathbf{u}$ and the tripotents $\mathbf{u}$ and $\mathbf{v}$ are algebraically orthogonal. Thus, we have shown that

$$\mathbf{u} \perp \mathbf{v} \;\Leftrightarrow\; \mathbf{u} \pm \mathbf{v} \text{ are tripotents.} \tag{5.99}$$

For two algebraically orthogonal tripotents $\mathbf{u}$ and $\mathbf{v}$ we have

$$D(\mathbf{u} \pm \mathbf{v}) = D(\mathbf{u}) + D(\mathbf{v}). \tag{5.100}$$

We can now introduce a partial order on the set of tripotents. We say that a tripotent $\mathbf{v}$ is *dominated* by a tripotent $\mathbf{u}$, written $\mathbf{u} \preceq \mathbf{v}$, if $\mathbf{v}$ and $\mathbf{u} - \mathbf{v}$ are algebraically orthogonal. It is obvious that

$$\mathbf{v} \preceq \mathbf{u} \text{ and } \mathbf{u} \preceq \mathbf{w}, \Rightarrow \mathbf{v} \preceq \mathbf{w}.$$

A tripotent $\mathbf{v}$ in a Hermitian Jordan triple system will be called *minimal* if it is minimal in the order $\preceq$. This means that a minimal tripotent can not be decomposed as a sum of non-zero algebraically orthogonal tripotents. We call a tripotent *maximal* if it is maximal with respect to this partial order. The *rank of a JB*-triple* is defined as the maximal number of mutually algebraically orthogonal tripotents.

5.4.6 The Jordan subalgebras

Let X be a Hermitian Jordan triple system, and let $\mathbf{a}$ be any element of X. We denote by $X^{(\mathbf{a})}$ the space X with the binary linear and symmetric product $\circ_a : \; X \times X \to X$ defined by

$$\mathbf{x} \circ_a \mathbf{y} = \{\mathbf{x}, \mathbf{a}, \mathbf{y}\}. \tag{5.101}$$

This definition makes $X^{(\mathbf{a})}$ into a commutative, but in general non-associative, algebra.

For any $\mathbf{x} \in X^{(\mathbf{a})}$ we can define a linear map (called the multiplication operator) $L(\mathbf{x})$ on $X^{(\mathbf{a})}$ by

$$L(\mathbf{x})\mathbf{y} = \mathbf{x} \circ_a \mathbf{y}$$

for any $\mathbf{y} \in X^{(\mathbf{a})}$. We will denote $\mathbf{x}^2 = \mathbf{x} \circ_a \mathbf{x}$. It is known that this product satisfies a weak associativity property, namely, for any element $\mathbf{x} \in X^{(\mathbf{a})}$, the multiplication operator $L(\mathbf{x}^2)$ commutes with the operator $L(\mathbf{x})$. This can be expressed as

$$L(\mathbf{x})L(\mathbf{x}^2) = L(\mathbf{x}^2)L(\mathbf{x}) \tag{5.102}$$

and directly verified from the main identity by appropriate substitutions, (see [23], p.452). This defines the Jordan identity for the product $\circ_a$, and the space $X^{(\mathbf{a})}$ with the *Jordan product* $\circ_a$ becomes a Jordan algebra.

An important case is when $\mathbf{a}$ is a tripotent, say $\mathbf{v}$. If we restrict ourselves to the Hermitian Jordan triple system $X_1(\mathbf{v})$, then the element $\mathbf{v}$ in the Jordan algebra $X_1^{(\mathbf{v})}$ is a unit, since

$$\mathbf{v} \circ_v \mathbf{b} = \{\mathbf{v}, \mathbf{v}, \mathbf{b}\} = \mathbf{b}$$

for any $\mathbf{b} \in X_1(\mathbf{v})$. In this case, we can also define a conjugate linear map

$$\mathbf{b}^* = \{\mathbf{v}, \mathbf{b}, \mathbf{v}\} = Q(\mathbf{v})\mathbf{b}. \tag{5.103}$$

Since by (5.83) we have $\mathbf{b}^{**} = Q(\mathbf{v})^2\mathbf{b} = P_1(\mathbf{v})\mathbf{b} = \mathbf{v}$, this map is a conjugation and is called the adjoint map on $X_1^{(\mathbf{v})}$. It can be shown that this map is a triple automorphism, meaning that

$$Q(\mathbf{v})\{\mathbf{a}, \mathbf{b}, \mathbf{c}\} = \{Q(\mathbf{v})\mathbf{a}, Q(\mathbf{v})\mathbf{b}, Q(\mathbf{v})\mathbf{c}\}. \tag{5.104}$$

This follows from the so-called *fundamental identity*

$$Q(Q(\mathbf{b})\mathbf{a}) = Q(\mathbf{b})Q(\mathbf{a})Q(\mathbf{b}), \tag{5.105}$$

which follows from the main identity, see [23] p. 440. Indeed, if we substitute in this identity $\mathbf{b} = \mathbf{v}$ and apply it on element $\mathbf{c}$, we obtain

$$\{Q(\mathbf{v})\mathbf{a}, \mathbf{c}, Q(\mathbf{v})\mathbf{a}\} = Q(\mathbf{v})\{\mathbf{a}, Q(\mathbf{v})\mathbf{c}, \mathbf{a}\}.$$

Substituting $\mathbf{d} = Q(\mathbf{v})\mathbf{c}$ and using polarization, we arrive at (5.104). We will say that an element $\mathbf{a} \in X_1^{(\mathbf{v})}$ is *self-adjoint* if $\mathbf{a} = \mathbf{a}^* = Q(\mathbf{v})\mathbf{a}$.

Note that from the main identity (5.76) we get, for any $\mathbf{a}, \mathbf{b}, \mathbf{c} \in X_1^{(\mathbf{v})}(\mathbf{v})$,

$$\{\mathbf{a}, \mathbf{b}^*, \mathbf{c}\} = \{\mathbf{a}, \{\mathbf{v}, \mathbf{b}, \mathbf{v}\}, \mathbf{c}\}$$

$$= \{\mathbf{b}, \mathbf{v}, \mathbf{a}\}, \mathbf{v}, \mathbf{c}\} + \{\mathbf{a}, \mathbf{v}, \{\mathbf{b}, \mathbf{v}, \mathbf{c}\}\} - \{\mathbf{b}, \mathbf{v}, \{\mathbf{a}, \mathbf{v}, \mathbf{c}\}\}$$

$$= (\mathbf{a} \circ_v \mathbf{b}) \circ_v \mathbf{c} + \mathbf{a} \circ_v (\mathbf{b} \circ_v \mathbf{c}) - (\mathbf{a} \circ_v \mathbf{c}) \circ_v \mathbf{b}). \tag{5.106}$$

This identity defines the connection between the Jordan triple product and the binary Jordan algebra product. It is used in Jordan algebras to make a Jordan algebra into a Jordan triple system.

We say that a Jordan algebra Z with product $a \circ b$ which is a complex linear space is a *JB*-algebra* if there is a norm $\|\cdot\|$ on Z such that for any two elements $\mathbf{a}, \mathbf{b} \in Z$, we have

$$\|\mathbf{a} \circ \mathbf{b}\| \leq \|\mathbf{a}\| \, \|\mathbf{b}\|$$

and

$$\|\{\mathbf{a}, \mathbf{a}^*, \mathbf{a}\}\| = \|\mathbf{a}\|^3,$$

where $\{\mathbf{a}, \mathbf{a}^*, \mathbf{a}\}$ is defined by (5.106) with respect to the product $a \circ b$. It is known that $X_1^{(\mathbf{v})}(\mathbf{v})$ is a JB^*-algebra. Note that for any tripotent $\mathbf{u}$ which is dominated by $\mathbf{v}$, $(\mathbf{u} \preceq \mathbf{v})$, we have

$$\{\mathbf{u}, \mathbf{v}, \mathbf{u}\} = \mathbf{u} = \mathbf{u} \circ_v \mathbf{u},$$

and

$$\{\mathbf{v}, \mathbf{u}, \mathbf{v}\} = \mathbf{v} = \mathbf{u}^*,$$

implying that $\mathbf{u}$ is a self-adjoint projection in the JB^*-algebra Z. So the spectral decomposition for any self-adjoint element in Z by self-adjoint projections becomes a decomposition for any self-adjoint element $\mathbf{a}$ in the Hermitian Jordan triple $X_1^{(\mathbf{v})}(\mathbf{v})$ by tripotents which are dominated by $\mathbf{v}$.

5.5 Bounded symmetric domains and JB^*-triples

In the previous sections, we constructed the triple product associated with a bounded symmetric domain D, in a Banach space E, from the Lie algebra of the group $Aut_a(D)$ of automorphisms on the domain. We have seen that the space E with this triple product becomes a Hermitian Jordan triple system. In order to study the geometry of the domain D, we need to establish a connection between the *norm* and the algebraic structure. This is done here. We describe the results connecting the bounded symmetric domains with JB^*-triples, which are Hermitian Jordan triple systems with additional norm axioms.

5.5.1 JB^*-triples and JBW^*-triples

A complex Banach space X is called a *JB^*-triple* if there is a triple product on X which makes it a Hermitian Jordan triple system and in addition the following holds:

1. the spectrum $\sigma(D(\mathbf{a}))$ is positive, and
2. $$\|\{\mathbf{a}, \mathbf{a}, \mathbf{a}\}\| = \|\mathbf{a}\|^3. \tag{5.107}$$

It can be shown that (5.107) is equivalent to

$$\|D(\mathbf{a})\| = \|\mathbf{a}\|^2. \tag{5.108}$$

It was shown by W. Kaup [46] that for any bounded symmetric domain D in a Banach space X, the triple product associated with this domain turns

X into a JB^*-triple if we use (5.108) for the definition of the norm. In this case, the domain D becomes the unit ball in this JB^*-triple. Conversely, if X is a JB^*-triple, then its unit ball D is a bounded symmetric domain. The homogeneity of the domain is given by the analytic map $\varphi_{\mathbf{a}}$, defined by (4.66) of Chapter 4, mapping the origin of the domain D to any point $\mathbf{a}$ in D. Thus, the categories of bounded symmetric domains and JB^*-triples are equivalent.

Let X be a JB^*-triple. If X is the dual of some Banach space X_*, then we can define on X a new type of convergence, called *w^*-convergence.* We say that the sequence $\mathbf{a}_n$ in X converges w^* to an element $\mathbf{a}$ if, for any $\mathbf{f} \in X_*$, we have

$$\lim_{n\to\infty} \mathbf{f}(\mathbf{a}_n) = \mathbf{f}(\mathbf{a}). \tag{5.109}$$

If for the JB^*-triple X with a predual X_*, the triple product is w^*-continuous, then we call X a *JBW^*-triple.*

It has been shown that the second dual of a JB^*-triple is a JBW^*-triple. We will show in Chapter 6 that any JBW^*-triple can be decomposed into an algebraically orthogonal sum of its *atomic* part, spanned by minimal tripotents, and its *non-atomic* part, containing no minimal tripotents. It has also been shown that a JB^*-triple can be isometrically embedded into the atomic part of its dual, which is a direct sum of Cartan factors.

5.5.2 Norm properties of Peirce projections

Let X be a JB^*-triple. We will show now that the Peirce projections are contractions. As we have shown in (5.75), for any element $\mathbf{a}$, the map $\exp(itD(\mathbf{a}))$, for real t, is a linear isometry of the unit ball D. Thus, for any tripotent $\mathbf{v}$, by using the spectral decomposition (5.86) of $D(\mathbf{v})$, we can define a family $S_t(\mathbf{v})$ of isometries by

$$S_t(\mathbf{v}) = \exp(itD(\mathbf{v})) = e^{it}P_1(\mathbf{v}) + e^{it/2}P_{1/2}(\mathbf{v}) + P_0(\mathbf{v}). \tag{5.110}$$

Since

$$P_1(\mathbf{v}) + P_0(\mathbf{v}) = \frac{S_{2\pi}(\mathbf{v}) + I}{2},$$

for any $\mathbf{x} \in X$ we have $\|(P_1(\mathbf{v}) + P_0(\mathbf{v}))\mathbf{x}\| \le \|\mathbf{x}\|$, implying that

$$\|P_1(\mathbf{v}) + P_0(\mathbf{v})\| \le 1. \tag{5.111}$$

Now, we also have

$$P_1(\mathbf{v}) = \frac{S_\pi(\mathbf{v}) + I}{2}(P_1(\mathbf{v}) + P_0(\mathbf{v})),\ P_0(\mathbf{v}) = \frac{S_\pi(\mathbf{v}) - I}{2}(P_1(\mathbf{v}) + P_0(\mathbf{v})),$$

implying that

$$\|P_1(\mathbf{v})\| \leq 1, \quad \|P_0(\mathbf{v})\| \leq 1. \tag{5.112}$$

Similarly,

$$P_{1/2}(\mathbf{v}) = \frac{S_{2\pi}(\mathbf{v}) - I}{2} \quad \Rightarrow \quad \|P_{1/2}(\mathbf{v})\| \leq 1. \tag{5.113}$$

Consider now any two algebraically orthogonal elements $\mathbf{a}, \mathbf{b}$ in (we are already assuming that X is a JB^*-triple) X such that $\mathbf{a} \in X_1(\mathbf{v})$ and $\mathbf{b} \in X_0(\mathbf{v})$. Using (5.97) and iterating (5.107) n times, we have

$$\|\mathbf{a}+\mathbf{b}\| = \|(\mathbf{a}+\mathbf{b})^{(3^n)}\|^{3^{-n}} = \|\mathbf{a}^{(3^n)} + \mathbf{b}^{(3^n)}\|^{3^{-n}} \leq (\|\mathbf{a}\|^{3^n} + \|\mathbf{b}\|^{3^n})^{3^{-n}}$$

$$\leq (2\max\{\|\mathbf{a}\|, \|\mathbf{b}\|\}^{3^n}) = 2^{3^{-n}} \max\{\|\mathbf{a}\|, \|\mathbf{b}\|\} \to \max\{\|\mathbf{a}\|, \|\mathbf{b}\|\}.$$

But from (5.112), we have

$$\|\mathbf{a}\| = \|P_1(\mathbf{v})(\mathbf{a}+\mathbf{b})\| \leq \|\mathbf{a}+\mathbf{b}\|$$

and, similarly,

$$\|\mathbf{b}\| = \|P_0(\mathbf{v})(\mathbf{a}+\mathbf{b})\| \leq \|\mathbf{a}+\mathbf{b}\|.$$

Thus,

$$\mathbf{a} \perp \mathbf{b} \quad \Rightarrow \quad \|\mathbf{a}+\mathbf{b}\| = \max\{\|\mathbf{a}\|, \|\mathbf{b}\|\}, \tag{5.114}$$

which is the same property as for the operator norm.

5.5.3 Spectral decomposition of elements of a *JBW**-triple

Let X be a JBW^*-triple with predual X_*, and let $\mathbf{a}$ be an element of X. Denote by $F(\mathbf{a})$ the set of elements $p_n(\mathbf{a}) \in X$, where p_n is an arbitrary odd polynomial. We define the JBW^*-triple singly-generated by $\mathbf{a}$ as the w^*-closure of the set $F(\mathbf{a})$ and denote it by $B(\mathbf{a})$. Recall that the spectral set of the operator $D(\mathbf{a})$ is a set of non-negative numbers, as mentioned earlier. We will define the singular set $\sigma(\mathbf{a})$ of $\mathbf{a}$ as the positive square roots of the spectral set of the operator $D(\mathbf{a})$. It can be shown that for any piecewise continuous function f on the spectral set $\sigma(\mathbf{a})$, we can define an element $f(\mathbf{a})$ in the singly-generated JBW^*-triple $B(\mathbf{a})$ as the w^*-limit of $p_n(\mathbf{a})$ for odd polynomials p_n converging w^* to f on $\sigma(\mathbf{a})$.

For instance, if we take $f(x) \equiv 1$ for all $x \in \sigma(\mathbf{a})$, then $f(\mathbf{a})$ is a tripotent which is denoted by $\mathbf{v}(\mathbf{a})$ and is called the *support tripotent* of $\mathbf{a}$. The support tripotent satisfies $Q(\mathbf{v}(\mathbf{a}))\mathbf{a} = \mathbf{a}$, implying that $\mathbf{a}$ is a self-adjoint element in the Jordan algebra $X_1^{(\mathbf{v}(\mathbf{a}))}$. Moreover $\mathbf{a}$ is positive in this Jordan algebra.

For any λ in the interval $(0, \|\mathbf{a}\|)$, we define $f_\lambda(x) = \chi(0,\lambda)(x),\ x \in \sigma(\mathbf{a})$, where $\chi(0,\lambda)$, called the characteristic function, is equal to 1 for $0 < x < \lambda$

and equal to 0 for all other x. Then $\mathbf{v}_\lambda = f_\lambda(\mathbf{a})$ is a tripotent dominated by $\mathbf{v}(\mathbf{a})$ and a projection in the Jordan algebra $X_1^{(\mathbf{v}(\mathbf{a}))}$. This defines the spectral decomposition

$$\mathbf{a} = \int_0^{\|\mathbf{a}\|} \lambda d\mathbf{v}_\lambda. \tag{5.115}$$

5.6 The dual of a JB^*-triple

In quantum mechanics, the observables are represented by self-adjoint operators on a Hilbert space, which form the real part of a JB^*-algebra. The dual, or predual, to this space is used to represent the states of a quantum system. Hence, in our model, we will use a JB^*-triple X to represent the observables and the dual or predual to represent the states of the systems. The measuring process defines filtering projections on the state space. Such projections are similar to the action of Peirce projections on the dual, representing the state space. In this section, we will study the properties of the *dual* X^* and the predual X_* of a JB^*-triple X and their behavior under the action of Peirce projections.

5.6.1 Norm of orthogonal states

Let X be a JB^*-triple. If X is the dual of some Banach space A, we will write $X_* = A$. We define the norm of elements of the dual X^* or the predual X_* by

$$\|\mathbf{f}\|_* = \sup_{\|\mathbf{x}\|=1} |\mathbf{f}(\mathbf{x})|, \text{ for } \mathbf{f} \in X^*(\text{or } X_*), \ \mathbf{x} \in X. \tag{5.116}$$

For any tripotent $\mathbf{v}$ in a JB^*-triple, we define the dual $P_k^*(\mathbf{v})$ of the Peirce projections $P_k(\mathbf{v})$, for $k \in \{1, 1/2, 0\}$, acting on the dual X^* (or predual X_*) by

$$(P_k^*(\mathbf{v})\mathbf{f})(\mathbf{x}) = \mathbf{f}(P_k(\mathbf{v})\mathbf{x}), \quad \text{for} \quad \mathbf{f} \in X^*, \ \mathbf{x} \in X. \tag{5.117}$$

From (5.111) and the definition of the dual norm, we have

$$\|(P_1^*(\mathbf{v}) + P_0^*(\mathbf{v}))\mathbf{f}\|_* = \sup_{\|\mathbf{x}\|=1} |((P_1^*(\mathbf{v}) + P_0^*(\mathbf{v}))\mathbf{f})(\mathbf{x})|$$

$$= \sup_{\|\mathbf{x}\|=1} |\mathbf{f}((P_1(\mathbf{v}) + P_0(\mathbf{v}))\mathbf{x})| \leq \|\mathbf{f}\|_*, \quad \text{for} \quad \mathbf{f} \in X^*, \ \mathbf{x} \in X.$$

Hence,

$$\|P_1^*(\mathbf{v}) + P_0^*(\mathbf{v})\| \leq 1. \tag{5.118}$$

Similarly, from (5.112) and (5.113), we have that $\|P_k^*(\mathbf{v}\| \leq 1$, for $k \in \{1, 1/2, 0\}$.

By use of the projections $P_k^*(\mathbf{v})$ we can define the notion of orthogonality for elements of X^*. We say that two elements $\mathbf{f}, \mathbf{g}$ of X^* are orthogonal if there is a tripotent $\mathbf{v}$ in X such that $P_1^*(\mathbf{v})\mathbf{f} = \mathbf{f}$ and $P_0^*(\mathbf{v})\mathbf{g} = \mathbf{g}$. We want to describe the norm of the sum of orthogonal elements $\mathbf{f}, \mathbf{g}$ in X^*. Since $P_1^*(\mathbf{v})\mathbf{f} = \mathbf{f}$, for any $\varepsilon > 0$ we can choose $\mathbf{x} \in X_1(\mathbf{v})$ with $\|\mathbf{x}\| = 1$ and $|\mathbf{f}(\mathbf{x})| \geq \|\mathbf{f}\|_* - \varepsilon$. Similarly, we can choose $\mathbf{y} \in X_0(\mathbf{v})$ with $\|\mathbf{y}\| = 1$ and $|\mathbf{g}(\mathbf{y})| \geq \|\mathbf{g}\|_* - \varepsilon$. Since by (5.114) the norm $\|\mathbf{x} + \mathbf{y}\| = 1$, we have

$$\|\mathbf{f} + \mathbf{g}\|_* \geq |(\mathbf{f} + \mathbf{g})(\mathbf{x} + \mathbf{y})| \geq \|\mathbf{f}\|_* + \|\mathbf{g}\|_* - 2\varepsilon.$$

Obviously, $\|\mathbf{f}+\mathbf{g}\|_* \leq \|\mathbf{f}\|_* + \|\mathbf{g}\|_*$, so by the arbitrariness of ε, for orthogonal elements $\mathbf{f}, \mathbf{g}$ in X^*, we have

$$\|\mathbf{f} + \mathbf{g}\|_* = \|\mathbf{f}\|_* + \|\mathbf{g}\|_*. \tag{5.119}$$

From this it follows that for any $\mathbf{f} \in X^*$, we have

$$\|(P_1^*(\mathbf{v}) + P_0^*(\mathbf{v}))\mathbf{f})\|_* = \|(P_1^*(\mathbf{v})\mathbf{f}\|_* + \|P_0^*(\mathbf{v})\mathbf{f}\|_*. \tag{5.120}$$

5.6.2 Neutrality of Peirce projections

We will now show that the projection $P_1^*(\mathbf{v})$ is neutral, meaning that

$$\|P_1^*(\mathbf{v})\mathbf{f}\|_* = \|\mathbf{f}\|_* \;\Rightarrow\; P_1^*(\mathbf{v})\mathbf{f} = \mathbf{f}, \text{ for } \; \mathbf{f} \in X^*. \tag{5.121}$$

Note that from (5.120), it follows that $P_0^*(\mathbf{v})\mathbf{f} = 0$. So it is enough to show that $P_{1/2}^*(\mathbf{v})\mathbf{f} = 0$. Without loss of generality, we will assume that $\|\mathbf{f}\|_* = 1$. Let $\mathbf{x} \in X_1(\mathbf{v})$ be as above, that is $\|\mathbf{x}\| = 1$ and $\mathbf{f}(\mathbf{x}) \geq 1 - \varepsilon$. Let $\mathbf{y}$ be any element in $X_{1/2}(\mathbf{v})$. By multiplying $\mathbf{y}$, if necessary, by a constant of absolute value 1, we may assume that $\mathbf{f}(\mathbf{y}) \geq 0$. Our result will follow if we can show that $\mathbf{f}(\mathbf{y}) = 0$.

We want first to estimate the norm of $\mathbf{x} + t\mathbf{y}$ for any small real positive t. By use of induction and (5.91), we have

$$(\mathbf{x} + t\mathbf{y})^{(3^n)} = \mathbf{x}^{(3^n)} + t2^n D(\mathbf{x}^{(3^{n-1})}) \cdots D(\mathbf{x}^{(3)})D(\mathbf{x})\mathbf{y} + O(t^2),$$

implying, by use of (5.107) and (5.108), that

$$\|\mathbf{x} + t\mathbf{y}\|^{3^n} = \|(\mathbf{x} + t\mathbf{y})^{(3^n)}\| \leq 1 + 2^n t\|\mathbf{y}\| + O(t^2).$$

Since we assumed that $\|\mathbf{f}\|_* = 1$, we have the following estimates:

$$\|\mathbf{x} + t\mathbf{y}\| \geq |\mathbf{f}(\mathbf{x} + t\mathbf{y})| \geq 1 - \varepsilon + t\mathbf{f}(\mathbf{y}).$$

From the last two equations, we have

$$(1-\varepsilon)^{3^n}+3^n t(1-\varepsilon)^{3^n-1}\mathbf{f}(\mathbf{y})+O(t^2)=(1-\varepsilon+t\mathbf{f}(\mathbf{y}))^{3^n}$$

$$\leq 1+2^n t\|\mathbf{y}\|+O(t^2).$$

If we take the limit as $\varepsilon \to 0$ and divide by t, we have

$$\mathbf{f}(\mathbf{y}) \leq (\frac{2}{3})^n \|\mathbf{y}\| + O(t).$$

Now take the limit as $n \to \infty$ and $t \to 0$ to get $\mathbf{f}(\mathbf{y}) = 0$, which proves (5.121).

The same proofs also establishes the *neutrality* of the Peirce projection $P_0^*(\mathbf{v})$, meaning that

$$\|P_0^*(\mathbf{v})\mathbf{f}\|_* = \|\mathbf{f}\|_* \;\Rightarrow\; P_0^*(\mathbf{v})\mathbf{f} = \mathbf{f}, \text{ for } \mathbf{f} \in X^*. \tag{5.122}$$

5.6.3 The facial structure of the predual of a JBW^*-triple

A non-empty convex subset G of a convex set K is called a *face* of K, if for any $\mathbf{g} \in G$ and $\mathbf{h}, \mathbf{k} \in K$ satisfying $\mathbf{g} = \lambda\mathbf{h} + (1-\lambda)\mathbf{k}$ for some $\lambda \in (0,1)$, then also $\mathbf{h}, \mathbf{k} \in G$. This means that if $\mathbf{g}$ belongs to a face G and belongs to some segment, meaning that the state $\mathbf{g}$ is a mixed state of two states $\mathbf{h}, \mathbf{k}$, then G also contains $\mathbf{h}, \mathbf{k}$. Thus faces can be considered as sets of states closed under the decomposition into mixed states.

A *norm-exposed face* of a unit ball Z_1 of a Banach space Z is a non-empty set (necessarily $\neq Z_1$) of the form

$$F_{\mathbf{x}} = \{\mathbf{f} \in Z_1 : \mathbf{f}(\mathbf{x}) = \|\mathbf{x}\|\}, \tag{5.123}$$

for some $\mathbf{x} \in Z^*$. We will say that the element $\mathbf{x}$ *exposes* the face $F_{\mathbf{x}}$. The physical interpretation of a norm-exposed face $F_{\mathbf{x}}$ is that it is the set of all states on which the observable $\mathbf{x}$ gets its maximal value. It is obvious that any norm-exposed face is norm closed.

Let X be a JBW^* triple with predual X_*. Denote by S the unit ball in X_*. It was shown in [18] that any norm-closed face in S is norm exposed by an element of X. Moreover, each norm-exposed face uniquely defines a tripotent $\mathbf{v}$ in X exposing the face, and the order on the norm-exposed faces, defined by inclusion, coincides with the order on the set of tripotents, as defined in section 5.4.4.

5.6.4 Polar decomposition

Let X be a JBW^*-triple with predual X_*. We can show that for any element $\mathbf{f} \in X_*$, there exists a unique tripotent $\mathbf{v} = \mathbf{v}(\mathbf{f})$, called the *support tripotent* of $\mathbf{f}$, such that

$$\mathbf{f} = P_1^*(\mathbf{v})\mathbf{f}$$

and $\mathbf{f}$ is a positive faithful functional on the Jordan algebra $X_1^{(\mathbf{v})}$. The support tripotent $\mathbf{v}(\mathbf{f})$ of a norm one element $\mathbf{f}$ exposes the minimal norm-exposed face containing $\mathbf{f}$.

A functional $\mathbf{f} \in X_*$ is called an *atom* if $\|\mathbf{f}\|_* = 1$ and its support tripotent $\mathbf{v} = \mathbf{v}(\mathbf{b})$ is a minimal tripotent. A functional $\mathbf{f} \in X_*$ is an atom if and only if it is an *extreme point* of the unit ball S in X_*. In this case, as in (3.106), we have

$$P_1(\mathbf{v})\mathbf{x} = \mathbf{f}(\mathbf{x})\mathbf{v}.$$

Such a functional corresponds to a pure state, since it is not a mixture of other states.

5.7 Facially symmetric spaces

In this section, we will review the results showing that a Banach space satisfying some physically significant geometric properties is the predual of a JBW^*-triple. We have already seen that a Banach space X is a JB^*-triple if and only if its unit ball is homogeneous under the group of holomorphic automorphisms. Thus, the results described in this section reveal how the facial structure of the predual X_* defines the homogeneity of the unit ball in X. If one considers the unit ball of X_* as the state space of a quantum system, these results show that the set of observables X, which is the dual to X_*, is equipped with a JBW^*-triple algebraic structure. This structure provides a basis for spectral theory and other tools for modelling the quantum mechanical measuring process.

The notion of a facially symmetric space for a two-state system was introduced in section 3.3.5. In this section, we describe the structure of *atomic* facially symmetric spaces. We will explain why an irreducible, neutral, strongly facially symmetric space is the predual of one of the Cartan factors of types I to VI.

5.7.1 The geometry of the quantum state space

The Jordan algebra of self-adjoint elements of a C^*-algebra A has long been used as a model for the bounded observables of a quantum mechanical system, and the states, consisting of norm 1 functionals in the predual A_*, as a model for the states of the system. The state space of this Jordan Banach algebra is the same as the state space of the C^*-algebra A and is a weak*-compact convex subset of the dual of A. With the development of both the structure theory of C^*-algebras and the representation theory of Jordan Banach algebras, the problem arose of determining which compact convex sets in locally convex spaces are affinely isomorphic to such a state space. In the context of ordered Banach spaces, such a characterization has been given for Jordan algebras by Alfsen and Shultz in [2].

Here we consider an analogous result without assuming an order structure on the Banach space. We will consider only the geometry of the state space that is implied by the measuring process for quantum systems. We may define the norm of a point representing the state of a quantum system to be 1. The states will then belong to the boundary ∂S of a unit ball S of a Banach space, which we will denote by X_*. Any element $\mathbf{f} \in S$ represents a state $\mathbf{f}/\|\mathbf{f}\| \in \partial S$ that is obtained with probability $\|\mathbf{f}\| \leq 1$. Recall that the state space consists of two types of points. The first type represents mixed states, which can be considered as a mixture of other states with certain probabilities. The second type represents pure states, those states which cannot be decomposed as a mixture of other states.

A physical quantity which can be measured by an experiment is called an *observable*. The observables can be represented as linear functionals $\mathbf{a} \in X$ on the state space. This representation is obtained by assigning to each state $\mathbf{f} \in S$ the expected value of the physical quantity $\mathbf{a}$, when the system is in state $\mathbf{f}$. A measurement causes the quantum system to move into an eigenstate of the observable that is being measured. Thus, the measuring process defines, for any set Δ of possible values of the observable $\mathbf{a}$, a projection $P_{\mathbf{a}}(\Delta)$ on the state space, called a *filtering projection*. The projection $P_{\mathbf{a}}(\Delta)$ represents a filtering device that will move any state to a state with the value of $\mathbf{a}$ in the set Δ. It is assumed that if the value of $\mathbf{a}$ on the state $\mathbf{f}$ was definitely in Δ, then the filtering does not change the state $\mathbf{f}$. Hence, applying the filtering a second time will not affect the output state of the first filtering, implying that $P_{\mathbf{a}}(\Delta)$ is a projection. Moreover, these projections are neutral in the sense defined in sections 3.3.5 and 5.6.2.

To each filtering projection $P_{\mathbf{a}}(\Delta)$, we can associate an element $\mathbf{u} \in X$ assigning for each state $\mathbf{f}$ the probability of passing through the filter. Obviously, $\mathbf{u}$ exposes a face $F_{\mathbf{u}}$, containing all the states on which the value of $\mathbf{a}$ is definitely in the set Δ. Thus, the image of $P_{\mathbf{a}}(\Delta)$ is the span of $F_{\mathbf{u}}$. Moreover, if we denote by Δ^c the complementary set to Δ in the set of all possible values of $\mathbf{a}$, then the projections $P_{\mathbf{a}}(\Delta)$ and $P_{\mathbf{a}}(\Delta^c)$ will be algebraically orthogonal. Thus, their sum $P_{\mathbf{a}}(\Delta) + P_{\mathbf{a}}(\Delta^c)$ will also be a projection whose image is spanned by $F_{\mathbf{u}}$ and those states that definitely do not pass the filter $P_{\mathbf{a}}(\Delta)$. This defines a symmetry

$$S_{\mathbf{u}} = 2(P_{\mathbf{a}}(\Delta) + P_{\mathbf{a}}(\Delta^c)) - I$$

on the state space S. Hence, to any norm-exposed face $F_{\mathbf{u}}$ in the state space, we can associate a symmetry $S_{\mathbf{u}}$, called the *facial symmetry* fixing the span of the face and its orthogonal complement. The exact meaning of orthogonality will be explained in the next section.

By definition, a pure state is an extreme point of the state space S. They also satisfy the *pure state properties*, which were introduced in section 3.3.5.

5.7.2 Norm-orthogonality in a Banach space

Here we explain the meaning of orthogonality in a Banach space, which corresponds to orthogonality of states. Let X_* be a Banach space. Two non-zero elements $\mathbf{f}, \mathbf{g} \in X_*$ are called *norm-orthogonal*, denoted by $\mathbf{f} \diamond \mathbf{g}$, if

$$\|\mathbf{f} + \mathbf{g}\| = \|\mathbf{f} - \mathbf{g}\| = \|\mathbf{f}\| + \|\mathbf{g}\|. \tag{5.124}$$

Note that if $\mathbf{f} \diamond \mathbf{g}$, then for any convex combination $\alpha\mathbf{f} + \beta\mathbf{g}$, with $\alpha, \beta \geq 0$, $\alpha + \beta = 1$, we have

$$\|\mathbf{f}\| + \|\mathbf{g}\| = \|\mathbf{f} \pm \mathbf{g}\| = \|(\alpha + \beta)\mathbf{f} \pm (\alpha + \beta)\mathbf{g}\| = \|(\alpha\mathbf{f} \pm \beta\mathbf{g}) + (\beta\mathbf{f} \pm \alpha\mathbf{g})\|$$

$$\leq \|\alpha\mathbf{f} \pm \beta\mathbf{g}\| + \|\beta\mathbf{f} \pm \alpha\mathbf{g}\| \leq \alpha\|\mathbf{f}\| + \beta\|\mathbf{g}\| + \beta\|\mathbf{f}\| + \alpha\|\mathbf{g}\| = \|\mathbf{f}\| + \|\mathbf{g}\|.$$

Since we now have equality in all of the above inequalities, we obtain that the norm, a convex combination of norm-orthogonal elements, is a convex combination of their norms, meaning that

$$\|\alpha\mathbf{f} \pm \beta\mathbf{g}\| = \alpha\|\mathbf{f}\| + \beta\|\mathbf{g}\|. \tag{5.125}$$

For any real linear combination $\alpha\mathbf{f} + \beta\mathbf{g}$, we define $\gamma = |\alpha| + |\beta|$ and $\widetilde{\mathbf{f}} = \pm\mathbf{f}$ and $\widetilde{\mathbf{g}} = \pm\mathbf{g}$. Then

$$\|\alpha\mathbf{f} + \beta\mathbf{g}\| = \gamma\|\gamma^{-1}|\alpha|\mathbf{f} + \gamma^{-1}|\beta|\mathbf{g}\| = |\alpha|\,\|\mathbf{f}\| + |\beta|\,\|\mathbf{g}\|. \tag{5.126}$$

From this it follows that for any real numbers α, β, we have

$$\mathbf{f} \diamond \mathbf{g} \Leftrightarrow \alpha\mathbf{f} \diamond \beta\mathbf{g}, \tag{5.127}$$

meaning that real multiples of norm-orthogonal elements are norm-orthogonal. This implies that the intersection of the unit ball of X_* with the plane generated by the real span of two norm-orthogonal elements is as shown in Figure 5.2.

There is an alternative characterization of norm-orthogonality. We will show now that $\mathbf{f}, \mathbf{g} \in X_*$ are norm-orthogonal if and only if there exist elements $\mathbf{u}, \mathbf{v}$ in X, the dual space to X_*, such that

$$\|\mathbf{u}\| = \|\mathbf{v}\| = 1 = \|\mathbf{u} \pm \mathbf{v}\| \tag{5.128}$$

and

$$\mathbf{f}(\mathbf{u}) = \|\mathbf{f}\|,\ \mathbf{g}(\mathbf{v}) = \|\mathbf{g}\|,\ \mathbf{f}(\mathbf{v}) = \mathbf{g}(\mathbf{u}) = 0. \tag{5.129}$$

Note that $\|\mathbf{u}\|$ denotes the norm in X.

Without loss of generality, we may assume that $\|\mathbf{f}\| = \|\mathbf{g}\| = 1$. By a well-known theorem, any element of a Banach space X_*, considered as a functional

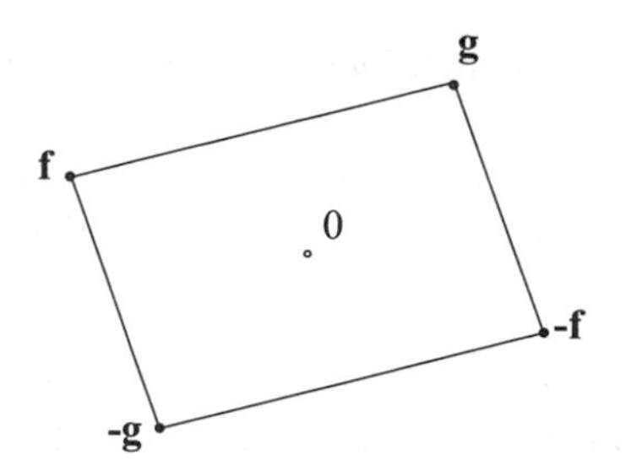

Fig. 5.2. The section of the unit ball with the real span of two norm-orthogonal elements $\mathbf{f}\diamond\mathbf{g}$.

on the dual space X, assumes its norm on the unit ball X_1 of the dual. Thus, there are norm 1 elements $\mathbf{x}, \mathbf{y} \in X$ such that

$$(\mathbf{f}+\mathbf{g})\mathbf{x} = 2 = (\mathbf{f}-\mathbf{g})\mathbf{y}. \tag{5.130}$$

Let $\mathbf{u} = \frac{1}{2}(\mathbf{x}+\mathbf{y})$ and $\mathbf{v} = \frac{1}{2}(\mathbf{x}-\mathbf{y})$. Obviously, $\|\mathbf{u} \pm \mathbf{v}\| = 1$, $\|\mathbf{u}\| \leq 1$, and $\|\mathbf{v}\| \leq 1$. From (5.130), it follows that

$$\mathbf{f}(\mathbf{x}) = \mathbf{g}(\mathbf{x}) = 1 = \mathbf{f}(\mathbf{y}) = -\mathbf{g}(\mathbf{y}),$$

implying (5.129). But since $\mathbf{f}(\mathbf{u}) = 1$ and $\|\mathbf{f}\| = 1$, we have $\|\mathbf{u}\| = 1$. Conversely, if $\mathbf{f}, \mathbf{g} \in X_*$ and $\mathbf{u}, \mathbf{v} \in X$ satisfy (5.128) and (5.129), then

$$\|\mathbf{f}\| + \|\mathbf{g}\| = \mathbf{f}(\mathbf{u}) + \mathbf{g}(\mathbf{v}) = \ (\mathbf{f} \pm \mathbf{g})(\mathbf{u} \pm \mathbf{v}) \leq \|\mathbf{f} \pm \mathbf{g}\| \leq \|\mathbf{f}\| + \|\mathbf{g}\|,$$

implying that $\mathbf{f}\diamond\mathbf{g}$.

5.7.3 Structure of facially symmetric spaces

A *norm-exposed face* of the unit ball S of X_* is a non-empty subset of S of the form

$$F_{\mathbf{x}} = \{\mathbf{f} \in Z_1 : \mathbf{f}(\mathbf{x}) = 1\}, \tag{5.131}$$

where $\mathbf{x} \in X$, $\|\mathbf{x}\| = 1$. Recall that a *face* G of a convex set K is a non-empty convex subset of K such that if $\mathbf{g} \in G$ and $\mathbf{h}, \mathbf{k} \in K$ satisfy $\mathbf{g} = \lambda\mathbf{h} + (1-\lambda)\mathbf{k}$ for some $\lambda \in (0,1)$, then $\mathbf{h}, \mathbf{k} \in G$. An element $\mathbf{u} \in X$ is called a *projective unit* if $\|\mathbf{u}\| = 1$ and $\mathbf{u}(F_{\mathbf{u}}^{\diamond}) = 0$. Here, for any subset A, $A^{\diamond}$ denotes the set of all elements orthogonal to each element of A.

Motivated by measuring processes in quantum mechanics, we define a *symmetric face* to be a norm-exposed face F in S with the following property: there is a linear isometry S_F of X_* onto X_* which is a symmetry, *i.e.*, $S_F^2 = I$, such that the fixed point set of S_F is $(\overline{\mathrm{sp}}F) \oplus F^{\diamond}$ (topological direct sum). The map S_F is called the facial symmetry associated to F. A complex normed space X_* is said to be *weakly facially symmetric* if every norm-exposed face in Z_1 is symmetric. For each symmetric face F, we define contractive projections $P_k(F)$, $k = 0, 1/2, 1$ on X_*, as follows. First, $P_{1/2}(F) = (I - S_F)/2$ is the projection on the -1 eigenspace of S_F. Next, we define $P_1(F)$ and $P_0(F)$ as the projections of X_* onto $\overline{\text{sp}}F$ and $F^\diamond$, respectively, so that $P_1(F)+P_0(F) = (I + S_F)/2$.

Define a *geometric tripotent* to be a projective unit $\mathbf{u}$ with the property that $F = F_\mathbf{u}$ is a symmetric face and $S_F^* \mathbf{u} = \mathbf{u}$. In [32] it was shown that in a weakly facially symmetric normed space X_*, every projective unit is a geometric tripotent. Moreover, there is a one-to-one correspondence between projective units and norm-exposed faces.

The projections $P_k(F_\mathbf{u})$, denoted also by $P_k(\mathbf{u})$, are called the *geometric Peirce projections* associated with a geometric tripotent $\mathbf{u}$. The symmetry corresponding to the symmetric face $F_\mathbf{u}$ will be denoted by $S_\mathbf{u}$. A contractive projection Q on a normed space X is said to be *neutral* if for each $\xi \in X$, $\|Q\xi\| = \|\xi\|$ implies $Q\xi = \xi$. A normed space X_* is called *neutral* if for every symmetric face F, the projection $P_1(F)$ is neutral, meaning that

$$\|P_1(F)\mathbf{f}\| = \|\mathbf{f}\| \Rightarrow P_1(F)\mathbf{f} = \mathbf{f}$$

for any $\mathbf{f} \in X_*$. Two geometric tripotents $\mathbf{u}$ and $\mathbf{v}$ are said to be *compatible* if their associated geometric Peirce projections commute, *i.e.*, $[P_k(\mathbf{u}), P_j(\mathbf{v})] = 0$ for $k, j \in \{0, 1/2, 1\}$. In [34] was shown that if X_* is a weakly facially symmetric neutral space and $\mathbf{u}, \mathbf{v}$ are geometric tripotents such that $S_\mathbf{v}(F_\mathbf{u}) = \pm F_\mathbf{u}$, then $\mathbf{u}$ and $\mathbf{v}$ are compatible. In particular, if $\mathbf{u} \in X_k(\mathbf{v})$ for some $k \in \{1, 1/2, 0\}$, then $\mathbf{u}$ and $\mathbf{v}$ are compatible.

A weakly facially symmetric space X_* is *strongly facially symmetric* if for every norm-exposed face F in S and every $\mathbf{y} \in X$ with $\|\mathbf{y}\| = 1$ and $F \subset F_\mathbf{y}$, we have $S_F^* \mathbf{y} = \mathbf{y}$. In [34] was shown that in a neutral strongly facially symmetric space X_*, for any non-zero element $\mathbf{f} \in X_*$, there exists a unique geometric tripotent $\mathbf{v} = \mathbf{v}(\mathbf{f})$, called the *support tripotent*, such that $\mathbf{f}(\mathbf{v}) = \|\mathbf{f}\|$ and $\mathbf{v}(\{\mathbf{f}\}^\diamond) = 0$. This is an analog of the *polar decomposition* for elements in the predual of a JBW^*-triple. The support tripotent $\mathbf{v}(\mathbf{f})$ is a minimal geometric tripotent if and only if $\mathbf{f}/\|\mathbf{f}\|$ is an extreme point of the unit ball of X_*.

Let $\mathbf{f}$ and $\mathbf{g}$ be extreme points of the unit ball of a neutral strongly facially symmetric space X_*. The *transition probability* of $\mathbf{f}$ and $\mathbf{g}$ is the number

$$\langle \mathbf{f} | \mathbf{g} \rangle := \mathbf{f}(\mathbf{v}(\mathbf{g})).$$

A neutral strongly facially symmetric space X_* is said to satisfy *"symmetry of transition probabilities"* (STP) if for every pair of extreme points $\mathbf{f}, \mathbf{g} \in S$, we have

$$\overline{\langle \mathbf{f} | \mathbf{g} \rangle} = \langle \mathbf{g} | \mathbf{f} \rangle,$$

where in the case of complex scalars, the bar denotes conjugation. We define the rank of a strongly facially symmetric space X_* to be the maximal number of orthogonal geometric tripotents. A normed space X_* is said to be *atomic* if every symmetric face of S has a norm-exposed extreme point. In [35] it was shown that if X_* is an atomic neutral strongly facially symmetric space satisfying STP, then, if X_* is of rank 1, it is linearly isometric to a Hilbert space, and if X_* is of rank 2, it is linearly isometric to the predual of a spin factor.

A neutral strongly facially symmetric space X_* is said to satisfy property FE if every norm-closed face of S is a norm-exposed face. A neutral strongly facially symmetric space X_* is said to satisfy the *"extreme rays property"* ERP if for every geometric tripotent $\mathbf{u}$ and every extreme point $\mathbf{f}$ of S, it follows that $P_1(\mathbf{u})\mathbf{f}$ is a scalar multiple of an extreme point in S. We also say that $P_1(\mathbf{u})$ preserves extreme rays. A strongly facially symmetric space X_* satisfies JP if for any pair $\mathbf{u}, \mathbf{v}$ of orthogonal minimal geometric tripotents, we have

$$S_{\mathbf{u}} S_{\mathbf{v}} = S_{\mathbf{u}+\mathbf{v}}.$$

The following is the main result of [36]: Let X_* be an atomic neutral strongly facially symmetric space satisfying FE, STP, ERP, and JP. Then $X_* = \oplus_\alpha J_\alpha$, where each J_α is isometric to the predual of a Cartan factor of one of the types I-VI. Thus, X is isometric to an atomic JBW^*-triple. If Z is irreducible, then X is isometric to a Cartan factor.

The proof of this theorem consists of contracting a grid basis, a technique we will use in the next chapter in order to classify JBW^*-triple factors. In [56] it was shown that there is an atomic decomposition for strongly facially symmetric spaces with the above properties.

5.8 Notes

The group of holomorphic automorphisms of a domain in C^n was studied by H. Cartan [13]. He also obtained the uniqueness theorems. Bounded symmetric domains of finite dimension were classified by E. Cartan [12]. M. Koecher [48] discovered the connection between bounded symmetric domains and the triple structure associated with them. L. Harris [40] extended the study of bounded symmetric domains to infinite-dimensional spaces consisting of operators between Hilbert spaces. The connection between bounded symmetric domains and Jordan pairs, which are generalizations of Jordan triple systems, was developed by O. Loos [52].

The extension of the above results to infinite-dimensional Banach spaces was done independently by J. P. Vigue [70] and H. Upmeier [68]. A short survey of the subject can be found in [69] and in [5]. W. Kaup [46] proved that the category of bounded symmetric domains D is equivalent to the category of JB^*-triples, where D is the unit ball. For a survey of new results for the Jordan triple structure, see [23].

The dual space was studied in [31]. The fact that the second dual of a JB^*-triple is a JBW^*-triple follows from [47], [16] and [9]. The structure of facially symmetric space was studied in [34], [35] and [36].

6 Classification of JBW^*-triple factors

In the previous chapters, we have shown that homogeneous balls and bounded symmetric domains can be realized as unit balls of JB^*-triples. Let X be a complex Banach space with the JB^*-triple product. The JB^*-triple X is called *atomic* if it is spanned by its minimal tripotents. It is known that the second dual X^{**} of X is a JBW^*-triple i.e., X^{**} is a JB^*-triple whose triple product is w^*-continuous. Moreover, X can be isometrically embedded into the atomic part of its second dual. Hence, any JB^*-triples can be considered as a subtriple of an atomic JBW^*-triple.

We will say that two atomic JBW^*-triples X, Y are orthogonal if any element $\mathbf{x}$ of X is orthogonal to any element $\mathbf{y}$ in Y. We say that X is a *factor* if it cannot be decomposed as an orthogonal sum of two non-zero JBW^*-triples. Any atomic JBW^*-triple can be decomposed (by Zorn's Lemma) into a sum (finite or infinite) of atomic factors. Thus, it is enough to classify the atomic JBW^*-triple factors.

The classification will be done by constructing a special basis, consisting of compatible tripotents, called a grid, in the JBW^*-triple factor. We will show that there are six different types of factors. This classification reveals the difference between the spin factor and other classical factors and sheds light on the occurrence of the exceptional triple factors in this theory.

At the end of the chapter, we obtain results on embeddings of JB^*-triples into direct sum of Cartan factors and Jordan algebras.

6.1 Building blocks of atomic JBW^*-triples

In this section, we will discuss the basic relations between two compatible tripotents with some condition of minimality. Algebraically orthogonal tripotents $\mathbf{u}, \mathbf{v}$, meaning $\mathbf{u} \in X_0(\mathbf{v})$, were studied in the previous chapter. Here we will concentrate on the case when $\mathbf{u} \in X_{1/2}(\mathbf{v})$. We will also study the subtriples generated by these tripotents.

6.1.1 General outline of the method

From now on, X will denote an *atomic* JBW^*-triple factor. The classification will be done by constructing a certain kind of basis, called a *grid*, for X.

This basis will consist of tripotents such that any pair $\mathbf{u}, \mathbf{v}$ of them satisfies one of the following relations: $\mathbf{u} \perp \mathbf{v}, \mathbf{u} \top \mathbf{v}, \mathbf{u} \vdash \mathbf{v}$, defined below. Each of the tripotents in the basis belongs to some Peirce component of the decomposition of X by any other tripotent in the basis. As shown (5.88) in the previous chapter, the Peirce projections of such $\mathbf{u}$ and $\mathbf{v}$ commute, and the tripotents are called compatible tripotents. Hence, we can use the joint Peirce decomposition and not worry about the order in the decomposition. In each case, we will obtain a commutative decomposition of X into one-dimensional subspaces with well-defined rules for the product between these spaces.

In JB^*-triples, all of the subspaces occurring in the Peirce decomposition remain JB^*-triples, and so the "analysis" can be carried out systematically.

Since a grid consists of tripotents, we take a closer look at the set of tripotents. As mentioned earlier, a tripotent $\mathbf{v}$ is *minimal* in a JBW^*-triple if it is not a sum of two orthogonal tripotents. If a tripotent $\mathbf{v}$ is minimal, then the JB^*-algebra $X_1^{(\mathbf{v})}(\mathbf{v})$, defined as in section 5.4.5, is one-dimensional.

We start our classification by choosing a minimal tripotent $\mathbf{v}$ in the atomic JBW^*-triple factor X. If $X_{\frac{1}{2}}(\mathbf{v}) = \{0\}$, then from the Peirce decomposition we have $X = X_1(\mathbf{v}) + X_0(\mathbf{v})$ and since X is assumed to be a factor, $X = X_1(\mathbf{v}) = C\mathbf{v}$ is a one-dimensional factor. Thus, we will consider only the case when $X_{\frac{1}{2}}(\mathbf{v}) \neq \{0\}$ and study the relations between $\mathbf{v}$ and the tripotents in $X_{\frac{1}{2}}(\mathbf{v})$.

6.1.2 Basic relations between pairs of tripotents in a grid

Let $\mathbf{v}$ be a minimal tripotent, and let $\mathbf{u}$ be any tripotent of the JB^*-triple $X_{\frac{1}{2}}(\mathbf{v})$, meaning that $D(\mathbf{v})\mathbf{u} = \frac{1}{2}\mathbf{u}$. For any $\mathbf{u} \in X_{\frac{1}{2}}(\mathbf{v})$, by (5.88) we have $[P_j(\mathbf{v}), P_k(\mathbf{u})] = 0$. Thus,

$$\mathbf{v} = (P_1(\mathbf{u}) + P_{\frac{1}{2}}(\mathbf{u}) + P_0(\mathbf{u}))\mathbf{v} = (P_1(\mathbf{u}) + P_{\frac{1}{2}}(\mathbf{u}) + P_0(\mathbf{u}))P_1(\mathbf{v})\mathbf{v}$$

$$= P_1(\mathbf{v})P_1(\mathbf{u})\mathbf{v} + P_1(\mathbf{v})P_{\frac{1}{2}}(\mathbf{u})\mathbf{v} + P_1(\mathbf{v})P_0(\mathbf{u})\mathbf{v}. \tag{6.1}$$

Since $\mathbf{v}$ is minimal, the image $X_1(\mathbf{v})$ of $P_1(\mathbf{v})$ is proportional to $\mathbf{v}$. Thus,

$$P_1(\mathbf{v})P_k(\mathbf{u})\mathbf{v} = P_k(\mathbf{u})P_1(\mathbf{v})\mathbf{v} = P_k(\mathbf{u})\mathbf{v} = \lambda_k\mathbf{v}, \ k \in \{1, 1/2, 0\}$$

for some constants λ_k satisfying $\lambda_1 + \lambda_2 + \lambda_3 = 1$. Since each $P_k(\mathbf{u})$ is a projection and λ_k is its eigenvalue, each $\lambda_k = 0$ or $\lambda_k = 1$, and so one and only one of $\lambda_k = 1$. If $\lambda_1 = 1$, then $\mathbf{v} \in X_1(\mathbf{u})$, and we say that $\mathbf{u}$ *governs* $\mathbf{v}$ and denote this by $\mathbf{u} \vdash \mathbf{v}$. If $\lambda_{1/2} = 1$, then $\mathbf{v} \in X_{\frac{1}{2}}(\mathbf{u})$ and we say that $\mathbf{u}$ is *co-orthogonal* to $\mathbf{v}$ and denote this by $\mathbf{u} \top \mathbf{v}$. If $\lambda_0 = 1$, then $\mathbf{v} \in X_0(\mathbf{u})$ and $\mathbf{v} \perp \mathbf{u}$. But then $\mathbf{u} \in X_0(\mathbf{v})$, and this contradicts $\mathbf{u} \in X_{\frac{1}{2}}(\mathbf{v})$.

Thus, we have shown that for any tripotent $\mathbf{u} \in X_{\frac{1}{2}}(\mathbf{v})$, with $\mathbf{v}$ a minimal tripotent, one of the following happens: either

$$\mathbf{v} \in X_1(\mathbf{u}), \tag{6.2}$$

and that $\mathbf{u}$ governs $\mathbf{v}$, which is denoted by $\mathbf{u} \vdash \mathbf{v}$, or

$$\mathbf{v} \in X_{\frac{1}{2}}(\mathbf{u}), \tag{6.3}$$

and that $\mathbf{u}$ is co-orthogonal to $\mathbf{v}$, which is denoted by $\mathbf{u} \top \mathbf{v}$.

Let $\mathbf{v}, \mathbf{u}$ be two co-orthogonal tripotents. We study first the norm of the linear subspace generated by $\mathbf{v}, \mathbf{u}$. Note that from the Peirce calculus, it follows that

$$\{\mathbf{v}, \mathbf{u}, \mathbf{v}\} \subset X_{1.5}(\mathbf{u}) = \{0\}. \tag{6.4}$$

By using this fact, for any complex numbers α, β, we get

$$(\alpha\mathbf{v} + \beta\mathbf{u})^{(3)} = \alpha|\alpha|^2\mathbf{v} + 2|\alpha|^2\beta D(\mathbf{v})\mathbf{u} + 2\alpha|\beta|^2 D(\mathbf{u})\mathbf{v} + \beta|\beta|^2\mathbf{u}$$

$$= (|\alpha|^2 + |\beta|^2)\alpha\mathbf{v} + (|\alpha|^2 + |\beta|^2)\beta\mathbf{u} = (|\alpha|^2 + |\beta|^2)(\alpha\mathbf{v} + \beta\mathbf{u}). \tag{6.5}$$

Thus, from the star identity (5.107), it follows that

$$\|\alpha\mathbf{v} + \beta\mathbf{u}\|^3 = (|\alpha|^2 + |\beta|^2)\|\alpha\mathbf{v} + \beta\mathbf{u}\|,$$

implying that

$$\|\alpha\mathbf{v} + \beta\mathbf{u}\| = \sqrt{|\alpha|^2 + |\beta|^2}. \tag{6.6}$$

Note that from (6.5), it follows that the span of $\mathbf{v}, \mathbf{u}$ is closed under the triple product and any element $\alpha\mathbf{v} + \beta\mathbf{u}$ is a tripotent when $\sqrt{|\alpha|^2 + |\beta|^2} = 1$.

Let $\mathbf{w} = \frac{\mathbf{v}+\mathbf{u}}{\sqrt{2}}$, which is a tripotent. Consider now the symmetry $S_{2\pi}(\mathbf{w}) = S_{\mathbf{w}} = P_1(\mathbf{w}) - P_{1/2}(\mathbf{w}) + P_0(\mathbf{w})$, defined by (5.110), for the pair of co-orthogonal tripotents. By use of (5.82) and (5.84) we have

$$S_{\mathbf{w}} = I - 8D(\mathbf{w}) + 8D(\mathbf{w})^2. \tag{6.7}$$

To calculate the action of $S_{\mathbf{w}}$ on $\mathbf{u}$, we will, by use of (6.4), calculate first

$$4D(\mathbf{w})\mathbf{u} = 2\{\mathbf{v} + \mathbf{u}, \mathbf{v} + \mathbf{u}, \mathbf{u}\} = \mathbf{u} + \mathbf{v} + 2\mathbf{u} = 3\mathbf{u} + \mathbf{v}$$

and

$$8D(\mathbf{w})^2\mathbf{u} = 4D(\mathbf{w})(2D(\mathbf{w})\mathbf{u}) = 4D(\mathbf{w})(1.5\mathbf{u} + 0.5\mathbf{v})$$

$$= 4.5\mathbf{u} + 1.5\mathbf{v} + 1.5\mathbf{v} + 0.5\mathbf{u} = 5\mathbf{u} + 3\mathbf{v}.$$

Thus, from (6.7), we have

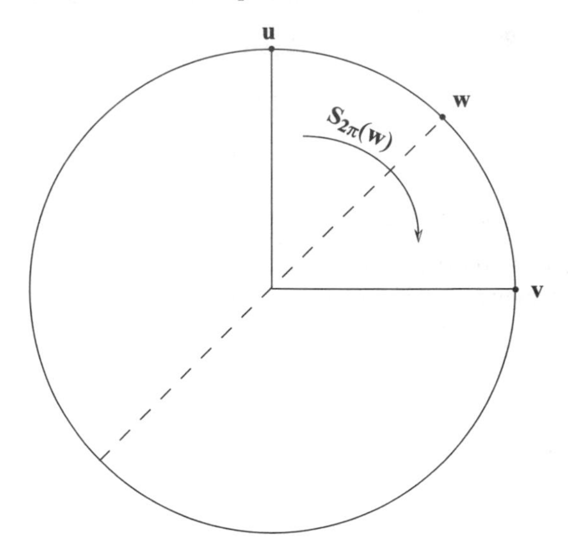

Fig. 6.1. The symmetry of co-orthogonal tripotents. The intersection of the real span of two co-orthogonal tripotents $\mathbf{v}$ and $\mathbf{u}$ with the unit ball is shown. The symmetry $S_{\mathbf{w}}$ exchanges $\mathbf{u}$ and $\mathbf{v}$.

$$S_{\mathbf{w}}\mathbf{u} = \mathbf{u} - 6\mathbf{u} - 2\mathbf{v} + 5\mathbf{u} + 3\mathbf{v} = \mathbf{v}, \tag{6.8}$$

implying that this symmetry exchanges the co-orthogonal tripotents. See Figure 6.1.

Thus, if $\mathbf{v}$ is a minimal tripotent and $\mathbf{u}$ is co-orthogonal to $\mathbf{v}$, then the tripotent $\mathbf{u}$ is also *minimal*. Since the sum of orthogonal tripotents cannot be minimal, the sum $\mathbf{u} = \mathbf{u}_1 + \mathbf{u}_2$ of any two mutually algebraically orthogonal tripotents $\mathbf{u}_1$ and $\mathbf{u}_2$, where each of them is co-orthogonal to a minimal tripotent $\mathbf{v}$, must govern $\mathbf{v}$, i.e.,

$$\mathbf{u}_1 \top \mathbf{v} \top \mathbf{u}_2,\ \mathbf{u}_1 \perp \mathbf{u}_2 \ \Rightarrow\ (\mathbf{u}_1 + \mathbf{u}_2) \vdash \mathbf{v}.$$

If $\mathbf{u}$ governs $\mathbf{v}$, then, using the fact that $\mathbf{v} \in X_1(\mathbf{u})$, any element $\mathbf{a} \in X_0(\mathbf{u})$ will be algebraically orthogonal to $\mathbf{v}$. This implies that $P_{1/2}(\mathbf{v})\mathbf{a} = 0$. Thus,

$$\mathbf{u} \vdash \mathbf{v} \ \Rightarrow\ X_0(\mathbf{u}) \cap X_{1/2}(\mathbf{v}) = \{0\}. \tag{6.9}$$

We can summarize the situation with tripotents in $X_{1/2}(\mathbf{v})$ for a minimal tripotent $\mathbf{v}$. Let $\mathbf{u}$ be a tripotent in $X_{1/2}(\mathbf{v})$. Then exactly one of the following three cases occurs:

(i) $\mathbf{u}$ is minimal in X. This occurs if and only if $\mathbf{u}$ and $\mathbf{v}$ are co-orthogonal.

(ii) $\mathbf{u}$ is not minimal in X but is minimal in $X_{1/2}(\mathbf{v})$. In this case, $\mathbf{u} \vdash \mathbf{v}$ and we say that $(\mathbf{v}, \mathbf{u})$ form a *pre-trangle*.

(iii) $\mathbf{u}$ is not minimal in $X_{1/2}(\mathbf{v})$. Thus, $\mathbf{u}$ is not minimal in X either. In this case we can decompose $\mathbf{u} = \mathbf{u}_1 + \widetilde{\mathbf{u}}_1$, where $\mathbf{u}_1, \widetilde{\mathbf{u}}_1$ are two mutually orthogonal tripotents of X, contained in $X_{1/2}(\mathbf{v})$. From (6.9), it follows that any one of $\mathbf{u}_1, \widetilde{\mathbf{u}}_1$ cannot be governing, so they must be co-orthogonal to $\mathbf{v}$ and thus minimal tripotents in X. We will call the triple of tripotents $(\mathbf{u}_1, \mathbf{v}, \widetilde{\mathbf{u}}_1)$ a *pre-quadrangle.*

These relations are summarized in Table 6.1.

Case	$\mathbf{v} \in X_k(\mathbf{u})$	minimal in X	minimal in $X_{1/2}(\mathbf{v})$
(i)	$\mathbf{v} \in X_{1/2}(\mathbf{u})$	yes	yes
(ii)	$\mathbf{v} \in X_1(\mathbf{u})$	no	yes
(iii)	$\mathbf{v} \in X_1(\mathbf{u})$	no	no

Table 6.1. Tripotent $\mathbf{u}$ in $X_{1/2}(\mathbf{v})$, with $\mathbf{v}$ as a minimal tripotent

This implies that if $\mathbf{v}$ is a minimal tripotent in X, then rank $X_{1/2}(\mathbf{v}) \leq 2$, that is, $X_{1/2}(\mathbf{v})$ cannot contain more than two mutually orthogonal tripotents. Moreover, if $\mathbf{v}, \mathbf{u}, \mathbf{w}$ are mutually co-orthogonal, then from (5.91), it follows that

$$\{\mathbf{v}, \mathbf{u}, \mathbf{w}\} = X_1(\mathbf{v}) \cap X_1(\mathbf{w}) = \{0\}. \tag{6.10}$$

We will say that two tripotents $\mathbf{v}, \mathbf{u}$ are *equivalent* and write $\mathbf{v} \approx \mathbf{u}$ if they generate the same Peirce decomposition, that is

$$\mathbf{v} \approx \mathbf{u} \Leftrightarrow P_k(\mathbf{v}) = P_k(\mathbf{u}) \text{ for } k = 1, 1/2, 0. \tag{6.11}$$

Note that only in case (i) is the span of $\mathbf{v}, \mathbf{u}$ closed under the triple product. In the remaining cases, we can complete these tripotents to a collection which will be closed under the triple product. We will do this now.

6.1.3 Trangles in JB^*-triples

Consider first the case $\mathbf{u} \vdash \mathbf{v}$ i.e., $(\mathbf{u}, \mathbf{v})$ is a pre-trangle. Denote $\widetilde{\mathbf{v}} = Q(\mathbf{u})\mathbf{v}$. Then the ordered triplet $(\mathbf{v}, \mathbf{u}, \widetilde{\mathbf{v}})$ is called a *trangle.* The elements of the trangle satisfy the following. From (5.91), it follows that $\widetilde{\mathbf{v}} = \{\mathbf{u}, \mathbf{v}, \mathbf{u}\} \in X_{1/2-1+1/2}(\mathbf{v}) = X_0(\mathbf{v})$ and $\widetilde{\mathbf{v}} = \{\mathbf{u}, \mathbf{v}, \mathbf{u}\} \in X_{1-1+1}(\mathbf{u}) = X_1(\mathbf{u})$. Thus $\mathbf{v} \perp \widetilde{\mathbf{v}}$ and $\mathbf{v}, \widetilde{\mathbf{v}} \in X_1(\mathbf{u})$. Moreover, from (5.83), we have $\mathbf{v} = Q^2(\mathbf{u})\mathbf{v} = Q(\mathbf{u})\widetilde{\mathbf{v}}$. By use of (5.100) we obtain $D(\mathbf{v} + \widetilde{\mathbf{v}})\mathbf{u} = D(\mathbf{v})\mathbf{u} + D(\widetilde{\mathbf{v}})\mathbf{u} = \frac{1}{2}\mathbf{u} + \frac{1}{2}\mathbf{u} = \mathbf{u}$, implying that $\mathbf{u} \in X_1(\mathbf{v} + \widetilde{\mathbf{v}})$. Thus,

$$\mathbf{u} \vdash \mathbf{v} \text{ and } \widetilde{\mathbf{v}} = Q(\mathbf{u})\mathbf{v} \Rightarrow \mathbf{v} \perp \widetilde{\mathbf{v}},\ \mathbf{u} \vdash \widetilde{\mathbf{v}},\ \mathbf{v} = Q(\mathbf{u})\widetilde{\mathbf{v}},\ \mathbf{u} \approx (\mathbf{v} + \widetilde{\mathbf{v}}). \tag{6.12}$$

Note also that from (5.76), we have

$$\{\mathbf{v}, \mathbf{u}, \widetilde{\mathbf{v}}\} = \{\mathbf{v}, \mathbf{u}, \{\mathbf{u}, \mathbf{v}, \mathbf{u}\}\} = \frac{1}{2}\mathbf{u} - \frac{1}{2}\mathbf{u} + \frac{1}{2}\mathbf{u} = \frac{1}{2}\mathbf{u}. \tag{6.13}$$

As an example of a trangle, consider the spin factor S^3 from Chapter 3 with the triple product defined by (3.4), and take $\mathbf{v} = (\frac{1}{2}, 0, \frac{i}{2})$ and $\mathbf{u} = (0, 1, 0)$. Then $\mathbf{v}$ is a minimal tripotent and $\mathbf{u} \vdash \mathbf{v}$. From (3.4), we get

$$\widetilde{\mathbf{v}} = Q(\mathbf{u})\mathbf{v} = (\frac{1}{2}, 0, -\frac{i}{2}).$$

The map $Q(\mathbf{u})$ is the conjugate linear map $Q(\mathbf{u})(x, y, iz) = (x, y, -iz)$ and is a reflection in the $x - y$ plane. Hence, the ball generated by the real span of these tripotents is as shown in Figure 6.2.

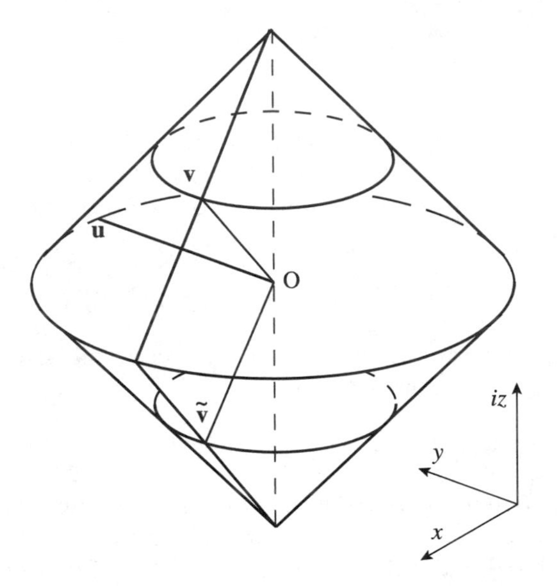

Fig. 6.2. The unit ball generated by the real span of a trangle $(\mathbf{v}, \mathbf{u}, \widetilde{\mathbf{v}})$, where $\mathbf{v} = (\frac{1}{2}, 0, \frac{i}{2})$, $\mathbf{u} = (0, 1, 0)$ and $\widetilde{\mathbf{v}} = (\frac{1}{2}, 0, -\frac{i}{2})$. $Q(\mathbf{u})$ acts as a reflection in the $x - y$ plane.

In addition to the conjugate linear symmetry given by $Q(\mathbf{u})$, we can also define a linear symmetry using $S_{\mathbf{w}}$ defined by (6.7). Let $\mathbf{w} = (\mathbf{v} + \mathbf{u} + \widetilde{\mathbf{v}})/2$. Then, by direct verification, we see that $\mathbf{w}$ is a tripotent. Moreover,

$$4D(\mathbf{w})\mathbf{v} = \{(\mathbf{v} + \widetilde{\mathbf{v}}) + \mathbf{u}, (\mathbf{v} + \widetilde{\mathbf{v}}) + \mathbf{u}, \mathbf{v}\} = 2\mathbf{v} + \mathbf{u},$$

$$2D(\mathbf{w})\mathbf{u} = \mathbf{v} + \widetilde{\mathbf{v}} + \mathbf{u} = 2\mathbf{w},$$

implying that

$$8D^2(\mathbf{w})\mathbf{v} = 4D(\mathbf{w})\mathbf{v} + 2D(\mathbf{w})\mathbf{v} = 3\mathbf{v} + \widetilde{\mathbf{v}} + 2\mathbf{u},\ D^2(\mathbf{w})\mathbf{u} = \mathbf{w}.$$

Thus,

$$S_{\mathbf{w}}\mathbf{v} = \mathbf{v} - 8D(\mathbf{w})\mathbf{v} + 8D^2(\mathbf{w})\mathbf{v} = \widetilde{\mathbf{v}},\ S_{\mathbf{w}}\mathbf{u} = \mathbf{u},$$

showing that the symmetry automorphism $S_{\mathbf{w}}$ of the triple structure acts on the trangle by

$$S_{\mathbf{w}}(\mathbf{v}, \mathbf{u}, \widetilde{\mathbf{v}}) = (\widetilde{\mathbf{v}}, \mathbf{u}, \mathbf{v}). \tag{6.14}$$

6.1.4 Quadrangles in JB^*-triples

An ordered quadruple $(\mathbf{u}_1, \mathbf{u}_2, \mathbf{u}_3, \mathbf{u}_4)$ of tripotents is called a *quadrangle* if $\mathbf{u}_1 \perp \mathbf{u}_3, \mathbf{u}_2 \perp \mathbf{u}_4, \mathbf{u}_1 \top \mathbf{u}_2 \top \mathbf{u}_3 \top \mathbf{u}_4 \top \mathbf{u}_1$ and

$$\mathbf{u}_4 \ = \ 2\{\mathbf{u}_1, \mathbf{u}_2, \mathbf{u}_3\}. \tag{6.15}$$

From (5.76) it follows that (6.15) is still true if the indices are permutated cyclically, *e.g.*, $\mathbf{u}_1 \ = \ 2\{\mathbf{u}_2, \mathbf{u}_3, \mathbf{u}_4\}, ...$ etc. Thus, cyclic permutations of a quadrangle remain quadrangles. We will represent a quadrangle by a diagram such as that in Figure 6.3 and denote it by $\mathbf{Q}$.

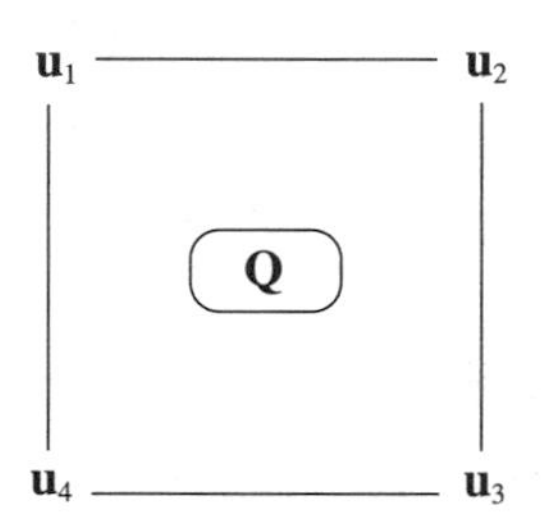

Fig. 6.3. In this diagram, two tripotents are orthogonal if they are on a diagonal, and are co-orthogonal if they are on an edge. The product of any three tripotents (on consecutive vertices) is equal to $\frac{1}{2}$ the remaining tripotent. For instance, $\{\mathbf{u}_1, \mathbf{u}_4, \mathbf{u}_3\} = \frac{1}{2}\mathbf{u}_2, \{\mathbf{u}_1, \mathbf{u}_1, \mathbf{u}_2\} = \frac{1}{2}\mathbf{u}_2, \{\mathbf{u}_1, \mathbf{u}_3, \mathbf{u}_2\} = 0, ...$etc.

Suppose $\mathbf{u}_1, \mathbf{u}_2, \mathbf{u}_3$ are tripotents with $\mathbf{u}_1 \perp \mathbf{u}_3$, and $\mathbf{u}_1 \top \mathbf{u}_2 \top \mathbf{u}_3$. We say that $(\mathbf{u}_1, \mathbf{u}_2, \mathbf{u}_3)$ forms a pre-quadrangle. Let $\mathbf{u}_4$ be defined by (6.15). Then $\mathbf{u}_4$ is a tripotent, and we will show that $\mathbf{Q} = (\mathbf{u}_1, \mathbf{u}_2, \mathbf{u}_3, \mathbf{u}_4)$ forms a quadrangle. Moreover,

$$\mathbf{u}_1 + \mathbf{u}_3 \approx \mathbf{u}_2 + \mathbf{u}_4. \tag{6.16}$$

This can be shown in the following way. Consider the map $Q(\mathbf{u}_1 + \mathbf{u}_3)$ on $X_1(\mathbf{u}_1 + \mathbf{u}_3)$, which is anti-linear and preserves the triple product. Obviously, $Q(\mathbf{u}_1 + \mathbf{u}_3)$ fixes both $\mathbf{u}_1$ and $\mathbf{u}_3$. Moreover, from (5.91), $\{\mathbf{u}_1, \mathbf{u}_2, \mathbf{u}_1\} \ = \ 0$ and $\{\mathbf{u}_3, \mathbf{u}_2, \mathbf{u}_3\} \ = \ 0$, implying that

$$Q(\mathbf{u}_1 + \mathbf{u}_3)(\mathbf{u}_2) \ = \ 2\{\mathbf{u}_1, \mathbf{u}_2, \mathbf{u}_3\} \ = \ \mathbf{u}_4. \tag{6.17}$$

Therefore, $\mathbf{u}_4$ is a tripotent satisfying $\mathbf{u}_1 \top \mathbf{u}_4 \top \mathbf{u}_3$. Moreover, (5.91) shows that $\mathbf{u}_4 = 2\{\mathbf{u}_1, \mathbf{u}_2, \mathbf{u}_3\} \in X_0(\mathbf{u}_2)$. Thus $(\mathbf{u}_1, \mathbf{u}_2, \mathbf{u}_3, \mathbf{u}_4)$ forms a quadrangle.

Since $\mathbf{u}_1$ and $\mathbf{u}_3$ are mutually orthogonal and $\mathbf{u}_2$ is co-orthogonal to each of them, it follows that $D(\mathbf{u}_1+\mathbf{u}_3)\mathbf{u}_2 = D(\mathbf{u}_1)\mathbf{u}_2+D(\mathbf{u}_3)\mathbf{u}_2 = \mathbf{u}_2$, implying that $\mathbf{u}_2 \in X_1(\mathbf{u}_1+\mathbf{u}_3)$. Similarly, $\mathbf{u}_4 \in X_1(\mathbf{u}_1+\mathbf{u}_3)$. Thus, $\mathbf{u}_2+\mathbf{u}_4 \in X_1(\mathbf{u}_1+\mathbf{u}_3)$. By the same argument, $\mathbf{u}_1 + \mathbf{u}_3 \in X_1(\mathbf{u}_2 + \mathbf{u}_4)$, showing that $\mathbf{u}_1 + \mathbf{u}_3$ and $\mathbf{u}_2 + \mathbf{u}_4$ are equivalent tripotents.

Also in this case, we can define a tripotent $\mathbf{w} = (\mathbf{u}_1 + \mathbf{u}_2 + \mathbf{u}_3 + \mathbf{u}_4)/2$ and a symmetry automorphism $S_{\mathbf{w}}$ of the triple structure satisfying

$$S_{\mathbf{w}}(\mathbf{u}_1, \mathbf{u}_2, \mathbf{u}_3, \mathbf{u}_4) = (\mathbf{u}_3, \mathbf{u}_4, \mathbf{u}_1, \mathbf{u}_2). \tag{6.18}$$

6.1.5 Joint Peirce decomposition of orthogonal tripotents

We finish this section with the following observation for the joint decomposition of two algebraically orthogonal tripotents. Let $\mathbf{v}$ and $\widetilde{\mathbf{v}}$ be orthogonal tripotents of a JB^*-triple X. Since the Peirce projections associated to these tripotents commute, also the operators $D(\mathbf{v})$ and $D(\widetilde{\mathbf{v}})$ are commuting. From (5.100), we have $D(\mathbf{v} + \widetilde{\mathbf{v}}) = D(\mathbf{v}) + D(\widetilde{\mathbf{v}})$. Thus, the 1/2-eigenspace $X_{1/2}(\mathbf{v} + \widetilde{\mathbf{v}})$ of this operator is a direct sum of two subspaces: the first one is the intersection of the 1/2-eigenspace of $D(\mathbf{v})$ with the 0-eigenspace of $D(\widetilde{\mathbf{v}})$, while the second one is the intersection of the 1/2-eigenspace of $D(\widetilde{\mathbf{v}})$ with the 0-eigenspace of $D(\mathbf{v})$. Thus,

$$X_{1/2}(\mathbf{v} + \widetilde{\mathbf{v}}) = [X_{1/2}(\mathbf{v}) \cap X_0(\widetilde{\mathbf{v}})] \oplus [X_{1/2}(\widetilde{\mathbf{v}}) \cap X_0(\mathbf{v})]. \tag{6.19}$$

Similarly, the 1-eigenspace $X_1(\mathbf{v}+\widetilde{\mathbf{v}})$ of $D(\mathbf{v}+\widetilde{\mathbf{v}})$ can be decomposed into three eigenspaces: the first one is the 1-eigenspace of $D(\mathbf{v})$ which belongs to the 0-eigenspace of $D(\widetilde{\mathbf{v}})$, the second one is the 1-eigenspace of $D(\widetilde{\mathbf{v}})$ which belongs to the 0-eigenspace of $D(\mathbf{v})$ and the third one is the intersection of the 1/2-eigenspace of $D(\mathbf{v})$ with the 1/2-eigenspace of $D(\widetilde{\mathbf{v}})$. Thus,

$$X_1(\mathbf{v} + \widetilde{\mathbf{v}}) = X_1(\mathbf{v}) \oplus X_1(\widetilde{\mathbf{v}}) \oplus [X_{1/2}(\mathbf{v}) \cap X_{1/2}(\widetilde{\mathbf{v}})]. \tag{6.20}$$

6.2 Methods of gluing quadrangles

By using both trangles and quadrangles as building blocks, we can construct Cartan factors of type III. The remaining Cartan factors are constructed primarily from quadrangles. How do we obtain the remaining five different types of factor from the same building blocks — quadrangles? The answer is that there are two construction techniques for gluing quadrangles together.

6.2.1 Gluing by diagonal

Two quadrangles $\mathbf{Q1} = (\mathbf{u}_1, \mathbf{u}_2, \mathbf{u}_3, \mathbf{u}_4)$ and $\mathbf{Q2} = (\mathbf{v}_1, \mathbf{v}_2, \mathbf{v}_3, \mathbf{v}_4)$ are said to be "glued" by diagonals if $\mathbf{u}_2 = \mathbf{v}_2$ and $\mathbf{u}_4 = \mathbf{v}_4$. The diagram of the

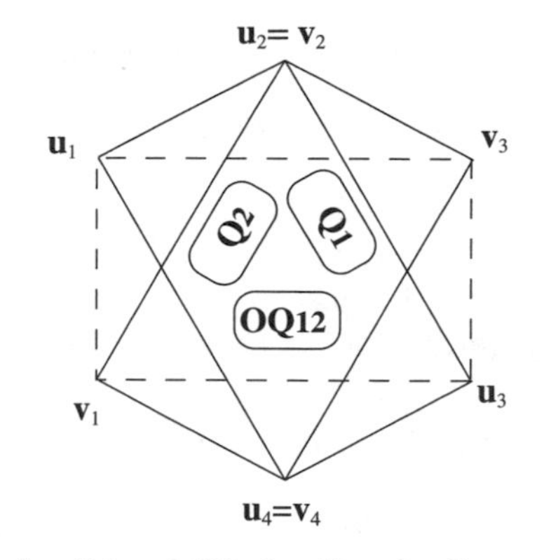

Fig. 6.4. Two quadrangles **Q**1 and **Q**2 glued by the diagonal. Here, $(\mathbf{u}_1, \mathbf{v}_1, \mathbf{u}_3, \mathbf{v}_3)$ forms an odd quadrangle, denoted **OQ**12.

construction is as follows: Let $\mathbf{Q}1 = (\mathbf{u}_1, \mathbf{u}_2, \mathbf{u}_3, \mathbf{u}_4)$ and $\mathbf{Q}2 = (\mathbf{v}_1, \mathbf{v}_2, \mathbf{v}_3, \mathbf{v}_4)$ be quadrangles glued together along the diagonal $(\mathbf{u}_2, \mathbf{u}_4)$. We assume that all of the vertices are minimal tripotents, and that one of $\mathbf{u}_1, \mathbf{u}_3$ is co-orthogonal to one of $\mathbf{v}_1, \mathbf{v}_3$. We will show now that $(\mathbf{u}_1, \mathbf{v}_1, \mathbf{u}_3, -\mathbf{v}_3)$ forms a quadrangle.

Without loss of generality, let us assume that $\mathbf{u}_1 \top \mathbf{v}_1$. To show that $(\mathbf{u}_1, \mathbf{v}_1, \mathbf{u}_3, -\mathbf{v}_3)$ forms a quadrangle, it suffices to show that $\mathbf{v}_1 \top \mathbf{u}_3$ and $\{\mathbf{u}_1, \mathbf{v}_1, \mathbf{u}_3\} = -\frac{1}{2}\mathbf{v}_3$. By applying (5.91) to $\mathbf{u}_3 = 2\{\mathbf{u}_4, \mathbf{u}_1, \mathbf{u}_2\}$, we see that $\mathbf{u}_3 \in X_{1/2}(\mathbf{v}_1)$. Since $\mathbf{v}_1$ is a minimal tripotent, $\mathbf{v}_1 \in X_{1/2}(\mathbf{u}_3)$ and $\mathbf{v}_1 \top \mathbf{u}_3$. To show that $\{\mathbf{u}_1, \mathbf{v}_1, \mathbf{u}_3\} = -\frac{1}{2}\mathbf{v}_3$, we have, from (5.76),

$$\begin{aligned}\{\mathbf{u}_1, \mathbf{v}_1, \mathbf{u}_3\} &= 2\{\mathbf{u}_1, \mathbf{v}_1, \{\mathbf{u}_4, \mathbf{u}_1, \mathbf{u}_2\}\} \\ &= 2(\{\{\mathbf{u}_1, \mathbf{v}_1, \mathbf{u}_4\}, \mathbf{u}_1, \mathbf{u}_2\} + \{\{\mathbf{u}_1, \mathbf{v}_1, \mathbf{u}_2\}, \mathbf{u}_1, \mathbf{u}_4\} \\ &\quad -\{\mathbf{u}_4, \{\mathbf{v}_1, \mathbf{u}_1, \mathbf{u}_1\}, \mathbf{u}_2\}). \end{aligned} \tag{6.21}$$

Since $\mathbf{u}_1, \mathbf{v}_1, \mathbf{u}_4$ are mutually co-orthogonal minimal tripotents, from (6.10) we have $\{\mathbf{u}_1, \mathbf{v}_1, \mathbf{u}_4\} = 0$. Similarly, $\{\mathbf{u}_1, \mathbf{v}_1, \mathbf{u}_2\} = 0$. Thus,

$$\{\mathbf{u}_1, \mathbf{v}_1, \mathbf{u}_3\} = -2\{\mathbf{u}_4, \{\mathbf{v}_1, \mathbf{u}_1, \mathbf{u}_1\}, \mathbf{u}_2\} = -\{\mathbf{u}_4, \mathbf{v}_1, \mathbf{u}_2\} = -\frac{1}{2}\mathbf{v}_3. \tag{6.22}$$

Let us call a quadruple $(\mathbf{u}_1, \mathbf{u}_2, \mathbf{u}_3, \mathbf{u}_4)$ an *odd quadrangle* and denote it by **OQ** if $(\mathbf{u}_1, \mathbf{u}_2, \mathbf{u}_3, -\mathbf{u}_4)$ is a quadrangle. We will denote the set of all odd quadrangles by OQ. Odd quadrangles have all the properties of quadrangles described earlier, except that the triple product of three consecutive tripotents will give $-\frac{1}{2}$ of the remaining one. In order to preserve the symmetry of our construction, we replace the original quadrangles by odd ones. Since $\mathbf{u}_3 = -2\{\mathbf{u}_4, \mathbf{u}_1, \mathbf{u}_2\}$, we will have a sign change in (6.21). Thus, (6.22) will become

$$\{\mathbf{u}_1, \mathbf{v}_1, \mathbf{u}_3\} = 2\{\mathbf{u}_4, \{\mathbf{v}_1, \mathbf{u}_1, \mathbf{u}_1\}, \mathbf{u}_2\} = \{\mathbf{u}_4, \mathbf{v}_1, \mathbf{u}_2\} = -\frac{1}{2}\mathbf{v}_3. \tag{6.23}$$

Consequently, assuming that $\mathbf{u}_1, \mathbf{u}_3$ are co-orthogonal, we obtain

$$(\mathbf{u}_1, \mathbf{u}_2, \mathbf{u}_3, \mathbf{u}_4), (\mathbf{v}_1, \mathbf{u}_2, \mathbf{u}_3, \mathbf{u}_4) \in OQ \Rightarrow (\mathbf{u}_1, \mathbf{v}_1, \mathbf{u}_3, \mathbf{v}_3) \in OQ. \quad (6.24)$$

Thus, as we shall see, if we begin with an appropriate family of odd quadrangles, we will obtain a spin grid consisting of odd quadrangles pairwise "glued" together diagonally. To such a grid we may add one trangle. These grids form bases for Cartan factors of type IV, called spin factors.

6.2.2 Side-by-side gluing

Two quadrangles $\mathbf{Q}1 = (\mathbf{u}_1, \mathbf{u}_2, \mathbf{u}_3, \mathbf{u}_4)$ and $\mathbf{Q}2 = (\mathbf{v}_1, \mathbf{v}_2, \mathbf{v}_3, \mathbf{v}_4)$ are glued side-by-side if $\mathbf{u}_2 = \mathbf{v}_2$ and $\mathbf{u}_3 = \mathbf{v}_3$. See Figure 6.5.

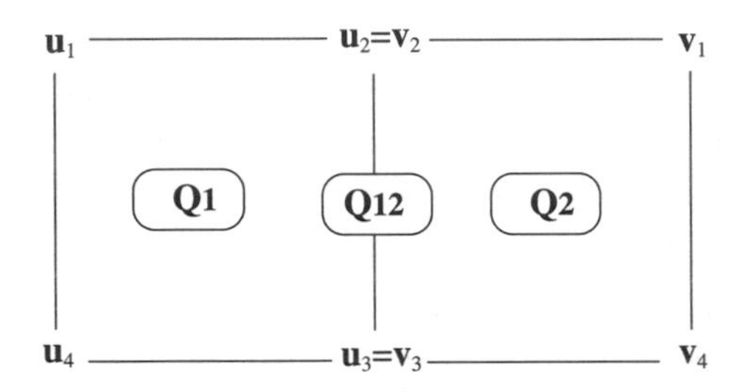

Fig. 6.5. Two quadrangles $\mathbf{Q}1$ and $\mathbf{Q}2$ glued side-by-side, generating a new quadrangle $\mathbf{Q}12 = (\mathbf{u}_1, \mathbf{v}_1, \mathbf{v}_4, \mathbf{u}_4)$.

Let $\mathbf{Q}1 = (\mathbf{u}_1, \mathbf{u}_2, \mathbf{u}_3, \mathbf{u}_4)$ and $\mathbf{Q}2 = (\mathbf{v}_1, \mathbf{u}_2, \mathbf{u}_3, \mathbf{v}_4)$ be quadrangles of minimal tripotents. Assume also that $\mathbf{u}_1 \top \mathbf{v}_1$. We will show that $\mathbf{Q}12 = (\mathbf{u}_1, \mathbf{v}_1, \mathbf{v}_4, \mathbf{u}_4)$ forms a quadrangle.

It suffices to show $\mathbf{u}_4 \perp \mathbf{v}_1$ and $\{\mathbf{v}_1, \mathbf{u}_1, \mathbf{u}_4\} = \frac{1}{2}\mathbf{v}_4$. From (5.91), it follows that $\mathbf{u}_4 = 2\{\mathbf{u}_1, \mathbf{u}_2, \mathbf{u}_3\} \in X_0(\mathbf{v}_1)$, that is, $\mathbf{u}_4 \perp \mathbf{v}_1$. From (5.76), it follows that

$$\{\mathbf{v}_1, \mathbf{u}_1, \mathbf{u}_4\} = 2\{\mathbf{v}_1, \mathbf{u}_1, \{\mathbf{u}_1, \mathbf{u}_2, \mathbf{u}_3\}\}$$

$$= 2(\{\{\mathbf{v}_1, \mathbf{u}_1, \mathbf{u}_1\}, \mathbf{u}_2, \mathbf{u}_3\} + \{\mathbf{u}_1, \mathbf{u}_2, \{\mathbf{v}_1, \mathbf{u}_1, \mathbf{u}_3\}\} - \{\mathbf{u}_1, \{\mathbf{u}_1, \mathbf{v}_1, \mathbf{u}_2\}, \mathbf{u}_3\}).$$

Since $\mathbf{u}_1 \perp \mathbf{u}_3$, from (5.94) we have $\{\mathbf{v}_1, \mathbf{u}_1, \mathbf{u}_3\} = 0$, and, since $\mathbf{u}_1, \mathbf{v}_1, \mathbf{u}_2$ are mutually co-orthogonal tripotents, from (6.10) we have $\{\mathbf{u}_1, \mathbf{v}_1, \mathbf{u}_2\} \in X_1(\mathbf{u}_1)$, implying that $\{\mathbf{u}_1, \{\mathbf{u}_1, \mathbf{v}_1, \mathbf{u}_2, \}, \mathbf{u}_2\} = 0$. Thus,

$$\{\mathbf{v}_1, \mathbf{u}_1, \mathbf{u}_4\} = 2\{\{\mathbf{v}_1, \mathbf{u}_1, \mathbf{u}_1\}, \mathbf{u}_2, \mathbf{u}_3\} = \{\mathbf{v}_1, \mathbf{u}_1, \mathbf{u}_3\} = \frac{1}{2}\mathbf{v}_4, \quad (6.25)$$

showing that $\mathbf{Q}12 = (\mathbf{u}_1, \mathbf{v}_1, \mathbf{v}_4, \mathbf{u}_4)$ forms a quadrangle.

Define a linear map ϕ from this grid to the natural basis $\mathbf{e}_{jk}$ of the 2×3 matrices, as follows:

$$\phi(\mathbf{u}_1) = \mathbf{e}_{11}, \quad \phi(\mathbf{v}_2) = \mathbf{e}_{12}, \quad \phi(\mathbf{v}_1) = \mathbf{e}_{13},$$

$$\phi(\mathbf{u}_4) = \mathbf{e}_{21}, \quad \phi(\mathbf{u}_3) = \mathbf{e}_{22}, \quad \phi(\mathbf{v}_4) = \mathbf{e}_{23}$$

(see Figure 6.5). The above observations imply that ϕ is a triple product automorphism.

If we take $\mathbf{w} = (\mathbf{u}_1 + \mathbf{u}_2 + \mathbf{u}_3 + \mathbf{u}_4)/2$, as above, then

$$\phi(\mathbf{w}) = \begin{pmatrix} 0.5 \; 0.5 \; 0 \\ 0.5 \; 0.5 \; 0 \end{pmatrix}.$$

By direct verification, using (6.7), we obtain

$$S_{\mathbf{w}}\mathbf{v}_1 = -\mathbf{v}_4, \quad S_{\mathbf{w}}\mathbf{v}_4 = -\mathbf{v}_1. \tag{6.26}$$

Note that if we had started out with odd quadrangles, we would still have obtained a quadrangle and not an odd one. Thus, it is more appropriate to apply side-by-side gluing to a family of quadrangles. Cartan factors of type I are obtained by applying side-by-side gluing to an appropriate family of quadrangles.

6.2.3 Construction of new factors

To construct the remaining three Cartan factors, we need to use both diagonal and side-by-side gluing. Basically, each of these factors consists of spin grids of the same dimension (partially) glued together side-by-side.

To justify this construction, we need the following observation. If a quadruple $(\mathbf{v}, \widetilde{\mathbf{v}}, \mathbf{u}, \widetilde{\mathbf{u}})$ forms a quadrangle and

$$X_{1/2}(\mathbf{v} + \widetilde{\mathbf{v}}) \neq 0, \tag{6.27}$$

then the *spin grid dimension*, defined to be $m = \dim X_1(\mathbf{v} + \widetilde{\mathbf{v}})$, is even and $4 \leq m \leq 10$.

Note that from (6.27) and (6.19), it follows that one of the summands in (6.19) is not zero. So, without loss of generality, we may assume that

$$X_{1/2}(\mathbf{v}) \cap X_0(\widetilde{\mathbf{v}}) = X_{1/2}(\mathbf{v}) \cap X_{1/2}(\mathbf{v} + \widetilde{\mathbf{v}}) \neq 0.$$

By applying (6.19) once more to the JBW^*-triple $X_{1/2}(\mathbf{v})$ and orthogonal tripotents $\mathbf{u}, \widetilde{\mathbf{u}}$, and using the fact that by (6.16) $\mathbf{v} + \widetilde{\mathbf{v}} \approx \mathbf{u} + \widetilde{\mathbf{u}}$, we may assume without loss of generality that

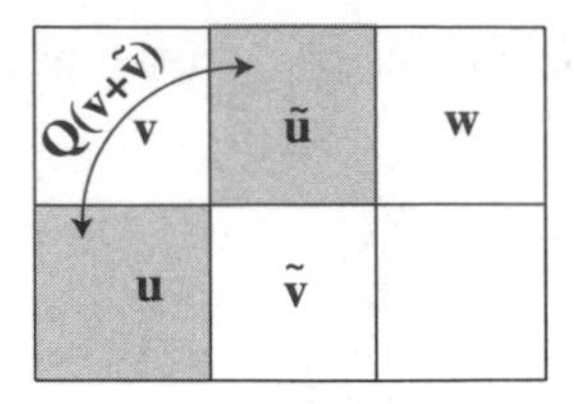

Fig. 6.6. The position of $\mathbf{w}$ in the grid relative to the decomposition by $\mathbf{v}, \widetilde{\mathbf{v}}, \mathbf{u}, \widetilde{\mathbf{u}}$.

$$X_{1/2}(\widetilde{\mathbf{u}}_1) \cap X_{1/2}(\mathbf{v} + \widetilde{\mathbf{v}}) \cap X_{1/2}(\mathbf{v}) \neq \{0\}.$$

Let $\mathbf{w}$ be a tripotent in this subspace. Since $\mathbf{w}$ and $\mathbf{u}$ are orthogonal tripotents in $X_{1/2}(\mathbf{v})$, from (6.9) it follows that $\mathbf{w}$ is a minimal tripotent of X. See Figure 6.6. Let $Y = X_{1/2}(\mathbf{v}) \cap X_{1/2}(\widetilde{\mathbf{v}})$, located in the shaded area of the figure. Since $\mathbf{w}$ is minimal, $P_1(\mathbf{w})Y = \{0\}$. Thus, denoting $P_k(\mathbf{w})Y$ by $Y_k(\mathbf{w})(k = 0, 1/2)$, we have $Y = Y_{1/2}(\mathbf{w}) + Y_0(\mathbf{w})$. Using (5.91) and the fact that $\mathbf{u} \top \mathbf{w} \perp \widetilde{\mathbf{v}}$, it is easy to check that the mapping $Q(\mathbf{v} + \widetilde{\mathbf{v}}) = 2Q(\mathbf{v}, \widetilde{\mathbf{v}})$ is a bijection from $Y_{1/2}(\mathbf{w})$ onto $Y_0(\mathbf{w})$. Therefore, $\dim Y_{1/2}(\mathbf{w}) = \dim Y_0(\mathbf{w})$ and, by (6.20), we see that the spin dimension

$$m = \dim X_1(\mathbf{v} + \widetilde{\mathbf{v}}) = 2 + \dim Y = 2 + 2\dim Y_0(\mathbf{w}) \tag{6.28}$$

is even.

To show that $m \leq 10$, let us assume that $m = \dim X_1(\mathbf{v} + \widetilde{\mathbf{v}}) \geq 12$. Since $Y_0(\mathbf{w})$ is a JBW^*-triple of rank 1, with $\dim Y_0(\mathbf{w}) \geq 5$, and $\mathbf{u} = \mathbf{u}_1 \in Y_0(\mathbf{w})$, we can choose $\mathbf{u}_2, \mathbf{u}_3, \mathbf{u}_4, \mathbf{u}_5$ in $Y_0(\mathbf{w})$ such that $\{\mathbf{u}_i\}_{i=1}^5$ is a co-orthogonal family of tripotents. Since $\mathbf{v}$ and $\widetilde{\mathbf{v}}$ are co-orthogonal, each triplet $(\mathbf{v}, \mathbf{u}_i, \widetilde{\mathbf{v}})$ forms a pre-quadrangle, which can be completed to a quadrangle $(\mathbf{v}, \mathbf{u}_i, \widetilde{\mathbf{v}}, \widetilde{\mathbf{u}}_i)$ with $\widetilde{\mathbf{u}}_i = 2\{\mathbf{u}, \mathbf{u}_i, \widetilde{\mathbf{v}}\}$. Using (5.91) and the fact that $\widetilde{\mathbf{u}}_i$ are minimal in X, we have $\mathbf{u}_i \top \widetilde{\mathbf{u}}_j$, for $i \neq j$, implying that $(\mathbf{w}, \widetilde{\mathbf{u}}_i, \mathbf{u}_j)$ forms a pre-quadrangle. Define

$$\mathbf{u}_{23} = 2\{\mathbf{w}, \widetilde{\mathbf{u}}_2, \mathbf{u}_3\}, \quad \mathbf{u}_{45} = 2\{\mathbf{w}, \widetilde{\mathbf{u}}_4, \mathbf{u}_5\}.$$

Let us show that the family of tripotents $\{\mathbf{u}_1, \mathbf{u}_{23}, \mathbf{u}_{45}\}$ is an orthogonal family. By (5.91), $\mathbf{u}_1$ is orthogonal to both $\mathbf{u}_{23}$ and $\mathbf{u}_{45}$, $\mathbf{u}_{23} \perp \mathbf{u}_5$ and $\mathbf{u}_{23} \in X_{1/2}(\widetilde{\mathbf{u}}_4)$. Since $\mathbf{u}_{23}$ is minimal in X, it follows that $\widetilde{\mathbf{u}}_4 \in X_{1/2}(\mathbf{u}_{23})$. Thus, by (5.91) again, $\mathbf{u}_{45} \in X_0(\mathbf{u}_{23})$, that is, $\mathbf{u}_{23}$ and $\mathbf{u}_{45}$ are orthogonal. The family of orthogonal tripotents $\{\mathbf{u}_1, \mathbf{u}_{23}, \mathbf{u}_{45}\}$ is in $X_{1/2}(\mathbf{v})$, contradicting that rank $X_{1/2}(\mathbf{v}) \leq 2$. Thus, $m = \dim X_1(\mathbf{v} + \widetilde{\mathbf{v}}) \leq 10$.

From the above considerations, we see that if a triple factor contains a quadrangle and is not a spin factor, then the dimension of its spin grids can only be 4,6,8 or 10. The dimension 4 corresponds to a type I factor. The other dimensions give rise to three additional Cartan factors of types II, V and VI, respectively. Moreover, for each minimal tripotent $\mathbf{v}$ in these factors, the space $X_{1/2}(\mathbf{v})$ is a JBW^*-triple constructed from spin grids of dimension two less than the dimension of the original grids. For factors involving spin grids of dimension 6, the space $X_{1/2}(\mathbf{v})$ is of rank 2 and the Cartan factors are of type I, which consist of all $2 \times k$ matrices, for arbitrary k. Therefore, we have Cartan factors of type II of arbitrary size.

However, if we wish to construct a factor from spin grids of dimension 8, $X_{1/2}(\mathbf{v})$ must be of rank 2 and Cartan type II, which is not a spin grid. Such a factor is unique — the 5×5 antisymmetric matrices. Thus, there is only one Cartan factor of type V of dimension 16. This factor, consisting of 5×5 antisymmetric matrices, paves the way from the special to the exceptional factors. By "special," we mean types I–IV, which exist for arbitrary dimension and can be represented as operators on a Hilbert space, while the two "exceptional" ones, types V and VI, exist only for a specific dimension and do not have an operator representation. Thus, any system modelled by an exceptional factor must consist of the 5×5 antisymmetric matrices. Note that this factor, as well as the Cartan factor of type V constructed from it, which is the first exceptional factor, are not Jordan algebras. On the other hand, the next factor, which is constructed from spin grids of dimension 10, is the exceptional Jordan algebra, the Cartan factor of type VI. Since this one is already of rank 3, no factor other than type IV can be constructed from spin grids of dimension greater than 10.

6.2.4 Extension of the triple product to the w^*-closure

Since we also want to consider infinite-dimensional JBW^*-triple factors, we have to clarify how to extend the triple automorphism from the span of the grid, which gives only finite rank elements of the triple, to the w^*-closure, which is needed for the classical domains of infinite rank. In order to extend a homomorphism to the w^*-closure, we will need the following observations. Let X be a JBW^*-triple. A functional $\mathbf{f}$ of X_* is said to be an *atom* if there is a minimal tripotent $\mathbf{v}$ of X such that

$$P_1(\mathbf{v})\mathbf{x} = \mathbf{f}(\mathbf{x})\mathbf{v} \tag{6.29}$$

for every $\mathbf{x}$ in X. In this case, we will write $\mathbf{f} = \mathbf{f}_{\mathbf{v}}$.

A JB^*-triple X is called *nuclear* if any $\mathbf{f}$ in X^* has a decomposition $f = \sum \alpha_i \mathbf{f}_{\mathbf{v}_i}$, where $\{\mathbf{v}_i\}$ is an orthogonal family of minimal tripotents in X and $\sum |\alpha_i| = ||\mathbf{f}||_* < \infty$. Note that if X is a JBW^*-triple and $\mathbf{v}$ is a minimal tripotent of X, then, since the left-hand side of (6.29) is w^*-continuous, we conclude that $\mathbf{f}_{\mathbf{v}}$ is in X_*.

In our construction of the triple factors, we will define a basis for these spaces consisting of a family F of compatible tripotents, called the grid of the space. Denote by X_1 the JB^*-triple which is the norm-closed span of elements of F. It can be shown that X_1 is w^*-dense in the JBW^*-triple $U = X_1^{**}$. Let M be a nuclear JB^*-triple corresponding to compact operators on a Hilbert space, such that M^{**} is a Cartan factor. Let $\phi : X_1 \to M$ be a triple isomorphism. We will show that this isomorphism extends to an isomorphism $\widetilde{\phi} : U \to M^{**}$.

The proof is as follows. Let $\mathbf{f_v}$ be an atom of M^*, and let $\mathbf{v} = \phi(\mathbf{w})$. Then, since ϕ is an isomorphism, we have

$$\langle \phi^* \mathbf{f_v}, \mathbf{x} \rangle \mathbf{v} = \langle \mathbf{f_v}, \phi(\mathbf{x}) \rangle \mathbf{v} = P_1(\mathbf{v})\phi(\mathbf{x}) = \phi(P_1(\mathbf{v})\mathbf{x}) = \langle \mathbf{f_v}, \mathbf{x} \rangle \phi(\mathbf{v}) \quad (6.30)$$

for any $\mathbf{x} \in X_1$, proving that $\phi^*(\mathbf{f_v})$ is an atom of X_1^*. It is known that each isomorphism is an isometry. Thus, ϕ^* is an isometry, implying that X_1 is nuclear.

Let $\mathbf{w}$ be a minimal tripotent of X_1. Since $P_1(\mathbf{w})$ is w^*-continuous, we conclude that $P_1(\mathbf{w})X_1 = C\mathbf{w}$ is w^*-dense in $P_1(\mathbf{w})U$, implying that $P_1(\mathbf{w})U = C\mathbf{w}$. Thus, every minimal tripotent of X_1 is also a minimal tripotent of U. Equation (6.30) defines $\mathbf{f}$ for all $\mathbf{x} \in X_1^{**} = U$. Thus, each atom of X_1^* can be identified with an atom of the predual U_*. This identification induces a natural embedding $i : X_1^* \to U_*$, with $i(\mathbf{f})|M = f$ for any f in X_1^*. Since X_1 is w^*-dense in $X_{1/2}$, we have $i(\mathbf{g}|M) = \mathbf{g}$ for any $\mathbf{g}$ in U_{1*}, implying that the map i is surjective. Thus, the map $\widetilde{\phi} = (i \circ \phi^*)^*$ is an isomorphism from U to M^{**}. A standard argument shows that $\widetilde{\phi}$ is a w^*-continuous extension of ϕ.

6.3 Classification of JBW^*-triple factors

6.3.1 Classification scheme

Let $\mathbf{v}$ be a minimal tripotent of a JBW^*-triple factor X. We will classify X according to the following scheme.

Classification scheme: rank $X_{1/2}(\mathbf{v})$ is either 0,1 or 2.

Case 0: Rank $X_{1/2}(\mathbf{v}) = 0$, that is, $X_{1/2}(\mathbf{v}) = \{0\}$. Then, obviously, $X = C\mathbf{v} \simeq C$.

If rank $X_{1/2}(\mathbf{v}) = 1$, let $\mathbf{u}$ be a minimal tripotent in $X_{1/2}(\mathbf{v})$. Then one of the following two cases occurs.

Case 1: The rank of $X_{1/2}(\mathbf{v})$ is 1 and $\mathbf{u} \top \mathbf{v}$. We will show that in this case, X is a Hilbert space, which is a special case of a Cartan factor of type I.

Case 2. The rank of $X_{1/2}(\mathbf{v})$ is 1 and $\mathbf{u} \vdash \mathbf{v}$. We will show that in this case, X is a Cartan factor of type III, the "symmetric matrices."

Suppose the rank of $X_{1/2}(\mathbf{v})$ is 2. Let $\mathbf{u}$ and $\widetilde{\mathbf{u}}$ be two orthogonal tripotents in $X_{1/2}(\mathbf{v})$. Then, from (6.9), it follows that neither of these tripotents governs $\mathbf{v}$, and, thus, from Table 6.1, it follows that they are co-orthogonal to $\mathbf{v}$ and minimal in X. Thus, $(\mathbf{u}, \mathbf{v}, \widetilde{\mathbf{u}})$ forms a pre-quadrangle. Let $\widetilde{\mathbf{v}} = \{\mathbf{u}, \mathbf{v}, \widetilde{\mathbf{u}}\}$. Then $\widetilde{\mathbf{v}}$ is a minimal tripotent of X, and $(\mathbf{u}, \mathbf{u}, \widetilde{\mathbf{v}}, \widetilde{\mathbf{u}})$ forms a quadrangle. Exactly one of the following five cases occurs.

Case 3. $X_{1/2}(\mathbf{v}+\widetilde{\mathbf{v}}) = \{0\}$. We will show that in this case, X is a Cartan factor of type IV, called the spin factor.

If $X_{1/2}(\mathbf{v}+\widetilde{\mathbf{v}}) \neq \{0\}$, then $m = \dim X_1(\mathbf{v}+\widetilde{\mathbf{v}})$ is even, and $4 \leq m \leq 10$. Thus, we have:

Case 4. $X_{1/2}(\mathbf{v}+\widetilde{\mathbf{v}}) \neq \{0\}$, $\dim X_1(v+\widetilde{\mathbf{v}}) = 4$. We will show that X is a Cartan factor of type I, the "full matrices."

Case 5. $X_{1/2}(\mathbf{v}+\widetilde{\mathbf{v}}) \neq \{0\}$, $\dim X_1(\mathbf{v}+\widetilde{\mathbf{v}}) = 6$. We will show that X is a Cartan factor of type II, the "anti-symmetric matrices."

Case 6. $X_{1/2}(\mathbf{v}+\widetilde{\mathbf{v}}) \neq \{0\}$, $\dim X_1(\mathbf{v}+\widetilde{\mathbf{v}}) = 8$. We will show that X is a Cartan factor of type V, the exceptional JB^*-triple factor of dimension 16.

Case 7. $X_{1/2}(\mathbf{v}+\widetilde{\mathbf{v}}) \neq \{0\}$, $\dim X_1(\mathbf{v}+\widetilde{\mathbf{v}}) = 10$. We will show that X is a Cartan factor of type VI, the exceptional JB^*-triple factor of dimension 27.

See the flowchart in Figure 6.7.

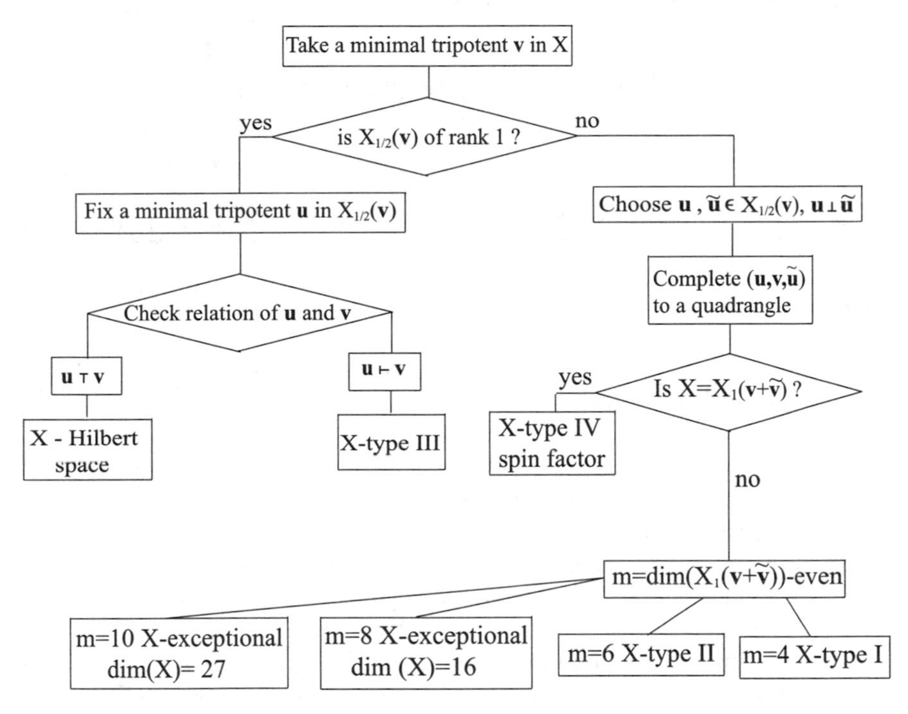

Fig. 6.7. The flowchart of the classification scheme.

We will now discuss each case in detail.

6.3.2 Case 1. The Hilbert space

Let $\mathbf{v}$ be a minimal tripotent in a JBW^*-triple factor X such that the rank of $X_{1/2}(\mathbf{v})$ is 1 and there is a tripotent $\mathbf{u}$ in $X_{1/2}(\mathbf{v})$, with $\mathbf{u}\top\mathbf{v}$. We will show that X is isometric to a Hilbert space.

We will show first that if $\{\mathbf{u}_i\}_{i\in I}$ is an arbitrary family of mutually co-orthogonal minimal tripotents in a JB^*-triple X, then the subspace $W = \overline{\text{span}}\{\mathbf{u}_i\}_{i\in I}$ is a JBW^*-triple which is isometric to a Hilbert space, with $\{\mathbf{u}_i\}_{i\in I}$ as an orthonormal basis.

To see this, let $\{\mathbf{u}_i\}_{i=1}^n$ be a finite set of mutually co-orthogonal minimal tripotents. We claim that for all $\alpha_i \in C$,

$$||\sum_{i=1}^n \alpha_i\mathbf{u}_i||^2 = \sum_{i=1}^n ||\alpha_i\mathbf{u}_i||^2. \tag{6.31}$$

From (6.10), it follows that $\{\mathbf{u}_i, \mathbf{u}_j, \mathbf{u}_k\} = 0$ unless it is of the form $\mathbf{u}_i^{(3)} = \mathbf{u}_i$, or $\{\mathbf{u}_i, \mathbf{u}_i, \mathbf{u}_k\} = \frac{1}{2}\mathbf{u}_k$, for $k \neq i$. Thus,

$$(\sum_{i=1}^n \alpha_i\mathbf{u}_i)^{(3)} = \sum_{i,j,k} \alpha_i\overline{\alpha}_j\alpha_k\{\mathbf{u}_i, \mathbf{u}_j, \mathbf{u}_k\}$$

$$= \sum_{i=1}^n |\alpha_i|^2\alpha_i\mathbf{u}_i + 2\sum_i\sum_{k\neq i} |\alpha_i|^2\alpha_k\{\mathbf{u}_i, \mathbf{u}_i, \mathbf{u}_k\} = \sum_i |\alpha_i|^2 \sum_k \alpha_k\mathbf{u}_k.$$

Therefore,

$$||\sum_{i=1}^n \alpha_i\mathbf{u}_i||^3 = ||(\sum_{i=1}^n \alpha_i\mathbf{u}_i)^{(3)}|| = (\sum_{i=1}^n |\alpha_i|^2)\ ||\sum_{k=1}^n a_k\mathbf{u}_k||, \tag{6.32}$$

implying (6.31). It follows from (6.31) that W is isometric to a Hilbert space with $\{\mathbf{u}_i\}_{i\in I}$ as an orthonormal basis. Moreover, since the family $\{\mathbf{u}_i\}_{i\in I}$ is closed under the triple product, W is a JBW^*-triple.

By Zorn's lemma, there are an index set I containing 1 and not containing 0 and a maximal family of co-orthogonal $\{\mathbf{u}_i\}_{i\in I}$ of tripotents in $X_{1/2}(\mathbf{v})$, with $\mathbf{u}_1 = \mathbf{u}$. Denote $\mathbf{u}_0 = \mathbf{v}$ and $J = I\bigcup\{0\}$. Since $\mathbf{u}_i \top \mathbf{u}_1$, it follows that each $\mathbf{u}_i$ is a minimal tripotent of X and $\mathbf{u}_i \top \mathbf{v}$. If $\bigcap_{i\in J} X_{1/2}(\mathbf{u}_i) \neq \{0\}$, let $\mathbf{w}$ be a tripotent in that space. Since every tripotent in the rank 1 JBW^*-triple $X_{1/2}(\mathbf{v})$ is minimal, we can deduce that $\mathbf{w} \top \mathbf{u}_i$ for all $i \in I$, contradicting the maximality of $\{\mathbf{u}_i\}_{i\in I}$. Thus $\bigcap_{i\in J} X_{1/2}(\mathbf{u}_i) = \{0\}$.

If $\mathbf{a}$ is an element of $X_{1/2}(\mathbf{v})$, then $\sum_{i\in F} P_1(\mathbf{u}_i)\mathbf{a}$ converges in norm by the above observation, implying that

$$\mathbf{a} - \sum_{i\in I} P_2(\mathbf{u}_i)z \in \bigcap_{i\in J} X_{1/2}(\mathbf{u}_i) = \{0\}. \tag{6.33}$$

Thus, $X_{1/2}(\mathbf{v}) = \overline{\text{span}}\{\mathbf{u}_i\}_{i\in I}$, and

$$X_1(\mathbf{v}) + X_{1/2}(\mathbf{v}) = \overline{\text{span}}\{\mathbf{u}_i\}_{i\in J} \tag{6.34}$$

is a JBW^*-triple isometric to a Hilbert space. Moreover, $X_1(\mathbf{v})+X_{1/2}(\mathbf{v})$ is clearly w^*-closed in X.

In order to show that $X = X_1(\mathbf{v}) + X_{1/2}(\mathbf{v})$, we only need to show that $X_0(\mathbf{v}) = \{0\}$. Obviously,

$$X_0(\mathbf{v}) = P_1(\mathbf{u}_i)X_0(\mathbf{v}) + P_{1/2}(\mathbf{u}_i)X_0(\mathbf{v}) + P_0(\mathbf{u}_i)X_0(\mathbf{v}). \tag{6.35}$$

Since $\mathbf{u}_i$ is minimal in X, $P_1(\mathbf{u}_i)X_0(\mathbf{v}) = \{0\}$. If $P_{1/2}(\mathbf{u}_i)X_0(\mathbf{v}) \neq \{0\}$, let $\widetilde{\mathbf{v}}$ be a tripotent in that space. Since $\mathbf{v}, \widetilde{\mathbf{v}}$ are two orthogonal tripotents in $X_{1/2}(\mathbf{u}_i)$, $(\mathbf{v}, \mathbf{u}_i, \widetilde{\mathbf{v}})$ form a pre-quadrangle which can be completed to a quadrangle with $\widetilde{\mathbf{u}}_i = 2\{\mathbf{v}, \mathbf{u}_i, \widetilde{\mathbf{v}}\}$, which is orthogonal to $\mathbf{u}_i$ and is in $X_{1/2}(\mathbf{v})$, contradicting the fact that the rank of $X_{1/2}(\mathbf{v}) = 1$. Thus, $X_0(\mathbf{v}) = P_0(\mathbf{u}_i)X_0(\mathbf{v})$, implying that $X_0(\mathbf{v})$ is orthogonal to $X_1(\mathbf{v})+X_{1/2}(\mathbf{v})$, and, since X is a factor, it must be equal to zero. So $X = X_1(\mathbf{v})+X_{1/2}(\mathbf{v})$ and is a JBW^*-triple isometric to a Hilbert space.

6.3.3 Case 2. The type III domain

Let $F = \{\mathbf{u}_{ij} | i,j \in I\}$ be a family of tripotents in a Jordan triple system X, for some index set I. We will say that F is a *Hermitian grid* if for every $i,j,k,l \in I$, we have $\mathbf{u}_{ij} = \mathbf{u}_{ji}$ and the tripotents satisfy the following relations:

$$\mathbf{u}_{ii} \dashv \mathbf{u}_{ij} \top \mathbf{u}_{jk}, \text{if } i,j,k \text{ are different}, \tag{6.36}$$

$$\mathbf{u}_{ij} \perp \mathbf{u}_{kl}, \text{ if } \{i,j\} \cap \{k,l\} = \varnothing, \tag{6.37}$$

every triple product among elements of F which cannot be brought to the form $\{\mathbf{u}_{ij}, \mathbf{u}_{jk}, \mathbf{u}_{kl}\}$ vanishes, and for arbitrary i,j,k,l, the triple products involving at least two different elements satisfy

$$\{\mathbf{u}_{ij}, \mathbf{u}_{jk}, \mathbf{u}_{kl}\} = \frac{1}{2}\mathbf{u}_{il}, \text{ for } i \neq l \tag{6.38}$$

and

$$\{\mathbf{u}_{ij}, \mathbf{u}_{jk}, \mathbf{u}_{ki}\} = \mathbf{u}_{ii}. \tag{6.39}$$

It is easy to see that the natural basis in a Cartan factor of type III is a hermitian grid whose span is w^*-dense.

We will show now that if $\mathbf{v}$ is a minimal tripotent of a JBW^*-triple factor X, rank $X_{1/2}(\mathbf{v}) = 1$, and there is a tripotent $\mathbf{u}$ in $X_{1/2}(\mathbf{v})$ such that $\mathbf{u} \vdash \mathbf{v}$, then X is a Cartan factor of type III.

By Zorn's lemma, there is in $X_{1/2}(\mathbf{v})$ a maximal co-orthogonal family of tripotents $\{\mathbf{u}_{1i}\}_{i\in I}$, which includes $\mathbf{u}$, where I is some index set not containing 1. If $\mathbf{v}\top\mathbf{u}_{1i}$, then $\mathbf{u}_{1i}$ is minimal in X. But since $\mathbf{u}\top\mathbf{u}_{1i}$, this would imply that also $\mathbf{u}$ is minimal in X, a contradiction. Thus, we can deduce that $\mathbf{u}_{1i} \vdash \mathbf{v}$ for all $i \in I$ and $(\mathbf{v}, \mathbf{u}_{1i})$ forms a pre-trangle. Let $J = I \cup \{1\}$ and let $\mathbf{u}_{11} = \mathbf{v}$. Define $\mathbf{u}_{i1} = \mathbf{u}_{1i}$ and

$$\mathbf{u}_{ii} = \{\mathbf{u}_{1i}, \mathbf{u}_{11}, \mathbf{u}_{1i}\}, \quad \mathbf{u}_{ij} = \mathbf{u}_{ji} = 2\{\mathbf{u}_{1j}, \mathbf{u}_{1i}, \mathbf{u}_{ii}\} \tag{6.40}$$

for any $i, j \in I$, where $i \neq j$. We will show that the family $F = \{\mathbf{u}_{ij}\}_{i,j} \in J$ forms a hermitian grid. First, we show that the family F satisfies properties (6.36) and (6.37).

Let $1, i, j$, be distinct indices. Since $\mathbf{u}_{11} \in X_1(\mathbf{u}_{1i})$, by (6.12), $\mathbf{u}_{ii} = Q(\mathbf{u}_{1i})\mathbf{u}_{11}$ is a minimal tripotent of X, and moreover, $\mathbf{u}_{1i}$ governs both $\mathbf{u}_{11}$ and $\mathbf{u}_{ii}$. On the other hand, by (5.91),

$$\mathbf{u}_{ii} \;=\; \{\mathbf{u}_{1i}, \mathbf{u}_{11}, \mathbf{u}_{1i}\} \in X_0(\mathbf{u}_{1j}), \tag{6.41}$$

i.e., the two tripotents $\mathbf{u}_{ii}$ and $\mathbf{u}_{1j}$ are orthogonal. Thus $\mathbf{w} = (\mathbf{u}_{1j} + \mathbf{u}_{ii})$ is a tripotent, and the map $Q(\mathbf{w})$ is an anti-linear homomorphism of order 2 on $X_1(\mathbf{w})$.

We claim that for distinct indices $1, i, j$, the map $Q(\mathbf{w})$ exchanges $\mathbf{u}_{11}$ with $\mathbf{u}_{jj}, \mathbf{u}_{1i}$ with $\mathbf{u}_{ij}$ and fixes $\mathbf{u}_{1j}$ as well as $\mathbf{u}_{ii}$. In other words, it exchanges 1 with j in the set of indices $\{1, i, j\}$. The action of this map can be represented as the symmetric reflection along the diagonal $\mathbf{u}_{1j} + \mathbf{u}_{ii}$ of the following table:

$\mathbf{u}_{11}$	$\mathbf{u}_{1i}$	$\mathbf{u}_{1j}$		$\mathbf{u}_{jj}$	$\mathbf{u}_{ij}$	$\mathbf{u}_{1j}$
$\mathbf{u}_{1i}$	$\mathbf{u}_{ii}$	$\mathbf{u}_{ij}$	$\overset{Q(\mathbf{w})}{\longleftrightarrow}$	$\mathbf{u}_{ij}$	$\mathbf{u}_{ii}$	$\mathbf{u}_{1i}$
$\mathbf{u}_{1j}$	$\mathbf{u}_{ij}$	$\mathbf{u}_{jj}$		$\mathbf{u}_{1j}$	$\mathbf{u}_{1i}$	$\mathbf{u}_{11}$

Table 6.2. Reflection of the grid for type III.

Let us check that $Q(\mathbf{w})$ exchanges $\mathbf{u}_{1i}$ with $\mathbf{u}_{ij}$. It suffices to show that $Q(\mathbf{u}_{1j} + Q(\mathbf{w})_{ii})\mathbf{u}_{1i} = \mathbf{u}_{ij}$. But

$$Q(\mathbf{u}_{1j} + \mathbf{u}_{ii})\mathbf{u}_{1i} = \{\mathbf{u}_{1j} + \mathbf{u}_{ii}, \mathbf{u}_{1i}, \mathbf{u}_{1j} + \mathbf{u}_{ii}\}$$

$$= \{\mathbf{u}_{ii}, \mathbf{u}_{1i}, \mathbf{u}_{ii}\} + \{\mathbf{u}_{1j}, \mathbf{u}_{1i}, \mathbf{u}_{1j}\} + 2\{\mathbf{u}_{1j}, \mathbf{u}_{1i}, \mathbf{u}_{ii}\}.$$

By (6.10), the first two terms are zero, implying that $Q(\mathbf{w})\mathbf{u}_{1i} = \mathbf{u}_{ij}$. The remaining statements of the claim are obvious.

Note that from this observation, it follows that

$$\mathbf{u}_{ii} \perp \mathbf{u}_{jj}, \; \mathbf{u}_{ij} \vdash \mathbf{u}_{ii}$$

and $\mathbf{u}_{ij}$ is a minimal tripotent of $X_{1/2}(\mathbf{u}_{jj})$. Moreover, $\{\mathbf{u}_{ii}\}$ form a family of mutually orthogonal minimal tripotents in X. From (5.91) and (6.40), it follows that $\mathbf{u}_{ij} \top \mathbf{u}_{jk}$ if i, j, k are distinct and (6.36) holds.

On the other hand, if $\{i,j\}\cap\{k,l\}=\varnothing$, then from (6.12), it follows that $\mathbf{u}_{ij}\approx\mathbf{u}_{ii}+\mathbf{u}_{jj}$ and $\mathbf{u}_{kl}\approx\mathbf{u}_{kk}+\mathbf{u}_{ll}$, implying (6.37).

Next, from (6.37) and (6.10), it follows that every non-vanishing triple product among elements of F must be of the form

$$\{\mathbf{u}_{ij},\mathbf{u}_{jk},\mathbf{u}_{kl}\}.$$

To show that (6.38) holds in case the four indices i,j,k,l are distinct, it is enough to show that $(\mathbf{u}_{ij},\mathbf{u}_{il},\mathbf{u}_{kl},\mathbf{u}_{kj})$ form a quadrangle. Consider first the case when one of indices, say j, is 1. By use of (6.40),(5.76) and (5.91), we have

$$\mathbf{u}_{kj}=2\{\mathbf{u}_{1k},\mathbf{u}_{1j},\mathbf{u}_{jj}\}=2\{\mathbf{u}_{1k},\mathbf{u}_{1j},\{\mathbf{u}_{1j},\mathbf{u}_{11},\mathbf{u}_{1j}\}\}=2\{\mathbf{u}_{1k},\mathbf{u}_{11},\mathbf{u}_{1j}\}. \tag{6.42}$$

Thus,

$$\{\mathbf{u}_{1i},\mathbf{u}_{1k},\mathbf{u}_{kl}\}=2\{\mathbf{u}_{1i},\mathbf{u}_{1k},\{\mathbf{u}_{1k},\mathbf{u}_{11},\mathbf{u}_{1l}\}\}=\{\mathbf{u}_{1i},\mathbf{u}_{11},\mathbf{u}_{1l}\}=\frac{1}{2}\mathbf{u}_{il},$$

verifying (6.38) for $j=1$. For the general case, we use the results of section 6.2.2 for side-by-side gluing applied to the quadrangles $(\mathbf{u}_{1j},\mathbf{u}_{1l},\mathbf{u}_{il},\mathbf{u}_{ij})$ and $(\mathbf{u}_{1j},\mathbf{u}_{1l},\mathbf{u}_{kl},\mathbf{u}_{kj})$ to obtain that $(\mathbf{u}_{ij},\mathbf{u}_{il},\mathbf{u}_{kl},\mathbf{u}_{kj})$ is a quadrangle. The above construction is depicted in Figure 6.8 below:

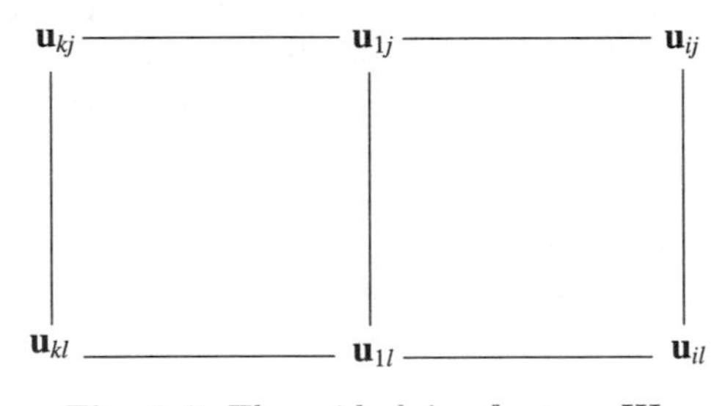

Fig. 6.8. The grid gluing for type III.

It remains to verify (6.39). Note that from (6.42) and (5.76), we have

$$\{\mathbf{u}_{1j},\mathbf{u}_{jk},\mathbf{u}_{k1}\}=\{\mathbf{u}_{1j},\{\mathbf{u}_{1j},\mathbf{u}_{11},\mathbf{u}_{k1}\},\mathbf{u}_{k1}\}=\mathbf{u}_{11}, \tag{6.43}$$

implying (6.39) for $i=1$. To get the result for arbitrary i, we use the symmetry $S_{\mathbf{w}}$ defined by (6.14), with $\mathbf{w}=(\mathbf{u}_{11}+\mathbf{u}_{1i}+\mathbf{u}_{ii})/2$. Note that this symmetry exchanges $\mathbf{u}_{11}$ with $\mathbf{u}_{ii}$. Since for $j\cap\{1,j\}=\varnothing$, $\mathbf{u}_{jj}$ is orthogonal to $\mathbf{w}$, it fixes all such $\mathbf{u}_{jj}$. Since $S_{\mathbf{w}}$ is a triple automorphism, it moves a trangle to a trangle and a quadrangle to a quadrangle, so any $\mathbf{u}_{kj}$ will be mapped to $\mathbf{u}_{\tilde{k}\tilde{j}}$, where the indexes $\tilde{k}\tilde{j}$ are obtained from k,j by exchanging 1 and k, if necessary. By applying $S_{\mathbf{w}}$ to equation (6.43), we get (6.39). Thus, the family $\{\mathbf{u}_{ij}\}$ defined by (6.40) forms a hermitian grid.

Let $\mathbf{p} = \sum_{i \in \widetilde{J}} \mathbf{u}_{ii}$. First, we show that $X = \overline{\text{span}}^{w^*}\{\mathbf{u}_{ij}\}$. It suffices to show $X_1(\mathbf{p}) = \overline{\text{span}}^{w^*}\{\mathbf{u}_{ij}\}$ and $X_{1/2}(\mathbf{p}) = \{0\}$. We know that

$$X_1(\mathbf{p}) \;=\; \overline{span}^{w^*}\;\{X_1(\mathbf{u}_{ii}), X_{1/2}(\mathbf{u}_{ii}) \cap X_{1/2}(\mathbf{u}_{jj}) : \; i, j \in J\}. \tag{6.44}$$

Since $\mathbf{u}_{ii}$ is minimal in X, $X_1(\mathbf{u}_{ii}) = C\mathbf{u}_{ii}$. On the other hand, since $\mathbf{u}_{ij}$ governs both $\mathbf{u}_{ii}$ and $\mathbf{u}_{jj}$, the tripotents $\mathbf{u}_{ij}$ and $\mathbf{u}_{ii} + \mathbf{u}_{jj}$ are equivalent, giving $X_1(\mathbf{u}_{ij}) = X_1(\mathbf{u}_{ii} + \mathbf{u}_{jj})$. By (6.20), we have

$$X_{1/2}(\mathbf{u}_{ii}) \cap X_{1/2}(\mathbf{u}_{jj}) \subseteq X_1(\mathbf{u}_{ij}). \tag{6.45}$$

Since $\mathbf{u}_{ij}$ is minimal in $X_{1/2}(\mathbf{u}_{ii})$, we conclude that $X_{1/2}(\mathbf{u}_{ii}) \cap X_{1/2}(\mathbf{u}_{jj}) = C\mathbf{u}_{ij}$. Thus,

$$X_1(\mathbf{p}) = \overline{span}^{w^*}\;\{\mathbf{u}_{ij}\}_{i,j \in J}. \tag{6.46}$$

For $X_{1/2}(\mathbf{p}) = \{0\}$, it suffices to show that $X_{1/2}(\mathbf{u}_{ii}) \subseteq X_1(p)$ for all $i \in J$. From the maximality of the family $\mathbf{u}_{1i}$ in the construction of the grid, it follows that $X_{1/2}(\mathbf{u}_{11}) = \overline{span}\{\mathbf{u}_{1i}\}_{i \in I} \subseteq X_1(p)$. For arbitrary i, we use the automorphism $S_{\mathbf{w}}$, with $\mathbf{w}$ as above, to exchange $X_{1/2}(\mathbf{u}_{11})$ and $X_{1/2}(\mathbf{u}_{ii})$, preserving the set $\mathbf{u}_{ij}$. Thus, $X_{1/2}(\mathbf{u}_{ii}) \subseteq X_1(\mathbf{p})$. Since X is a factor, this implies that $X = X_1(\mathbf{p}) = \overline{span}^{w^*}\;\{\mathbf{u}_{ij}\}_{i,j \in J}$.

Finally, we show that X is a Cartan factor of type III in the sense defined in Chapter 4. Let H be a Hilbert space with an orthonormal basis $\{\xi_i : i \in J\}$. Equip H with a conjugation Q defined by $Q(\sum \lambda_i \xi_i) = \sum \overline{\lambda}_i \xi_i$. For $i, j \in J$, let $S_{ii} = e_{ii}$ and $S_{ij} = e_{ij} + e_{ji}$, for $i \neq j$, where e_{ij} and e_{ji} are matrix units corresponding to the basis $\{\xi_i\}$. The family $\{S_{ij}\}$ is then a Hermitian gird. Let $X_2 = \overline{span}\;\{S_{ij}\}_{i,j \in J}$ be its norm-closed span, consisting of all compact symmetric operators on H. Then X_2 is a JB^*-triple. Moreover, X_2^{**} consists of all compact symmetric operators on H and is a Cartan factor of type III. Let $X_1 = \overline{span}^{w^*}\{\mathbf{u}_{ij}\}_{i,j \in J}$ and define a linear map $\phi : \; X_1 \to X_2$ by $\phi(\mathbf{u}_{ij}) = S_{ij}$. Obviously, ϕ is a triple isomorphism. Thus, it extends to an isomorphism $\widetilde{\phi} : \; X \to X_2^{**}$, showing that X is a Cartan factor of type III.

6.3.4 Case 3. Type IV factor

Let $\mathbf{v}$ and $\widetilde{\mathbf{v}}$ be minimal tripotents of a JBW^*-triple factor X, such that rank $X_{1/2}(\mathbf{v}) = 2$ and $X_{1/2}(\mathbf{v} + \widetilde{\mathbf{v}}) = \{0\}$. We will show that in this case, X is a Cartan factor of type IV, called the spin factor. We studied the properties of spin factors in Chapter 3.

Let $\mathbf{u}_1 = \mathbf{v}$ and $\widetilde{\mathbf{u}}_1 = \widetilde{\mathbf{v}}$. By Zorn's Lemma, there is a maximal co-orthogonal family $\{\mathbf{u}_i\}_{i \in I}$ of minimal tripotents of X which includes $\mathbf{u}_1$. For $i \in I$, with $i \neq 1$, let $\widetilde{\mathbf{u}}_i = -2\{\mathbf{u}_1, \mathbf{u}_i, \widetilde{\mathbf{u}}_1\}$. Then the quadruples $(\mathbf{u}_1, \mathbf{u}_i, \widetilde{\mathbf{u}}_1, \widetilde{\mathbf{u}}_i)$

form odd quadrangles glued together along the diagonal $(\mathbf{u}_1, \widetilde{\mathbf{u}}_1)$. It follows from (6.24) that $(\mathbf{u}_j, \mathbf{u}_i, \widetilde{\mathbf{u}}_j, \widetilde{\mathbf{u}}_i)$ form odd quadrangles for $i \neq j$.

If $\bigcap_{i\in I} X_{1/2}(\mathbf{u}_i) \neq \{0\}$, let $\mathbf{u}_0$ be a tripotent in that JBW^*-triple. From Table 6.1 and the maximality of the family $\{\mathbf{u}_i\}_{i\in I}$, it follows that $\mathbf{u}_0$ is a minimal tripotent of $\bigcap_{i\in i} X_{1/2}(\mathbf{u}_i)$ and that $\mathbf{u}_0$ governs both $\mathbf{u}_i, \widetilde{\mathbf{u}}_i$ for all $i \in I$. Hence,

$$\bigcap_{i\in I} X_{1/2}(\mathbf{u}_i) \;=\; P_2(\mathbf{u}_0)(\bigcap_{i} X_{1/2}(\mathbf{u}_i)) \;=\; C\mathbf{u}_0. \tag{6.47}$$

By multiplying $\mathbf{u}_0$ by an appropriate scalar λ if necessary, we can assume without loss of generality that $\mathbf{u}_0$ is a tripotent satisfying $\{\mathbf{u}_0, \mathbf{u}_1, \mathbf{u}_0\} = -\widetilde{\mathbf{u}}_1$. From (5.76) and (5.91), it follows that $\{\mathbf{u}_0, \mathbf{u}_i, \mathbf{u}_0\} = -\mathbf{u}_i$ for all $i \in I$. Since, from (6.31), both $\{\mathbf{u}_i\}$ and $\{\widetilde{\mathbf{u}}_i\}$ span a Hilbert space, the summations $\sum_{i\in I} P_2(\mathbf{u}_i)\mathbf{a}$ and $\sum_{i\in I} P_2(\widetilde{\mathbf{u}}_i)\mathbf{a}$ converge in norm for all $\mathbf{a} \in X$. Thus, X is the norm-closed span of either $\{\mathbf{u}_i, \widetilde{\mathbf{u}}_i\}_{i\in I}$ or $\{\mathbf{u}_i, \widetilde{\mathbf{u}}_i, \mathbf{u}_0\}_{i\in I}$.

We can define an inner product on X by

$$\langle \sum_{i\in I} \alpha_i \mathbf{u}_i + \widetilde{\alpha}_i \widetilde{\mathbf{u}}_i + \beta \mathbf{u}_0 | \sum_{i\in I} \lambda_i \mathbf{u}_i + \widetilde{\lambda}_i \widetilde{\mathbf{u}}_i + \delta \mathbf{u}_0 \rangle = \sum_{i\in I} \alpha_i \overline{\lambda}_i + \widetilde{\alpha}_i \overline{\widetilde{\lambda}}_i + 2\beta\overline{\delta}.$$

We can define a conjugation on X by $(\mathbf{u}_i)^* = \widetilde{\mathbf{u}}_i$, $(\mathbf{u}_0)^* = \mathbf{u}_0$. Then the triple product on X can be expressed in terms of the inner product by

$$2\{\mathbf{a}, \mathbf{b}, \mathbf{c}\} \;=\; \langle \mathbf{a}|\mathbf{b}\rangle \mathbf{c} + \langle \mathbf{c}|\mathbf{b}\rangle \mathbf{a} - \langle \mathbf{a}|\mathbf{c}^*\rangle \mathbf{b}^*. \tag{6.48}$$

A straightforward verification shows that this identity holds on the basis elements $\{\mathbf{u}_i, \widetilde{\mathbf{u}}_i, \mathbf{u}_0\}_{i\in I}$ of X. Then, by passing to the limit, we conclude that it holds for arbitrary $\mathbf{a}, \mathbf{b}, \mathbf{c}$ in X.

Since X is a factor, $X = X_1(\mathbf{v} + \widetilde{\mathbf{v}})$ and is the norm-closed span of a family consisting of minimal tripotents $\{\mathbf{u}_i, \widetilde{\mathbf{u}}_i\}_{i\in I}$ and possibly a tripotent $\mathbf{u}_0$ such that

(i) $\mathbf{u}_0 \vdash \mathbf{u}_i, \mathbf{u}_0 \vdash \widetilde{\mathbf{u}}_i, Q(\mathbf{u}_0)\mathbf{u}_i = -\widetilde{\mathbf{u}}_i$ for all $i \in I$,
(ii) $(\mathbf{u}_i, \mathbf{u}_j, \widetilde{\mathbf{u}}_i, \widetilde{\mathbf{u}}_j)$ are odd quadrangles for $i \neq j$,
(iii) $X = X_1(\mathbf{u}_i + \widetilde{\mathbf{u}}_i)$ for all $i \in I$.

Such a family $\{\mathbf{u}_i, \widetilde{\mathbf{u}}_i, \mathbf{u}_0\}$ will be called a *spin grid*. By use of transformation (3.114) of Chapter 3, we can replace the spin grid constructed from minimal tripotents with a basis of maximal tripotents, called a TCAR basis. In this basis, the triple product, defined by (6.48), coincides with the product defined by (3.4).

6.3.5 Case 4. Type I factor

Let I, J be index sets. A family of minimal tripotents $\{\mathbf{u}_{ij}\}_{i\in I, j\in J}$ is said to form a *rectangular grid* if

(i) $\mathbf{u}_{jk}, \mathbf{u}_{il}$ are co-orthogonal when they share a common row index ($j = i$) or a column index ($k = l$), and are algebraically orthogonal otherwise,

(ii) for any $j \neq i, l \neq k$ the quadruple $(\mathbf{u}_{jk}, \mathbf{u}_{jl}, \mathbf{u}_{il}, \mathbf{u}_{ik})$ forms a quadrangle, and

(iii) all products vanish unless they are of the form $D(\mathbf{a})\mathbf{b}$ or $\{\mathbf{a}, \mathbf{b}, \mathbf{c}\}$ when $(\mathbf{a}, \mathbf{b}, \mathbf{c})$ forms a pre-quadrangle.

It is easy to see that each Cartan factor of type I has a rectangular grid whose span is w^*-dense.

Let $\mathbf{v}$ be a minimal tripotent in a JBW^*-triple factor X. Suppose there are minimal tripotents $\mathbf{u}, \widetilde{\mathbf{u}}, \widetilde{\mathbf{v}}$ such that $(\mathbf{v}, \mathbf{u}, \widetilde{\mathbf{v}}, \widetilde{\mathbf{u}})$ forms a quadrangle, and $\dim X_1(\mathbf{v} + \widetilde{\mathbf{v}}) = 4$. We will show now that X is a Cartan factor of type I.

We will consider the structure of $X_{1/2}(\mathbf{v})$. Since $\mathbf{u} + \widetilde{\mathbf{u}} \approx \mathbf{v} + \widetilde{\mathbf{v}}$, we have $X_0(\mathbf{u} + \widetilde{\mathbf{u}}) \subseteq X_0(\mathbf{v})$, and, thus,

$$X_{1/2}(\mathbf{v}) = (X_1(\mathbf{u} + \widetilde{\mathbf{u}}) + X_{1/2}(\mathbf{u} + \widetilde{\mathbf{u}})) \cap X_{1/2}(\mathbf{v}).$$

From our assumption that $\dim X_1(\mathbf{v} + \widetilde{\mathbf{v}}) = 4$, it follows that $X_1(\mathbf{v} + \widetilde{\mathbf{v}}) = X_1(\mathbf{u} + \widetilde{\mathbf{u}}) = \operatorname{span}\{\mathbf{v}, \mathbf{u}, \widetilde{\mathbf{v}}, \widetilde{\mathbf{u}}\}$. Hence,

$$X_1(\mathbf{u} + \widetilde{\mathbf{u}}) \cap X_{1/2}(\mathbf{v}) = C\mathbf{u} \oplus C\widetilde{\mathbf{u}}.$$

By use of (6.20), we get

$$X_{1/2}(\mathbf{u} + \widetilde{\mathbf{u}}) \cap X_{1/2}(\mathbf{v})$$

$$= X_{1/2}(\mathbf{u}) \cap X_0(\widetilde{\mathbf{u}}) \cap X_{1/2}(\mathbf{v}) \oplus X_0(\mathbf{u}) \cap X_{1/2}(\widetilde{\mathbf{u}}) \cap X_{1/2}(\mathbf{v}) := M_1 \oplus M_2.$$

Note that any tripotent $\mathbf{w} \in M_1 = X_{1/2}(\mathbf{u}) \cap X_0(\widetilde{\mathbf{u}}) \cap X_{1/2}(\mathbf{v})$ is orthogonal to $\widetilde{\mathbf{u}}$ in $X_{1/2}(\mathbf{v})$, and, thus, by (6.9), it cannot govern $\mathbf{u}$ and must be co-orthogonal to $\mathbf{u}$ and a minimal tripotent. Hence, M_1 is of rank 1. By applying the classification scheme for the JBW^*-triple $X_{1/2}(\mathbf{v})$ and the minimal tripotent $\mathbf{u}$, from Case 1 of our classification (see section 6.3.2), $H_1 = C\mathbf{u} \oplus M_1$ is isometric to a Hilbert space and any orthonormal basis in it is given by minimal tripotents of X and is the JBW^*-triple factor generated by $\mathbf{u}$. So the subspace $H_2 = C\widetilde{\mathbf{u}} \oplus M_2$, which is linearly independent of H_1, must be orthogonal to H_1. Moreover, the subspace H_2 is also isometric to a Hilbert space and

$$X_{1/2}(\mathbf{v}) = H_1 \oplus H_2 \tag{6.49}$$

is a direct sum of two orthogonal Hilbert spaces.

Let us denote $\mathbf{u}_{11} = \mathbf{v}$, $\mathbf{u}_{21} = \mathbf{u}$ and $\mathbf{u}_{12} = \widetilde{\mathbf{u}}$. Complete $\{\mathbf{u}_{21}\}$ via elements $\{\mathbf{u}_{i1}\}_{i \in I}$ to an orthonormal basis in H_1. Note that $\{\mathbf{u}_{i1}\}_{i \in I \cup \{2\}}$ forms a maximal family of co-orthogonal minimal tripotents in H_1. Similarly, complete $\{\mathbf{u}_{12}\}$ via elements $\{\mathbf{u}_{1j}\}_{j \in J}$ to an orthonormal basis in H_2, which forms a maximal family of co-orthogonal minimal tripotents in H_2. It is obvious that for $(i,j) \in I \times J$, the triple of tripotents $(\mathbf{u}_{i1}, \mathbf{u}_{11}, \mathbf{u}_{1j})$ forms a pre-quadrangle. Thus, one can define

$$\mathbf{u}_{ij} = 2\{\mathbf{u}_{i1}, \mathbf{u}_{11}, \mathbf{u}_{1j}\}$$

and complete them to a quadrangle $(\mathbf{u}_{i1}, \mathbf{u}_{11}, \mathbf{u}_{1j}, \mathbf{u}_{ij})$. Since $\mathbf{u}_{ij}$ is co-orthogonal to a minimal tripotent $\mathbf{u}_{i1}$, it is also minimal.

Let $\widetilde{I} = I \cap \{1\}$, $\widetilde{J} = J \cap \{1\}$. We will show that the family

$$F = \{\mathbf{u}_{ij} : (i,j) \in \widetilde{I} \times \widetilde{J}\}$$

forms a rectangular grid. Properties (i) and (iii) of the definition of a rectangular grid follow from (5.91). To show (iii), i.e., that $(\mathbf{u}_{jk}, \mathbf{u}_{jl}, \mathbf{u}_{il}, \mathbf{u}_{ik})$ forms a quadrangle (for $i \neq j$ and $k \neq l$), we apply the side-by-side gluing of section 6.2.2 to

a) the pair of quadrangles $(\mathbf{u}_{j1}, \mathbf{u}_{11}, \mathbf{u}_{1k}, \mathbf{u}_{jk})$, $(\mathbf{u}_{j1}, \mathbf{u}_{11}, \mathbf{u}_{1l}, \mathbf{u}_{jl})$ to yield the quadrangle $(\mathbf{u}_{1k}, \mathbf{u}_{1l}, \mathbf{u}_{jl}, \mathbf{u}_{jk})$,

b) the pair of quadrangles $(\mathbf{u}_{i1}, \mathbf{u}_{11}, \mathbf{u}_{1l}, \mathbf{u}_{il})$, $(\mathbf{u}_{i1}, \mathbf{u}_{11}, \mathbf{u}_{1k}, \mathbf{u}_{ik})$ to yield the quadrangle $(\mathbf{u}_{1k}, \mathbf{u}_{1l}, \mathbf{u}_{il}, \mathbf{u}_{ik})$, and

c) the pair of quadrangles $(\mathbf{u}_{1k}, \mathbf{u}_{1l}, \mathbf{u}_{jl}, \mathbf{u}_{jk})$, $(\mathbf{u}_{1k}, \mathbf{u}_{1l}, \mathbf{u}_{il}, \mathbf{u}_{ik})$, to yield the quadrangle $(\mathbf{u}_{jk}, \mathbf{u}_{jl}, \mathbf{u}_{il}, \mathbf{u}_{ik})$. This process is illustrated in Figure 6.9 and is similar to the one in Case 2.

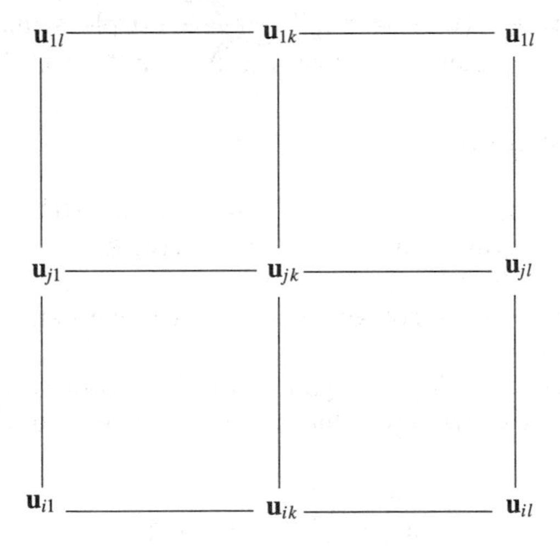

Fig. 6.9. The grid gluing for type I.

From (6.49) and case 1 of section 6.3.2, we have

$$X_{1/2}(\mathbf{u}_{11}) = \overline{\text{span}}\{\mathbf{u}_{i1}, \mathbf{u}_{1j}\}_{(i,j)\in I\times J} \subseteq \overline{\text{span}} F.$$

For any $(i, j) \in I \times J$, let $\mathbf{w} = (\mathbf{u}_{1l} + \mathbf{u}_{il} + \mathbf{u}_{1j} + \mathbf{u}_{ij})/2$. Then the symmetry $S_{\mathbf{w}}$, by use of (6.18), exchanges $\mathbf{u}_{11}$ with $\mathbf{u}_{ij}$, and, since it is a triple automorphism, it maps $X_{1/2}(\mathbf{u}_{11})$ onto $X_{1/2}(\mathbf{u}_{ij})$. Moreover, $S_{\mathbf{w}}$ maps the elements of the family F into itself, with a possible change of sign. Thus, we also have

$$X_{1/2}(\mathbf{u}_{ij}) \subseteq \overline{\operatorname{span}} F.$$

Denote by $J(\mathbf{v}) = \overline{\operatorname{span}}^{w^*} F$, which is the w^*-closed ideal generated by $\mathbf{v}$. If some element $\mathbf{a} \in X$ has the property $P_1(\mathbf{u}_{ij})\mathbf{a} = 0$ for any $(i, j) \in \widetilde{I} \times \widetilde{J}$, then from the above observation, we have

$$\mathbf{a} = P_1(\mathbf{u}_{ij})\mathbf{a} + P_{1/2}(\mathbf{u}_{ij})\mathbf{a} + P_0(\mathbf{u}_{ij})\mathbf{a} = P_0(\mathbf{u}_{ij})\mathbf{a}.$$

Hence, $\mathbf{a}$ is algebraically orthogonal to $J(\mathbf{v})$, and, since X is a factor, $\mathbf{a} = 0$, implying that

$$X = J(\mathbf{v}) = \overline{\operatorname{span}}^{w^*} F = \overline{\operatorname{span}}^{w^*} \{\mathbf{u}_{ij}\}_{(i,j) \in \widetilde{I} \times \widetilde{J}},$$

where $\mathbf{u}_{ij}$ form a rectangular grid.

To show that X is a Cartan factor of type I, let H, K be Hilbert spaces of appropriate dimensions, and let $\{e_{ij} : (i, j) \in \widetilde{I} \times \widetilde{J}\}$ be a system of matrix units of $L(H, K)$. The family $\{e_{ij}\}$ is a rectangular grid. Let E_c be the norm-closed span of $\{e_{ij}\}$. Then E_c is a JB^*-triple and $E_c^{**} = L(H, K)$ is a Cartan factor of type I. Let X_c be the norm-closed span of $\{\mathbf{u}_{ij}\}$ and define a linear map $\phi : X_c \to E_c$ by $\phi(\mathbf{u}_{ij}) = e_{ij}$. Then ϕ is a triple isomorphism and thus has an extension to an automorphism $\widetilde{\phi} : X \to E_c^{**} = L(H, K)$.

6.3.6 Case 5. Type II factor

A family of tripotents $F = \{\mathbf{u}_{ij} : i, j \in I\}$ in a JBW^*-triple is called a *symplectic grid* if $\mathbf{u}_{ij}$ are minimal tripotents with $\mathbf{u}_{ij} = -\mathbf{u}_{ji}$ for all $i, j \in I$, implying that $\mathbf{u}_{ii} = 0$ and

(i) $\mathbf{u}_{ij}$ and $\mathbf{u}_{kl}$ are co-orthogonal when they share an index and are orthogonal otherwise,

(ii) $(\mathbf{u}_{ij}, \mathbf{u}_{il}, \mathbf{u}_{kl}, \mathbf{u}_{kj})$ forms a quadrangle for distinct i, j, k, l.

(iii) Non-vanishing triple products among elements of the family are of the forms

$$\mathbf{u}_{ij} = \frac{1}{2}\{\mathbf{u}_{il}, \mathbf{u}_{kl}, \mathbf{u}_{kj}\}$$

and

$$\mathbf{u}_{kj} = \frac{1}{2}\{\mathbf{u}_{ij}, \mathbf{u}_{ij}, \mathbf{u}_{kj}\}.$$

Each Cartan factor of type II is the w^*-closed span of a symplectic grid.

Let $\mathbf{v}$ be a minimal tripotent in a JBW^*-triple X such that rank $X_{1/2}(\mathbf{v}) = 2$. Suppose that there is a minimal tripotent $\widetilde{\mathbf{v}}$ orthogonal to $\mathbf{v}$ such that $\dim X_1(\mathbf{v} + \widetilde{\mathbf{v}}) = 6$. We will show that X is a Cartan factor of type II.

Since $X_1(\mathbf{v}+\widetilde{\mathbf{v}})$ is a 6-dimensional spin factor, from the analysis of Case 3, it follows that $X_{1/2}(\mathbf{v}) \cap X_1(\mathbf{v}+\widetilde{\mathbf{v}})$ contains a quadrangle consisting of minimal tripotents, which we will denote by $(\mathbf{u}_{13}, \mathbf{u}_{14}, \mathbf{u}_{24}, \mathbf{u}_{23})$. Since $\dim X_{1/2}(\mathbf{v}) \cap X_1(\mathbf{u}_{13} + \mathbf{u}_{24}) = 4$, the JBW^*-triple $X_{1/2}(\mathbf{v})$ satisfies the conditions of case 4 from the previous section. Let $J(\mathbf{u}_{13})$ be the w^*-closed ideal generated by $\mathbf{u}_{13}$. It follows from Case 4 that $J(\mathbf{u}_{13})$ is isomorphic to a Cartan factor of type I, and since it is of rank 2, it is the norm-closed span of a rectangular grid $\{\mathbf{u}_{1i}, \mathbf{u}_{2i}\}_{i\in I}$, for some index set I not containing $\{1, 2\}$. Since $J(\mathbf{u}_{13})$ is a summand of rank 2 in the rank 2 JBW^*-triple $X_{1/2}(\mathbf{v})$, we have $X_{1/2}(\mathbf{v}) = J(\mathbf{u}_{13})$. Let $\mathbf{u}_{12} = \mathbf{v}$. Then

$$X_{1/2}(\mathbf{u}_{12}) = \overline{\operatorname{span}}\{\mathbf{u}_{1i}, \mathbf{u}_{2i}\}_{i\in I}.$$

Let $\widetilde{I} = I \cup \{1, 2\}$. For distinct $i, j \in I$, let $\mathbf{u}_{j1} = -\mathbf{u}_{1j}, \mathbf{u}_{i2} = -\mathbf{u}_{2i}$. From the definition of a rectangular grid, we have $\mathbf{u}_{2i} \perp \mathbf{u}_{1j}$, implying that $(\mathbf{u}_{i2}, \mathbf{u}_{12}, \mathbf{u}_{1j})$ forms a pre-quadrangle. Define

$$\begin{cases} \mathbf{u}_{ij} &= 2\{\mathbf{u}_{12}, \mathbf{u}_{12}, \mathbf{u}_{1j}\} & \text{for distinct } i, j \in I, \\ \mathbf{u}_{ii} &= 0 & \text{for all } i \in \widetilde{I}. \end{cases} \tag{6.50}$$

We will show that

$$F = \{\mathbf{u}_{ij} \ : i, j \in \widetilde{I}\}$$

forms a symplectic grid. To show that $\mathbf{u}_{ij} = -\mathbf{u}_{ji}$ for $i, j \in \widetilde{I}$, it suffices to consider the cases $i, j \in I$. By diagonal gluing of quadrangles (section 6.2.1), the two quadrangles $(\mathbf{u}_{1i}, \mathbf{u}_{i2}, \mathbf{u}_{j2}, \mathbf{u}_{1j})$ and $(\mathbf{u}_{ij}, \mathbf{u}_{i2}, \mathbf{u}_{12}, \mathbf{u}_{1j})$ can be glued together along the diagonal $(\mathbf{u}_{i2}, \mathbf{u}_{1j})$ to give the quadrangle $(\mathbf{u}_{1i}, \mathbf{u}_{ij}, \mathbf{u}_{j2}, -\mathbf{u}_{12})$, i.e., $\mathbf{u}_{ij} = -2\{\mathbf{u}_{j2}, \mathbf{u}_{12}, \mathbf{u}_{1i}\} = -\mathbf{u}_{ji}$. This construction is illustrated in Figure 6.10.

Property (i) of the definition of a symplectic grid can be verified easily with (5.91). Property (iii) follows from (5.91) and the fact that any two distinct non-zero elements of the family $\{\mathbf{u}_{ij} \ : i, j \in \widetilde{I}\}$ are either orthogonal or co-orthogonal minimal tripotents. To verify (ii), note that $(\mathbf{u}_{ij}, \mathbf{u}_{il}, \mathbf{u}_{kl}, \mathbf{u}_{kj})$ forms a quadrangle. We apply side-by-side gluing (section 6.2.2) to

a) the pair of quadrangles $(\mathbf{u}_{i2}, \mathbf{u}_{12}, \mathbf{u}_{1j}, \mathbf{u}_{ij})$, $(\mathbf{u}_{i2}, \mathbf{u}_{12}, \mathbf{u}_{1l}, \mathbf{u}_{il})$ to yield the quadrangle $(\mathbf{u}_{ij}, \mathbf{u}_{1j}, \mathbf{u}_{ll}, \mathbf{u}_{il})$,

b) the pair of quadrangles $(\mathbf{u}_{k2}, \mathbf{u}_{12}, \mathbf{u}_{1l}, \mathbf{u}_{kl})$, $(\mathbf{u}_{k2}, \mathbf{u}_{12}, \mathbf{u}_{1j}, \mathbf{u}_{kj})$ to yield the quadrangle $(\mathbf{u}_{kj}, \mathbf{u}_{1j}, \mathbf{u}_{ll}, \mathbf{u}_{kl})$, and

c) the pair of quadrangles $(\mathbf{u}_{ij}, \mathbf{u}_{1j}, \mathbf{u}_{ll}, \mathbf{u}_{il})$, $(\mathbf{u}_{kj}, \mathbf{u}_{1j}, \mathbf{u}_{ll}, \mathbf{u}_{kl})$, to yield the quadrangle $(\mathbf{u}_{ij}, \mathbf{u}_{il}, \mathbf{u}_{kl}, \mathbf{u}_{kj})$.

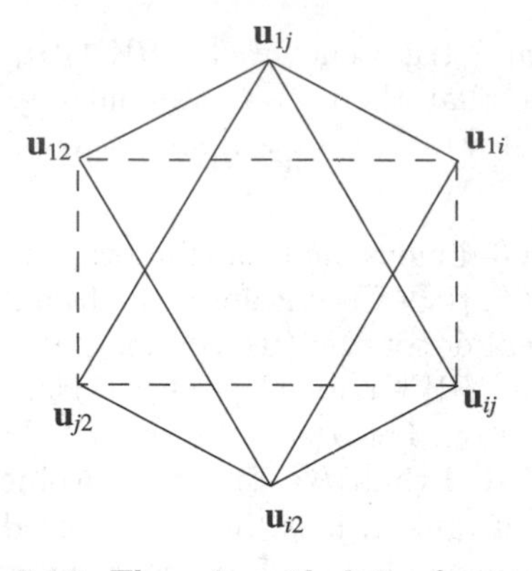

Fig. 6.10. The spin grid gluing for type II.

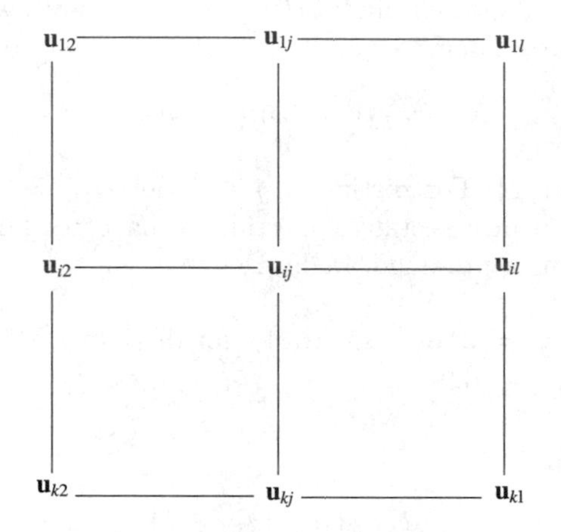

Fig. 6.11. The grid gluing for type II.

This gluing is shown in Figure 6.11.

For any $(i,j) \in I \times I$, $i \neq j$, let $\mathbf{w} = (\mathbf{u}_{12} + \mathbf{u}_{i2} + \mathbf{u}_{1j} + \mathbf{u}_{ij})/2$. Then the symmetry $S_{\mathbf{w}}$, by use of (6.18), exchanges $\mathbf{u}_{12}$ with $\mathbf{u}_{ij}$, and, since it is a triple automorphism, it maps $X_{1/2}(\mathbf{u}_{12})$ onto $X_{1/2}(\mathbf{u}_{ij})$. Moreover, $S_{\mathbf{w}}$ maps the elements of the family F into itself with a possible change of sign. Thus, we also have

$$X_{1/2}(\mathbf{u}_{ij}) = \subseteq \overline{\operatorname{span}} F.$$

Denote by $J(\mathbf{v}) = \overline{\operatorname{span}}^{w^*} F$, which is the w^*-closed ideal generated by $\mathbf{v}$. If some element $\mathbf{a} \in X$ has the property $P_1(\mathbf{u}_{ij})\mathbf{a} = 0$ for any $(i,j) \in \widetilde{I} \times \widetilde{J}$, then from the above observation, we have

$$\mathbf{a} = P_1(\mathbf{u}_{ij})\mathbf{a} + P_{1/2}(\mathbf{u}_{ij})\mathbf{a} + P_0(\mathbf{u}_{ij})\mathbf{a} = P_0(\mathbf{u}_{ij})\mathbf{a}.$$

Hence, $\mathbf{a}$ is algebraically orthogonal to $J(\mathbf{v})$, and, since X is a factor, $\mathbf{a} = 0$, implying that

$$X = J(\mathbf{v}) = \overline{\text{span}}^{w^*} F = \overline{\text{span}}^{w^*} \{\mathbf{u}_{ij}\}_{(i,j)\in\widetilde{I}\times\widetilde{I}},$$

where $\mathbf{u}_{ij}$ form a symplectic grid.

To show that X is a Cartan factor of type II, let H be a Hilbert space with an orthonormal basis $\{\xi_i\}_{i\in\widetilde{I}}$ and a conjugation Q defined by $Q(\sum \lambda_i \xi_i) = \sum \overline{\lambda_i}\xi_i$. For $i, j \in \widetilde{I}, i < j$, let $a_{ii} = 0$ and $a_{ij} = e_{ij} - e_{ji}$, where the e_{ij} are matrix units corresponding to the basis $\{\xi_i\}$. The family $\{a_{ij} : i, j \in \widetilde{I}\}$ is a symplectic grid. Let X_1 be its norm-closed span, consisting of compact antisymmetric operators on H. Then X_1 is a JB^*-triple, and, moreover, X_1^{**} is the space of bounded antisymmetric operators on H and is a Cartan factor of type II.

Let F_c be the norm-closed span of $\{\mathbf{u}_{ij}\}$ and define a linear map $\phi : F_c \to X_1$ by $\phi(\mathbf{u}_{ij}) = a_{ij}$. Obviously, ϕ is a triple isomorphism. It extends to an isomorphism $\widetilde{\phi}: J(\mathbf{v}) \to X_1^{**}$, verifying that X is a Cartan factor of type II.

6.3.7 Case 6. Exceptional factor of dimension 16

Throughout this case and case 7, we will use the following notation. Let $I = \{0, 1, 2, 3, 4, 5\}$, and for any $i \in I$, denote $I_i = I - \{i\}$. If ϕ is a permutation on I, then the sign of the permutation is denoted by $\text{sign}(\phi)$ and is equal to 1 or -1, depending on whether ϕ is even or odd, respectively.

For any fixed $i \in I$, a family of minimal tripotents

$$F = \{\mathbf{u}_i, \mathbf{u}_{jk}, \mathbf{u}_n \ : \ j, k, l \in I_i\}$$

is called an *exceptional grid of the first type* if:

(i) The family of tripotents $\{\mathbf{u}_{jk} : \ j, k \in I_i\}$ form a symplectic grid.

(ii) For any $j, k, l \in I_i$, we have

$$\begin{cases} \mathbf{u}_i \top \mathbf{u}_{jk}, \ \mathbf{u}_i \perp \mathbf{u}_k & \\ \mathbf{u}_l \perp \mathbf{u}_{jk} & \text{if } l \in \{jk\}, \\ \mathbf{u}_l \top \mathbf{u}_{jk} & \text{if } l, j, k \text{ are distinct}, \\ \mathbf{u}_l \top \mathbf{u}_k & \text{if } k \neq l. \end{cases} \tag{6.51}$$

(iii) The elements of F generate three types of quadrangles:

(a)$(\mathbf{u}_{jk}, \mathbf{u}_{jl}, \mathbf{u}_{ml}, \mathbf{u}_{mk})$, for $(j, k, l, m \in I_i)$,

(b) $(\text{sign}\,(\varphi)\mathbf{u}_i, \mathbf{u}_{jk}, \mathbf{u}_l, \mathbf{u}_{mn})$, where $\varphi = (i, j, k, l, m, n)$ is a permutation of I, and

(c) $(\mathbf{u}_j, \mathbf{u}_k, -\mathbf{u}_{jl}, \mathbf{u}_{kl})$, for j, k, l distinct in I_i.

(iv) The non-vanishing products among elements of F are of the form $D(\mathbf{a})\mathbf{b}$ or $\{\mathbf{a}, \mathbf{b}, \mathbf{c}\}$ where $(\mathbf{a}, \mathbf{b}, \mathbf{c})$ forms a pre-quadrangle of one of the quadrangles in (iii).

Let $\mathbf{v}$ and $\widetilde{\mathbf{v}}$ be two orthogonal minimal tripotents in a JBW^*-triple factor X. We will show that if $\dim X_1(\mathbf{v} + \widetilde{\mathbf{v}}) = 8$ and $X_{1/2}(\mathbf{v} + \widetilde{\mathbf{v}}) \neq \{0\}$, then X

is 16 dimensional and spanned by an exceptional grid of the first type. X is called a Cartan factor of type V.

Choose two algebraically orthogonal tripotents $\mathbf{u}, \widetilde{\mathbf{u}}$ in $X_{1/2}(\mathbf{v}) \cap X_1(\mathbf{v}+\widetilde{\mathbf{v}})$. Then $\dim(X_{1/2}(\mathbf{v}) \cap X_1(\mathbf{u}+\widetilde{\mathbf{u}})) = 6$. By use of (6.19), we may assume that $X_{1/2}(\mathbf{v}) \cap X_{1/2}(\mathbf{u}+\widetilde{\mathbf{u}}) \neq \{0\}$. Otherwise, we replace $\mathbf{v}$ by $\widetilde{\mathbf{v}}$. By the previous case, the ideal generated by $\mathbf{u}$ in $X_{1/2}(\mathbf{v})$ is isomorphic to a Cartan factor of type II. Since $\operatorname{rank} X_{1/2}(v) = 2$, this ideal and $X_{1/2}(\mathbf{v})$ must be isomorphic to the 5×5 antisymmetric matrices, which is the only type II factor of rank 2 which is not a spin factor. Let $\{\mathbf{u}_{jk} : j, k \in I_0\}$ be a symplectic grid spanning $X_{1/2}(\mathbf{v})$.

Let $\mathbf{u}_0 = \mathbf{v}$. For any integer $l \in I_0$, let φ_l be the unique permutation

$$\varphi_l = (0, j, k, l, m, n) \tag{6.52}$$

such that $j < k < m < n$. Note that $(\mathbf{u}_{jk}, \mathbf{u}_0, \mathbf{u}_{mn})$ forms a pre-quadrangle. Thus, we can define a new tripotent in X by

$$\mathbf{u}_l = 2 \operatorname{sign}(\varphi_l)\{\mathbf{u}_{mn}, \mathbf{u}_0, \mathbf{u}_{jk}\}, \tag{6.53}$$

completing the pre-quadrangle to a quadrangle

$$(\operatorname{sign}(\varphi_l)\mathbf{u}_0, \mathbf{u}_{jk}, \mathbf{u}_l, \mathbf{u}_{mn}). \tag{6.54}$$

From the quadrangle construction, each $\mathbf{u}_l$ is a minimal tripotent in X.

We will show now that the family $F = \{\mathbf{u}_0, \mathbf{u}_{jk}, \mathbf{u}_n : \ j, k, n \in I_0\}$ is an exceptional grid of the first type.

Assume that $i = 0$ in the definition of an exceptional grid of the first type. We claim that if $\varphi = (j, k, l, m, n)$ is any permutation on I_0, then

$$(\operatorname{sign}(\varphi)\mathbf{u}_0, \mathbf{u}_{jk}, \mathbf{u}_l, \mathbf{u}_{mn})$$

is also a quadrangle, weakening the restriction in (6.52).

Indeed, since every permutation is a product of transpositions, i.e., permutations that exchange only two elements of the set and leave the other elements fixed, we can assume that φ is a transposition of φ_l. Moreover, if φ exchanges k with j, or n with m, then the claim holds, since $\mathbf{u}_{kj} = -\mathbf{u}_{jk}$ and $\mathbf{u}_{nm} = -\mathbf{u}_{mn}$. Thus, we can assume further that φ exchanges j with m and is $\varphi = (0, m, k, l, j, n)$, and, thus, $\operatorname{sign} \varphi = -\operatorname{sign} \varphi_l$. By applying the diagonal gluing to the two quadrangles $(\operatorname{sign} \varphi_l \mathbf{u}_0, \mathbf{u}_{jk}, \mathbf{u}_l, \mathbf{u}_{mn})$ and $(\mathbf{u}_{mk}, \mathbf{u}_{jk}, \mathbf{u}_{jn}, \mathbf{u}_{mn})$ which are glued together along the diagonal $(\mathbf{u}_{jk}, \mathbf{u}_{mn})$, we obtain the quadrangle $(-\operatorname{sign} \varphi_l \mathbf{u}_0, \mathbf{u}_{mk}, \mathbf{u}_l, \mathbf{u}_{jn})$, which is equal to

$$(\operatorname{sign} \varphi \mathbf{u}_0, \mathbf{u}_{mk}, \mathbf{u}_l, \mathbf{u}_{jn}), \tag{6.55}$$

verifying our claim. This construction is shown in Figure 6.12.

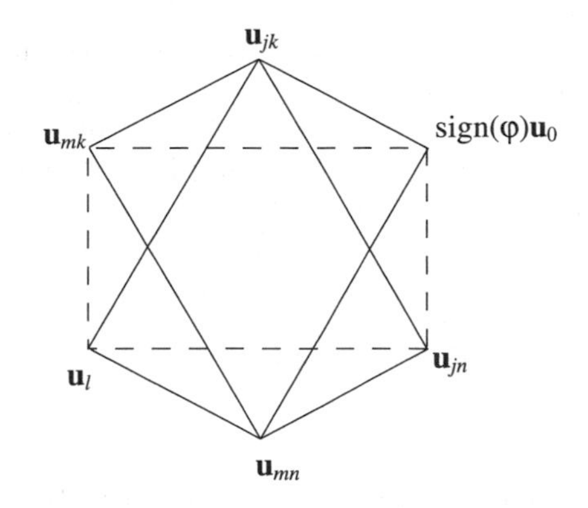

Fig. 6.12. The grid diagonal gluing for type V.

Property (i) of the definition of an exceptional grid of the first type follows from the construction of the family F. Type (a) quadrangles follow from the construction and type (b) from (6.55).

To obtain quadrangles of type (c), let $\varphi = (0, k, l, j, m, n)$ and $\varphi' = (0, j, l, k, m, n)$. Then $\operatorname{sign}\varphi = -\operatorname{sign}\varphi'$. Thus, we have two quadrangles $(\operatorname{sign}\varphi\, \mathbf{u}_i, \mathbf{u}_{kl}, \mathbf{u}_j, \mathbf{u}_{mn})$ and

$$(-\operatorname{sign}\varphi\mathbf{u}_i, \mathbf{u}_{jl}, \mathbf{u}_k, \mathbf{u}_{mn}) = (\operatorname{sign}(\varphi)\mathbf{u}_i, -\mathbf{u}_{jl}, \mathbf{u}_k, \mathbf{u}_{mn}) \tag{6.56}$$

which can be glued together along the side $(\operatorname{sign}\varphi\, \mathbf{u}_i, \mathbf{u}_{mn})$, using side-by-side gluing, to obtain the quadrangle

$$(\mathbf{u}_j, \mathbf{u}_k, -\mathbf{u}_{jl}, \mathbf{u}_{kl}). \tag{6.57}$$

See Figure 6.13.

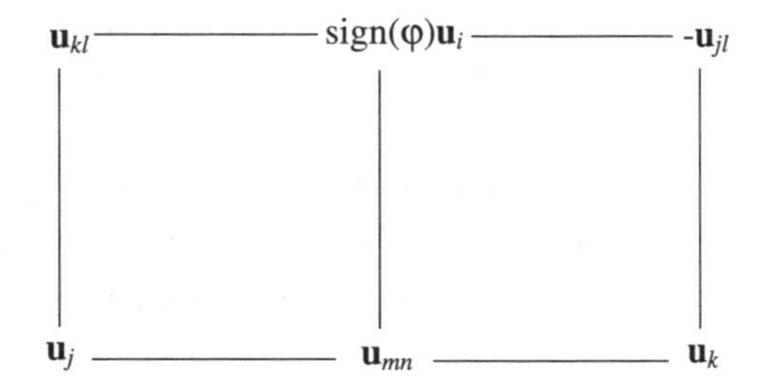

Fig. 6.13. The side-by-side grid gluing for type V.

Property (ii) follows from (6.55), (6.57) and the construction, by use of (5.91). So F is an *ortho-co-orthogonal* family of minimal tripotents, meaning that every pair of elements of F is either orthogonal or co-orthogonal. Therefore, (iv) holds as well, showing that F is an *exceptional grid of the first type.*

Next, we show that $J(\mathbf{v}) = \text{span } F$ and is a summand. Now $\mathbf{u}_0$ can be exchanged by an automorphism $S_\mathbf{w}$ with any $\mathbf{u}_{jk}$ and any $\mathbf{u}_k$, for any distinct indices j, k. Since $S_\mathbf{w}$ moves any element of F to a multiple of F, we get $X_{1/2}(\mathbf{u}_{jk})J(\mathbf{v}) \subset J(\mathbf{v})$ and $X_{1/2}(\mathbf{u}_k)J(\mathbf{v}) \subset J(\mathbf{v})$. Thus, $J(v)$ is a summand, and, since X is a factor, we have

$$X = J(\mathbf{v}) = \text{span } F,$$

where F is an exceptional grid of the first type.

6.3.8 Case 7. Exceptional factor of dimension 27

Let $\varepsilon = \pm 1$. When ε occurs as an index, we will use only the sign $+$ or $-$. A family of tripotents $F = \{\mathbf{u}_j^\varepsilon, \mathbf{u}_{ij} : i, j \in I, \ \varepsilon = \pm 1\}$ is called an *exceptional grid of the second type* if

(i) $\{\mathbf{u}_{ij}\}_{i,j \in I}$ is a 6×6 symplectic grid,

(ii) for any $i \in I$ and $\varepsilon = \pm 1$, the family $\{-\varepsilon \mathbf{u}_i^\varepsilon, \mathbf{u}_{jk}, \mathbf{u}_j^\varepsilon : j, k \in I_i\}$ is an exceptional grid of the first type,

(iii) the quadrangles of the family F are those determined by parts (i) and (ii) above or of the form

$$(\mathbf{u}_j^\varepsilon, \mathbf{u}_j^{-\varepsilon}, \mathbf{u}_k^{-\varepsilon}, \mathbf{u}_k^\varepsilon) \text{ for distinct } j, k \in I, \tag{6.58}$$

(iv) the family F is ortho-co-orthogonal, and, therefore, all non-vanishing products among elements of F are either of the form $D(\mathbf{a})\mathbf{b}$, or $\{\mathbf{a}, \mathbf{b}, \mathbf{c}\}$, where $(\mathbf{a}, \mathbf{b}, \mathbf{c})$ forms a pre-quadrangle.

Let $\mathbf{v}$ be a minimal tripotent in JBW^*-triple X. If there is a minimal tripotent $\widetilde{\mathbf{v}}$ such that $\dim X_1(\mathbf{v} + \widetilde{\mathbf{v}}) = 10$ and $X_{1/2}(\mathbf{v} + \widetilde{\mathbf{v}}) \neq \{0\}$, then X is spanned by an exceptional grid of the second type and is isomorphic to the Jordan algebra $H_3(\mathcal{O})$ consisting of 3×3 hermitian matrices over the octonians.

Let $\mathbf{u}_0^-$ be a minimal tripotent in $X_{1/2}(\mathbf{v})$, and let $J(\mathbf{u}_0^-)$ be the ideal in $X_{1/2}(\mathbf{v})$ generated by $\mathbf{u}_0^-$. The same argument as in case 6 shows that $J(\mathbf{u}_0^-) = X_{1/2}(\mathbf{v})$ and is spanned by an exceptional grid $\{\mathbf{u}_0^-, \mathbf{u}_{jk}, \mathbf{u}_m^+ : j, k, m \in I_0\}$ of the first type.

Let $\mathbf{v} = -\mathbf{u}_0^+$, and, as above, for any $l \in I_0$, let $\varphi_l = (0, j, k, l, m, n)$, where $j < k < m < n$. Note that $(\text{sign}(\varphi_l)\mathbf{u}_{jk}, -\mathbf{u}_0^+, \mathbf{u}_{mn})$ defines a pre-quadrangle which can be completed to a quadrangle by

$$\mathbf{u}_l^- = \text{sign}\,(\varphi_l)\, 2\{\mathbf{u}_{jk}, -\mathbf{u}_0^+, \mathbf{u}_{mn}\}.$$

Similarly, the pre-quadrangle $(\mathbf{u}_{ij}, -\mathbf{u}_0^+, \mathbf{u}_j^+)$ can be completed to a quadrangle by

$$\mathbf{u}_{i0} = 2\{\mathbf{u}_{ij}, -\mathbf{u}_0^+, \mathbf{u}_j^+\}, \tag{6.59}$$

where $i \in I_0$, and j is some element of $I_0 \cap I_i$.

Note that for $i, l \in I_0, \mathbf{u}_l{}^-$ and $\mathbf{u}_{i0}$ are also minimal tripotents of X and that the family $\{-\mathbf{u}_0{}^+, \mathbf{u}_{jk}, \mathbf{u}_n{}^- : j, k, n \in I_0\}$ is also an exceptional grid of the first type, by construction. Moreover, the definition of $\mathbf{u}_{i0}$ in (6.59) does not depend on the choice of j.

To see this, let l be another element of $I_0 \cap I_i \cap I_j$. Since $\{\mathbf{u}_0{}^-, \mathbf{u}_{jk}, \mathbf{u}_n{}^+\}$ is an exceptional grid of the first type,

$$(\mathbf{u}_j{}^+, \mathbf{u}_l{}^+, \mathbf{u}_{ij}, \mathbf{u}_{li}) = (\mathbf{u}_j{}^+, \mathbf{u}_l{}^+, -\mathbf{u}_{ji}, \mathbf{u}_{li}) \tag{6.60}$$

is a quadrangle. Thus, by use of diagonal gluing, it can be glued to the quadrangle $(\mathbf{u}_{i0}, \mathbf{u}_{ij}, -\mathbf{u}_0{}^+, \mathbf{u}_j{}^+)$ to obtain that

$$(\mathbf{u}_{i0}, \mathbf{u}_{li}, -\mathbf{u}_0{}^+, -\mathbf{u}_l{}^+) = (\mathbf{u}_{i0}, \mathbf{u}_{il}, -\mathbf{u}_0{}^+, \mathbf{u}_l{}^+) \tag{6.61}$$

is a quadrangle, implying that $\mathbf{u}_{i0} = 2\{\mathbf{u}_{il}, -\mathbf{u}_0{}^+, \mathbf{u}_l{}^+\}$. This construction is shown in Figure 6.14.

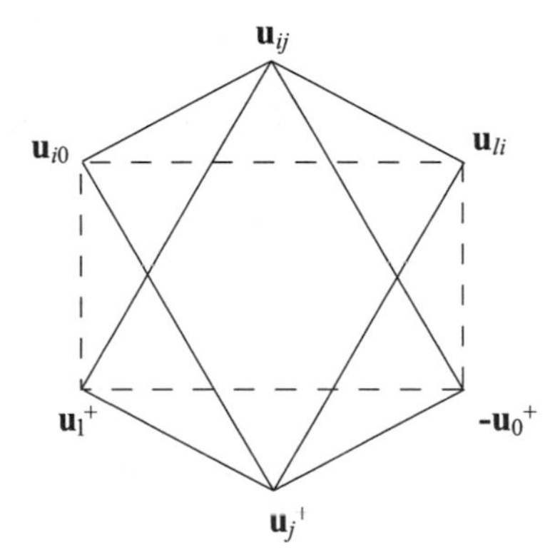

Fig. 6.14. The grid gluing for type VI.

Let $\mathbf{u}_{00} = 0$ and $\mathbf{u}_{0i} = -\mathbf{u}_{i0}$. Let i, j, k be arbitrary elements of I_0, with $j \neq k$. From (5.91), it follows that $\mathbf{u}_{i0} \top \mathbf{u}_{jk}$, if $i \in \{j, k\}$, and $\mathbf{u}_{i0} \perp \mathbf{u}_{jk}$, if i, j, k are distinct. Moreover, if i, j, k are distinct, we can glue the quadrangles $(\mathbf{u}_{i0}, \mathbf{u}_{ij}, -\mathbf{u}_0{}^+, \mathbf{u}_j{}^+)$ and $(\mathbf{u}_{k0}, \mathbf{u}_{kj}, -\mathbf{u}_0{}^+, \mathbf{u}_j{}^+)$ side-by-side to obtain that $(\mathbf{u}_{i0}, \mathbf{u}_{jk}, \mathbf{u}_{kj}, \mathbf{u}_{k0})$ is a quadrangle. Thus, since the family $\{\mathbf{u}_{ij}\}_{i,j \in I_0}$ is a 5×5 symplectic grid, the family $\{\mathbf{u}_{ij}\}_{i,j \in I}$ is a 6×6 symplectic grid, showing that (i) holds.

Let $\varphi = (0, i, j, k, l, m)$. Then we have two quadrangles

$$(-\operatorname{sign}(\psi)\mathbf{u}_0{}^-, \mathbf{u}_{ij}, \mathbf{u}_k{}^-, \mathbf{u}_{lm}), \; (\operatorname{sign}(\psi)\mathbf{u}_0{}^-, \mathbf{u}_{ij}, \mathbf{u}_k{}^+, \mathbf{u}_{lm})$$

which can be glued along the diagonals $(\mathbf{u}_{ij}, \mathbf{u}_{lm})$ to obtain the quadrangle $(\mathbf{u}_0{}^+, \mathbf{u}_0{}^-, \mathbf{u}_k{}^-, \mathbf{u}_k{}^+)$, for arbitrary $k \neq 0$. Now, if $j \neq 0$ and $j \neq k$, then the two quadrangles $(\mathbf{u}_0{}^+, \mathbf{u}_0{}^-, \mathbf{u}_j{}^-, \mathbf{u}_j{}^+)$ and $(\mathbf{u}_0{}^+, \mathbf{u}_0{}^-, \mathbf{u}_k{}^-, \mathbf{u}_k{}^+)$ can be glued side-by-side to obtain the quadrangle $(\mathbf{u}_j{}^+, \mathbf{u}_j{}^-, \mathbf{u}_k{}^-, \mathbf{u}_k{}^+)$. Thus, (6.58) are quadrangles for any distinct $j, k \in I$.

Next, we show that

$$(\mathbf{u}_0{}^\varepsilon, \mathbf{u}_i{}^\varepsilon, \mathbf{u}_{jk}, \mathbf{u}_{ij}) \tag{6.62}$$

are quadrangles, for any $i \in I$ and any ε. From (6.59), we obtain the quadrangle $(\mathbf{u}_0{}^+, \mathbf{u}_i{}^+, \mathbf{u}_{j0}, \mathbf{u}_{ij})$. By gluing this quadrangle with the quadrangle $(\mathbf{u}_0{}^+, \mathbf{u}_0{}^-, \mathbf{u}_i{}^-, \mathbf{u}_i{}^+)$, obtained from (6.58), we obtain the quadrangle $(\mathbf{u}_0{}^+, \mathbf{u}_0{}^-, \mathbf{u}_{j0}, \mathbf{u}_{ij})$. Thus, (6.62) are quadrangles.

As noted previously, part (ii) holds if $i = 0$. For $i \neq 0$, we can apply (5.91) to verify that $\mathbf{u}_i^\varepsilon \top \mathbf{u}_{jk}$ for all distinct $j, k \in I_i$. Thus, it suffices to show that

$$(-\mathrm{sign}(\psi)\varepsilon \mathbf{u}_i^\varepsilon), \mathbf{u}_{jk}, \mathbf{u}_l, \mathbf{u}_{mn}) \tag{6.63}$$

are quadrangles for any $\psi = (i, j, k, l, m, n)$. Since $\{-\varepsilon \mathbf{u}_0^\varepsilon, \mathbf{u}_{jk}, \mathbf{u}_n : j, k, n \in I_0\}$ are exceptional grids of the first type, elementary properties of permutations imply that (6.63) are quadrangles if $0 \in \{i, l\}$. Thus, it remains to consider the case when $0 \in \{j, k, m, n\}$. By applying a transposition, if necessary, we may assume that $k = 0$.

Let $\sigma = (0, i, j, l, m, n)$ and $\varphi = (i, j, 0, l, m, n)$. From (6.62), we obtain the quadrangle $(\{\varepsilon \mathbf{u}_i^\varepsilon, \varepsilon \mathbf{u}_0^\varepsilon, \mathbf{u}_{ij}, \mathbf{u}_{j0})$. Gluing to it the quadrangle $(\varepsilon \mathbf{u}_0^\varepsilon, \mathbf{u}_{ij}, \mathbf{u}_l{}^{-\varepsilon}, -\mathrm{sign}(\sigma)\mathbf{u}_{mn})$, we obtain the quadrangle

$$(\varepsilon \mathbf{u}_i^\varepsilon, \mathbf{u}_{j0}, \mathbf{u}_l{}^{-\varepsilon}, -\mathrm{sign}(\sigma)\mathbf{u}_{mn}).$$

It is easy to verify that sign(σ)=sign(φ), implying that $(-\mathrm{sign}(\varphi)\varepsilon \mathbf{u}_i^\varepsilon)$, $\mathbf{u}_{j0}, \mathbf{u}_l^{-\varepsilon}, \mathbf{u}_{mn})$ is a quadrangle for $\varphi = (i, j, 0, l, m, n)$. Thus, (ii) is verified.

We have already verified that (6.58) are quadrangles. The remaining claim of (iii) can be verified by inspection, using (i) and the relations (6.51). Part (iv) follows from (i) and (ii). Therefore, we have shown that F is an exceptional grid of the second type and consists of minimal tripotents. Next, observe that each element in F is co-orthogonal to exactly 16 other elements of F. From property (i), the family F contains three minimal orthogonal tripotents $\mathbf{u}_{01}, \mathbf{u}_{23}$ and $\mathbf{u}_{45}$. If we let $\mathbf{e} = \mathbf{u}_{01} + \mathbf{u}_{23} + \mathbf{u}_{45}$, then $\mathbf{e}$ is a tripotent and $X = X_1(\mathbf{e})$ is a rank 3 Jordan algebra with $\mathbf{e}$ as an identity. The JB^*-triple X can be decomposed with the joint Peirce decomposition with respect to the three minimal tripotents as

$$X = X_1(\mathbf{u}_{01}) + X_1(\mathbf{u}_{23}) + X_1(\mathbf{u}_{45})$$

$$+X_{1/2}(\mathbf{u}_{01}) \cap X_{1/2}(\mathbf{u}_{23}) + X_{1/2}(\mathbf{u}_{01}) \cap X_{1/2}(\mathbf{u}_{45}) + X_{1/2}(\mathbf{u}_{23}) \cap X_{1/2}(\mathbf{u}_{45}).$$

The subspaces $X_1(\mathbf{u}_{jk}) = C\mathbf{u}_{jk}$ are one-dimensional, while the dimension of $X_{1/2}(\mathbf{u}_{jk}) \cap X_{1/2}(\mathbf{u}_{lm})$ is 8, and this space can be identified with the octonians. Thus, the space X can be considered as the space of 3×3 hermitian matrices over the octonians, known as the exceptional Jordan algebra.

6.4 Structure and representation of JB^*-triples

The first corollary of our Classification Theorem is the decomposition of a JBW^*-triple into atomic and non-atomic parts. For every JBW^*-triple X, we can define two w^*-closed ideals A and N, where A is an ideal generated by minimal tripotents and is called the *atomic part* of X, and N, which contains no minimal tripotents, is called the *non-atomic part* of X.

Let $\{J_i\}_{i\in I}$ be the set of w^*-closed ideals generated by minimal tripotents of X. As we have shown in the previous section, each of them is a Cartan factor and a summand. Denote the orthogonal complement of J_i in X by $J_i^\perp$. Let

$$A = \oplus_{i\in I} J_i, \quad N = \bigcap_{i\in I} J_i^\perp. \tag{6.64}$$

A and N are orthogonal w^*-closed ideals in X.

For each $i \in I$, let P_i be the contractive projection from X to J_i with kernel $J_i^\perp$. Moreover, since their ranges, the ideals J_i, are mutually orthogonal, $\sum_{i\in I} P_i(x)$ converges w^* for each $\mathbf{x} \in U$. For $j \in I$,

$$P_j(x - \sum_i P_i(x)) = P_j(x) - \sum_i P_j P_i(x) = P_j(x) - P_j(x) = 0, \tag{6.65}$$

implying that $\mathbf{x} - \sum_i P_i(x) \in N$. Thus, any JBW^*-triple decomposes into an orthogonal sum as

$$X = A \oplus N, \tag{6.66}$$

which is called the *atomic decomposition.* Moreover, A can be decomposed as an orthogonal sum of Cartan factors.

Recall that in our constructions, each Cartan factor other than those of type III is spanned by a grid consisting of minimal tripotents. On the other hand, the Cartan factors of type III are spanned by hermitian grids $\{\mathbf{u}_{ij}\}$ with $\mathbf{u}_{ii}$ minimal, and $\mathbf{u}_{ij}$ not minimal for $i \neq j$. However, it is easy to verify that such $\mathbf{u}_{ij} (i \neq j)$ can be written as the sum of two minimal tripotents:

$$\mathbf{u}_{ij} = \frac{1}{2}(\mathbf{u}_{ij} + \mathbf{u}_{ii} + \mathbf{u}_{jj}) + \frac{1}{2}(\mathbf{u}_{ij} - \mathbf{u}_{ii} - \mathbf{u}_{jj}). \tag{6.67}$$

Thus, A is the w^*-closed span of the minimal tripotents.

We are now ready to verify the so-called Gelfand–Naimark Theorem for JB^*-triples. The original Gelfand–Naimark theorem showed that any C^*-algebra can be realized (isometrically embedded) as a subalgebra of operators on a Hilbert space. For JB^*-triples, we get the following: Every JB^*-triple can be isometrically embedded (as a triple system) into a direct sum of Cartan factors.

To see this, let X be a JB^*-triple. Then its bidual X^{**} is a JBW^*-triple in which X can be isometrically embedded. Denote this embedding of X into X^{**} by π. Since X^{**} is a JBW^*-triple, it has a decomposition into atomic and non-atomic parts: $X^{**} = A \oplus N$, as described above. Denote the projection from X^{**} onto A with kernel N by P. Let $T = P \circ \pi : M \to A$. Obviously, P is a triple homomorphism and $||T\mathbf{x}|| \leqq ||\mathbf{x}||$ for all $\mathbf{x} \in M$. We will show T is isometric.

Let $\mathbf{x} \in M$, with $||\mathbf{x}|| = 1$. The set $\{\varphi \in M^* : \varphi(\mathbf{x}) = 1 = ||\varphi||\}$ is closed and convex and, hence, contains an extreme point ψ. An elementary argument shows that ψ is also an extreme point of the unit ball of X^*. Thus, there is a minimal tripotent $\mathbf{v} \in M^{**}$ such that $\psi(\mathbf{v}) = 1$ and $\psi(\mathbf{a}) = \psi(P_1(\mathbf{v})(\mathbf{a}))$ for all $\mathbf{a} \in M^{**}$. In particular,

$$\psi(T(\mathbf{x})) = \psi(P_1(\mathbf{v})T(\mathbf{x})) = \psi(P_1(\mathbf{v}) \circ P \circ \pi(\mathbf{x})) = \psi(P_1(\mathbf{v}) \circ \pi(\mathbf{x}))$$

$$= \psi(\pi(\mathbf{x})) = \psi(\mathbf{x}) = 1.$$

Thus $||T(\mathbf{x})|| = 1$.

Finally, from our construction, it follows that any Cartan factor either has a basis consisting of an ortho-co-orthogonal family of tripotents or can be embedded as an invariant subspace of a symmetry acting on a space spanned by an ortho-co-orthogonal family. This follows immediately from the fact that a triangle can be embedded into a quadrangle in a natural way.

Moreover, any $JB*$-triple can be isometrically embedded (as a triple system) into a Jordan algebra with the triple product on it. First of all, the classical domains are embedded into $L(H)$, which is a Jordan algebra. Next, the spin factor itself is a Jordan algebra. Finally, every type V domain is a subspace of a domain of type VI, which is itself a Jordan algebra.

6.5 Notes

This chapter is based mainly on the work in [15]. The Gelfand–Naimark Theorem for $JB*$-triples was proven in [33]. The classification of grids for Cartan factors was also done in [53] and [57].

For possible applications of the exceptional domains to physics, see [58] and the references therein. We want to thank Prof. N. Kamiya for bringing this reference to our attention.

References

1. L. V. Ahlfors, Möbius transformations and Clifford numbers, in: *Differential Geometry and Complex Analysis*, Springer-Verlag, Berlin, 1985.
2. E. Alfsen and F. Shultz, State spaces of Jordan algebras, *Acta Math.* **140** (1978), 155–190.
3. E. Alfsen and F. Shultz, *State Spaces of Operator Algebras, Basic Theory, Orientation and C^*-product*, Birkhäuser Boston, Cambridge, MA, 2001.
4. J. Arazy, The isometries of C_p, *Israel J. Math.***22** (1975), 247–256.
5. J. Arazy, An application of infinite-dimensional holomorphy to the geometry of Banach spaces, *Geometrical Aspects of Functional Analysis*, Lecture notes in Mathematics, Springer-Verlag, 1987.
6. J. Arazy, T. Barton, and Y. Friedman, Operator differentiable functions, *Journal of Operator Integral Theory* **13**(4) (1990), 462–487.
7. J. Arazy and Y. Friedman, Contractive projections in C_1 and C_∞, *Memoirs of the American Mathematical Society*, **200**, 1978.
8. J. Arazy and Y. Friedman, Contractive projections in C_p, *Memoir Amer. Math. Soc.* **459**, 1992.
9. T.J. Barton and R.M. Timoney, Weak*-continuity of Jordan triple products and applications, *Math. Scand.* **59** (1986), 177–191.
10. W.E. Baylis (ed.), *Clifford (Geometric) Algebras with Applications to Physics, Mathematics and Engineering*, Birkhäuser, Boston, 1996.
11. M. Born, *Einstein's Theory of Relativity*, Dover Publications, New York, 1965.
12. E. Cartan, Sur les domains bornés homogènes de l'espace de n variables complexes, *Abh. Math. Sem. Univ. Hamburg* **11** (1935), 116–162.
13. H. Cartan, Les functions de deux variables complexes et le probleme de la representation analytique, *J. Math. Pures Appl.* **10** (1931), 1–114.
14. J.L. Daletski and S.G. Krein, Integration and differentiation of functions of Hermitian operators and application to theory of perturbations, *A. M. S. Translations* **47**(2)(1965), 1–30.
15. T. Dang and Y. Friedman, Classification of JBW^*-triples and applications, *Math. Scand.* **61** (1987), 292–330.
16. S. Dineen, Complete holomorphic vector fields on the second dual of a Banach space, *Math. Scand.* **59** (1986), 131–142.
17. P.A.M. Dirac, *Spinors in Hilbert Space*, Plenum Press, New York, 1974.
18. C.M. Edwards and G.T. Rüttiman, On the facial structure of the unit balls in a JBW^*-triple and its predual, *J. London Math. Soc.* **38** (1988), 317–332.
19. E. Effros, Order ideals in a C^*-algebra and its dual, *Duke Math. Journal* **30** (1963), 391–412.
20. A. Einstein, Zur Elektrodynamik bewegter Körper, *Ann. Phys.* **17** (1905), 891.

21. A. Einstein, *The Meaning of Relativity*, Princeton University Press, Princeton, NJ, 1955.
22. L.J. Eisenberg, Necessity of the linearity of relativistic transformations between inertial sytems, *Am. J. Phys.* **35** (1967), 649.
23. J. Faraut, S. Kaneyuki, A. Koranyi, Q. Lu, G. Ross, *Analysis and Geometry on Complex Homogeneous Domains,* Progress in Mathematics **185**, Birkhäuser, Boston, Cambridge, MA, 2000.
24. J.H. Field, Space-time exchange invariance: Special relativity as a symmetry principle, *Am. J. Phys.* **69**(5)(2001), 569–575.
25. Y. Friedman, Bounded symmetric domains and the JB^*-triple structure in physics, in: *Jordan Algebras, Proceedings of the Oberwolfach Conference 1992*, W. Kaup, K. McCrimmon, H. Petersson, eds., de Gruyter, Berlin (1994), 61–82.
26. Y. Friedman, Analytic solution of relativistic dynamic equation in a constant uniform electromagnetic field, to appear.
27. Y. Friedman and Yu. Gofman, Why Does the Geometric Product Simplify the Equations of Physics?, *International Journal of Theoretical Physics* **41** (2002), 1861–1875.
28. Y. Friedman and Yu. Gofman, Relativistic Linear Spacetime Transformations Based on Symmetry, *Foundations of Physics* **32** (2002), 1717–1736.
29. Y. Friedman and Y. Gofman, Transformation between Accelerated Systems Based on Symmetry, *Gravitation, cosmology and relativistic astrophysics,* 2nd Gravitation Conference, June 23–27, 2003, Kharkov National University, Kharkov, Ukraine.
30. Y. Friedman and A. Naimark, The homogeneity of the ball in R^3 and special relativity, *Foundation of Physics Letters* **5** (1992), 337–354.
31. Y. Friedman and B. Russo, Structure of the predual of a JBW^*-triple, *J. Reine Angew. Math.* **356** (1985), 67–89.
32. Y. Friedman and B. Russo, A geometric spectral theorem, *Quart. J. Math. Oxford* **37** (1986), 263–277.
33. Y. Friedman and B. Russo, The Gelfand-Naimark Theorem for JB^* triples, *Duke Math. J.* **53** (1986), 139–148.
34. Y. Friedman and B. Russo, Affine structure of facially symmetric spaces, *Math. Proc. Camb. Philos. Soc.* **106** (1989), 107–124.
35. Y. Friedman and B. Russo, Geometry of the dual ball of the spin factor, *Proc. Lon. Math. Soc.* **65**(1992), 142–174.
36. Y. Friedman and B. Russo, Classification of atomic facially symmetric spaces, *Canad. J. Math* **45** (1) (1993), 33–87.
37. Y. Friedman and B. Russo, A new approach to spinors and some representation of the Lorentz group on them, *Foundations of Physics* **31**(12) (2001), 1733–1766.
38. I.C. Gohberg and M.G. Krein, Introduction to the theory of linear nonselfadjoint operators, *Translations of the AMS,* **18**, Providence, 1969.
39. M. Günaydin, Quadratic Jordan formulation of quantum mechanics and construction of Lie (super)algebras from Jordan (super)algebras, VIII Internat. Coloq. on Group Theoretical Methods in Physics, Kiryat Anavim, Israel,1979, *Ann. Israel Phys. Soc.* **3**(1980), 279–296.
40. L.A. Harris, Bounded symmetric domains in infinite dimensional spaces, in: *Infinite dimensional holomorphy,* T. L.Hayden and T. J. Suffridge eds., Lecture Notes in Math. **364**, 13–40.

41. D. Hestenes and G. Sobczyk, *Clifford Algebra to Geometric Calculus,* Kluwer Academic, Dordrecht, 1984.
42. J.D. Jackson, *Classical electrodynamics*, John Wiley and Sons, Inc., New York, 1999.
43. P. Jordan, J. von Neumann, and E. Wigner, On an algebraic generalization of the quantum mechanical formalism, *Ann. of Math.* **35** (1934), 29–64.
44. R.V. Kadison and J.R. Ringrose, *Fundamentals of the Theory of Operator Algebras*, Academic Press, New York, 1986.
45. I.L. Kantor, Transitive differential groups and invariant connections on homogeneous spaces, *Trudy Sem. Vektor. Tenzor. Anal.* **13** (1966), 310–398.
46. W. Kaup, A Riemann mapping theorem for bounded symmetric domains in complex Banach spaces, *Math. Zeit.* **183** (1983), 503–529.
47. W. Kaup, Contractive projections on Jordan C^*-algebras and generalizations, *Math. Scand.* **54** (1984), 95–100.
48. M. Koecher, *An elementary approach to bounded symmetric domains*, Rice University, Houston, 1969.
49. J. Lasenby, A.N. Lasenby, and C.J.L. Doran, A unified mathematical language for physics and engineering in the 21st century, *Phil. Trans. R. Soc. Lond.*, **A 358** (2000), 21–39.
50. A. Lee and T.M. Kalotas, Lorentz transformation from the first postulate, *Am. J. of Physics* **43**(5) (1975).
51. L.D. Landau and E.M. Lifshitz, *The Classical Theory of Fields*, Course of Theoretical Physics, Volume **2**, Pergamon Press, Addison-Wesley, 1971.
52. O. Loos, *Bounded symmetric domains and Jordan pairs*, University of California, Irvine, 1977.
53. K. McCrimmon and K. Mayberg, Coordination of Jordan triple systems, *Comm. Algebra* **9** (1981), 1495–1542.
54. C. Marchal, Henri Poincaré: a Decisive Contribution to Special Relativity, *IHEP Preprint 99–21, Protvino*, 1999.
55. N.D. Mermin, Relativity without light, *Am. J. Phys.* **52**(2) (1984), 119–124.
56. M. Neal and B. Russo, State space of JB^*-triples, *Math. Ann.* **328**(4)(2004), 585–624.
57. E. Neher, *Jordan triple systems by the grid approach*, Lecture notes in Mathematics **1280**, Springer-Verlag, Berlin-New-York, 1987.
58. S. Okubo, *Introduction to Octonion and Other Non-Associative Algebras in Physics*, Cambridge University Press, Cambridge, U.K., 1995.
59. W. Pauli, *Theory of Relativity*, Pergamon Press, London, 1958.
60. R. Prosser, On ideal structure of operator algebras, *Mem. Amer. Math. Soc.* **45** (1963).
61. S. Sakai, *C^*-algebras and W^*-algebras,* Ergebnisse Math. **60,** Springer-Verlag, New York, Berlin, Heidelberg, 1971.
62. I. Satake, *Algebraic Structures of Symmetric Domains,* Princeton University Press, 1980.
63. H.M. Schwartz, Deduction of the general Lorentz transformations from a set of necessary assumptions, *Am. J. Phys.* **52**(4) (1984), 346–350.
64. L.L. Stacho, A projection principle concerning biholomorphic automorphisms, *Acta Sci. Math.* **44** (1982), 99–124.
65. S. Takeuchi, Relativistic $E \times B$ acceleration, *Physical Review E* **66**, 37402 (2002).

66. Y.P. Terletskii, *Paradoxes in the Theory of Relativity*, Plenum, NY, 1968.
67. A.A. Ungar, *Beyond the Einstein Addition Law and its Gyroscopic Thomas Precession,* Fundamental Theories of Physics, Volume 117, Kluwer Academic Publisher, 2001.
68. H. Upmeier, *Symmetric Banach Manifolds and Jordan C^*-algebras,* North Holland, 1985.
69. H. Upmeier, *Jordan Algebras in Analysis, Operator Theory, and Quantum Mechanics,* AMS-CBMS Regional Conference Series no. 67, 1987.
70. J.P. Vigué, Le groupe des automorphismes analytiques d'un domaine borné d'un espace de Banach complexe. Application aux domains bornés symmétriques, *Ann. Sci. Ec. Norm. Sup.*, 4e serie **9** (1976), 203–282.
71. S. Weinberg, *The Quantum Theory of Fields*, Cambridge University Press, 2000.

Index